Marine Glycomics

Marine Glycomics

Editors

Yuki Fujii
Marco Gerdol
Yasuhiro Ozeki

MDPI • Basel • Beijing • Wuhan • Barcelona • Belgrade • Manchester • Tokyo • Cluj • Tianjin

Editors
Yuki Fujii
Nagasaki International University
Japan

Marco Gerdol
University of Trieste
Italy

Yasuhiro Ozeki
Yokohama City University
Japan

Editorial Office
MDPI
St. Alban-Anlage 66
4052 Basel, Switzerland

This is a reprint of articles from the Special Issue published online in the open access journal *Marine Drugs* (ISSN 1660-3397) (available at: https://www.mdpi.com/journal/marinedrugs/special_issues/Marine_Glycomics).

For citation purposes, cite each article independently as indicated on the article page online and as indicated below:

LastName, A.A.; LastName, B.B.; LastName, C.C. Article Title. *Journal Name* **Year**, *Volume Number*, Page Range.

ISBN 978-3-0365-5821-9 (Hbk)
ISBN 978-3-0365-5822-6 (PDF)

Contents

About the Editors

Yuki Fujii

Dr. Yuki Fujii has been Associate Professor at Nagasaki International University (Sasebo, Nagasaki, Japan) since 2022, and was Senior Assistant Professor (2017–2022) and Assistant Professor (2012–2016). He obtained a PhD in 2010 at Yokohama City University under the direction of Prof. Y. Ozeki (see below), an expert on marine lectins. He was a postdoctoral Visiting Scientist in 2006 at The Biomembrane Institute (Seattle, WA, USA; PI: Prof. Sen-itiroh Hakomori), where he studied cancer cell surface expression of Lex glycosphingolipid. At Nagasaki International University, he has identified a series of glycan-binding specificities of marine lectins using Frontal Affinity Chromatography Technology (FACT), leading to the discovery of lectins that cause cancer cell apoptosis. His ongoing collaborations with overseas researchers are focused on the development of these lectins as post-antibody drugs using next-generation techniques.

Marco Gerdol

Dr. Marco Gerdol has been Assistant Professor (RtdB) at University of Trieste (Italy) since 2015. He obtained a BA in Functional Genomics at the same university in 2008, and a PhD in Environmental Biology in 2012, and continued as a postdoctoral fellow in the Applied and Comparative Genomics group (Dept. of Life Sciences) from 2012 to 2014. His research has been focused mainly on genomics of non-model marine invertebrate species, with an emphasis on bivalve mollusks, which are important "sentinel" organisms for biomonitoring, and useful models for comparative immunological studies. His recent (2019–2022) publications have been focused on the study of the mussel pan-genome, characterized by massive gene presence-absence variation, which he further investigated in a number of follow-up studies targeting different families of immune-related genes. His group uses genomic and transcriptomic tools to study the evolution of immune systems in marine invertebrates.

Yasuhiro Ozeki

Prof. Dr. Yasuhiro Ozeki is Professor at Yokohama City University, Japan. He is a lectinologist and a graduate of Meijo University (Nagoya, Japan). Dr. Ozeki obtained his PhD at Fujita Health University (Toyoake, Japan) in 1992, and was an NIH Fellow in 1994–95 at The Biomembrane Institute (Director: Prof. Sen-itiroh Hakomori) and University of Washington, Seattle, WA, conducting glycobiological studies. From 2005 through to the present, he has supervised five Ph.D. students from Japan and Bangladesh. He has been invited as an external examiner of five Ph.D. dissertations from different Univs in India since 2008. He served as Guest Editor of Special Issues ("Marine Glycobiology, Glycomics, and Lectins", "Marine Glycomics", and "Marine Glycomics 2.0" of the journal *Marine Drugs*; and "Recent Advances in Carbohydrates", as well as "Glycosides, and Nucleosides for Biological and Computational Research" of the journal *Molecules*).

marine drugs

Editorial

Coming New Age of Marine Glycomics: The Fundamental, Medical, and Ecological Aspects

Yuki Fujii [1,*], Marco Gerdol [2,*] and Yasuhiro Ozeki [3,*]

1 Graduate School of Pharmaceutical Sciences, Nagasaki International University, 2825-7 Huis Ten Bosch, Sasebo 859-3298, Japan
2 Department of Life Sciences, University of Trieste, Via Licio Giorgieri 5, 34127 Trieste, Italy
3 Graduate School of NanoBio Sciences, Yokohama City University, 22-2 Seto, Kanazawa-ku, Yokohama 236-0027, Japan
* Correspondence: yfujii@niu.ac.jp (Y.F.); mgerdol@units.it (M.G.); ozeki@yokohama-cu.ac.jp (Y.O.); Tel.: +81-956205642 (Y.F.); +39-0405588627 (M.G.); +81-457872221 (Y.O.)

Citation: Fujii, Y.; Gerdol, M.; Ozeki, Y. Coming New Age of Marine Glycomics: The Fundamental, Medical, and Ecological Aspects. *Mar. Drugs* **2022**, *20*, 613. https://doi.org/10.3390/md20100613

Received: 12 September 2022
Accepted: 21 September 2022
Published: 28 September 2022

This Special Issue "Marine Glycomics" (https://www.mdpi.com/journal/marinedrugs/special_issues/Marine_Glycomics, accessed on 12 September 2022) provided new approaches and information on bioactive compounds, such as glycans and lectins from marine animals, seaweeds, and microorganisms for the application of clinical therapy and elucidation of the physiological functions of marine organisms.

Marine glycomics is a discipline which concerns the study of glycans and their binding partners, i.e., lectins, in marine species. First, we would like to introduce the articles and reviews related to glycans that were published in this Special Issue.

A review by Fonseca and Mourao discusses the pharmacological activity of sulfated fucose-rich polysaccharides, which can be administered orally [1]. Many reports of polysaccharides' pharmacological effects were examined using experimental animal models. These polysaccharides are available from echinoderms (sea urchins and sea cucumbers) and algae. This review describes several functions of polysaccharides, such as their antilipidemic, anticancer, and immunomodulatory activities, as well as their effects on diabetes, thrombosis, and hemostasis. Finally, this review summarizes the potential of oral treatment of various diseases by utilizing sulfated polysaccharides.

Lin, Li, Gao et al. introduce the anticoagulation function of fucosylated glycosaminoglycan (FG) derived from two species of sea cucumber, *Thelenota bananas* and *Holothuria fuscopunctata* [2]. Both high and low (depolymerized) molecular weight of FG showed anticoagulation properties. Nevertheless, the injection of high molecular weight FG caused the death of rats due to the induction of hypotension through the activation of rat plasma kallikrein and an increase in the level of the vasoactive peptide bradykinin. On the other hand, low molecular weight FG did not show any side effects. The authors argue that depolymerized FG will be able to specifically target iXase, allowing its use as a safe anticoagulant compared to native FG.

A review by Besednova, Zaporozhets, Andryukov et al. discusses the antiparasitic activity of sulfated polysaccharides (SPS) obtained from marine algae and invertebrates [3]. SPS has potential for the treatment and prevention of infections caused by several types of protozoa, as well as the prevention of helminthiasis. The combination treatment of silver nanoparticles and fucoidans from brown algae showed an antitrypanosomal effect. Some SPS, such as fucoidans or their derivatives, have also reported trypanocidal activity. Cryptosporidiosis and trichomoniasis are also potential therapeutic targets of SPS. These findings will not only contribute to the design of novel SPS-based drugs, but also to the development of biologically active food additives and functional food products with antiparasitic activity.

A review by Qi, Zhou, Shen et al. discusses the considerable potential of bioactive compounds, including peptides and polysaccharides derived from peanut worms (Annelida, Sipuncula), which may find application both in food and pharmaceutical industry [4]. Peanut worms (e.g., *Sipunculus nudus*) are used in traditional Chinese medicine for the treatment of different medical conditions, such as hypertension, neurosis, and cough, and for many other purposes. The 162 known species of peanut worms, all living in shallow water, have a significant potential for the discovery of novel bioactive compounds. Peanut worm-based molecules endowed with antioxidant, anti-inflammatory, antihypertensive, anticancer, and wound healing activities have already been reported. Based on these observations, the authors discuss the great potential that peanut worms might have for the discovery of new peptides and polysaccharides with antidiabetic, antimicrobial, and many other biologically relevant properties.

Ren, Gao, Dai et al. report the bioactivity of light-yellow pigmented bacterial strains, which were newly isolated from the strongly toxic and dinoflagellate *Alexandrium catenella*, associated with harmful algal blooms [5]. Taxonomic and phenotypic analyses allowed the identification of this bacteria as a *Mameliella alba* strain. Nevertheless, the authors found an extraordinary production of bioflocculanting exopolysaccharides. Interestingly, the portion of exopolysaccharides containing glucose and fucose positively contributed to this flocculant ability. These findings revealed the potential pharmaceutical, environmental, and biotechnological implications of active EPS produced by this new *Mameliella alba* strain. It also highlighted the importance of studying bacteria associated with harmful algae due to the presence of unique compounds that still remain to be uncovered.

Next, we would like to introduce advanced findings published in this species issue related to the identification and functional characterization of marine lectins.

Gerdol provides a first insight into the diversity of secretory glycan-binding lectin gene from Rotifera, a phylum of small aquatic animals belonging to Lophotrochozoa [6]. While a high number of lectins have been described in lophotrochozoans, no information was previously available about rotifer lectins. The author identified secretory lectin-like sequences from the genomes of from six rotifer species, *Brachionus calyciflorus*, *Brachionus plicatilis*, *Proales similis* (class Monogononta), *Adineta ricciae*, *Didymodactylos carnosus*, and *Rotaria sordida* (class Bdelloidea). At least nine lectin families were identified, including fibrinogen-related domain-containing proteins (FreDs), C-type lectins, C1q domain-containing proteins, galectins, R-type lectins, F-type lectins, SUEL-type lectins, H-type lectins, and jacalin-type lectins. Each of the lectin families is characterized by a different structural fold. Interestingly, several unique features were identified in rotifer lectins, supporting the importance of expanding the range of organisms targeted by lectins research in order to elucidate their role in the future.

Takeuchi, Jimbo, Tanimoto et al. describe the symbiosis between the coral *Acropora tenuis* and Symbiodiniaceae cells, discussing the involvement of a coral lectin in this interaction [7]. The coral N-acetyl-d-glucosamine-binding lectin ActL plays a fundamental role in attracting the Symbiodiniaceae cells to establish symbiosis. ActL is a unique coral lectin because it does not attract all Symbiodiniaceae cells, but selectively allows an interaction with only some species, including *Symbiodinium tridacnidorum*, *Symbiodinium tridacnidorum*, *Durusdinium trenchii*, and *Breviolum* sp. The finding that ActL is a main contributor in the acquisition of Symbiodiniaceae cell in the coral *A. tenuis* is valuable, since it supports the role of coral lectins in regulating the attraction and phagocytosis of Symbiodiniaceae cells. This exciting model, supported by strong experimental evidence, encourages us to keep investigating the physiological endogenous role played by lectins in marine organisms.

Wang, Zhou, Liu et al. demonstrate the construction of a oncoVV-WCL recombinant virus, obtained though the insertion of the white-spotted charr lectin (WCL) gene into an oncolytic vaccinia virus (oncoVV) vector [8]. WCL is a rhamnose-binding lectin belonging to the SUEL-type lectin family, identified from white-spotted Charr (*Salvelinus leucomaenis*) eggs based on its primary structure. The viewpoint of this study is quite new, because it is the first to report the combination between a rhamnose-binding lectin and an oncoVV vector.

OncoVV-WCL showed an anticancer effect against hepatocellular carcinoma Huh-7cells through activation of caspase-9 and -3. In addition, oncoVV-WCL stimulates the production of IFN-alpha and beta in Huh-7 cells. The anticancer activity of oncoVV-WCL was further demonstrated by inoculating Huh-7 cells into hepatocellular carcinoma tumor-bearing Balb/c nude mice to investigate its function in vivo.

Swarna, Asaduzzaman, Kabir et al. elucidated the diverse immune functions of the sea hare *Aplysia kurodai* egg lectin, AKL-40 [9]. AKL-40 showed cytotoxic activity against human erythroleukemia cells through the activation of signal transduction molecules, but not in human B-lymphoma cells or rat basophilic leukemia cells. The cytotoxic effect was also demonstrated though an in vivo study. AKL-40 was administered to Ehrlich ascites carcinoma cells injected into Swiss albino mice, inducing the reduction of cell viability and cell aggregation. AKL-40 also showed strong antifungal activity against *Talaromyces verruculosus* and antibacterial activity against some bacteria, including *Staphylococcus aureus*. The present study suggests that AKL-40 may contribute to the embryo defense of sea hares.

Grinchenko, Kriegsheim, Shved et al., have reported the biochemical properties of the novel C1q domain containing protein, MkC1qDC, from the hemolymph of the bivalve mollusk *Modiolus kurilensis* [10]. MkC1qDC showed a lectin-like activity by binding galactose and mannose with a requirement of Ca2+. Moreover, it showed antibacterial activity by aggregating both gram positive and negative strains. Among the many compounds tested, alginate, κ-carrageenan, fucoidan, and pectin showed the highest inhibition of MkC1qDC activity. Interestingly, this study implies the potential of biomedical application of MkC1qDC by demonstrating the anticancer effect it exerts on human adenocarcinoma HeLa cells in a dose dependent manner.

Andreeva, Budenkova, Babich et al. report unique and original research which supports the use of carbohydrate additives to increase the biomass and carbohydrate content of cultured microalgae [11]. This finding is highly relevant, since it demonstrates that the nutrient medium can stimulate carbohydrate accumulation in microalgae utilized for biofuel production. In this study, a nutrient medium containing carbohydrate additives was supplied to several species of microalgae, including *Arthrospira platensis*, *Chlorella vulgaris*, and *Dunaliella salina*. The authors have already reported stirring during cultivation as a key factor in reducing carbohydrate production in microalgae. This study provides novel information for improving the culture of microalgae with high carbohydrate content, which may be used for biofuel production.

We are delighted to share these new publications with the scientific community, since we have taken all the necessary steps to ensure their high quality and novelty. We hope that many scientists will enjoy these articles and contribute to start the collaboration with these authors and experts in other scientific areas to develop research fields. Fortunately, the efforts of this Special Issue will be followed by a new Special Issue concerning the same themes, entitled "Marine Glycomics 2.0" (https://www.mdpi.com/journal/marinedrugs/special_issues/marine_glycomics_2, accessed on 12 September 2022).

We invite you to consider submitting manuscripts focused on your recent work for our consideration for a possible publication in this new, upcoming Special Issue. Yuki Fujii, Marco Gerdol, Yasuhiro Ozeki.

Funding: This research received no external funding.

Conflicts of Interest: The authors declare no conflict of interest.

References

1. Fonseca, R.J.C.; Mourão, P.A.S. Pharmacological activities of sulfated fucose-rich polysaccharides after oral administration: Perspectives for the development of new carbohydrate-based drugs. *Mar. Drugs* **2021**, *19*, 425. [CrossRef] [PubMed]
2. Lin, L.; Li, S.; Gao, N.; Wang, W.; Zhang, T.; Yant, L.; Yang, X.; Luo, D.; Ji, X.; Zhao, J. The toxicology of native fucosylated glycosaminoglycans and the safety of their depolymerized products as anticoagulants. *Mar. Drugs* **2021**, *19*, 487. [CrossRef] [PubMed]

3. Besednova, N.N.; Zaporozhets, T.S.; Andryukov, B.G.; Kryzhanovsky, S.P.; Ermakova, S.P.; Kuznetsova, T.A.; Voronova, A.N.; Shchelkanov, M.Y. Antiparasitic effects of sulfated polysaccharides from marine hydrobionts. *Mar. Drugs* **2021**, *19*, 637. [CrossRef] [PubMed]
4. Qi, Y.; Zhou, J.; Shen, X.; Chalamaiah, M.; Lv, S.; Luo, H.; Chen, L. Bioactive properties of peptides and polysaccharides derived from peanut worms: A review. *Mar. Drugs* **2022**, *20*, 10. [CrossRef] [PubMed]
5. Ren, C.-Z.; Gao, H.-M.; Dai, J.; Zhu, W.-Z.; Xu, F.-F.; Ye, Y.; Zhang, X.-L.; Yang, Q. Taxonomic and bioactivity characterizations of Mameliella alba strain LZ-28 isolated from highly toxic marine dinoflagellate *Alexandrium catenella* LZT09. *Mar. Drugs* **2022**, *20*, 321. [CrossRef] [PubMed]
6. Gerdol, M. First insights into the repertoire of secretory lectins in rotifers. *Mar. Drugs* **2022**, *20*, 130. [CrossRef] [PubMed]
7. Takeuchi, R.; Jimbo, M.; Tanimoto, F.; Iijima, M.; Yamashita, H.; Suzuki, G.; Harii, S.; Nakano, Y.; Yasumoto, K.; Watabe, S. N-Acetyl-d-glucosamine-binding lectin in *Acropora tenuis* attracts specific Symbiodiniaceae cell culture strains. *Mar. Drugs* **2021**, *19*, 146. [CrossRef] [PubMed]
8. Wang, X.; Zhou, N.; Liu, T.; Jia, X.; Ye, T.; Chen, K.; Li, G. Oncolytic vaccinia virus expressing white-spotted charr lectin regulates antiviral response in tumor cells and inhibits tumor growth in vitro and in vivo. *Mar. Drugs* **2021**, *19*, 292. [CrossRef] [PubMed]
9. Swarna, R.R.; Asaduzzaman, A.K.M.; Kabir, S.R.; Arfin, N.; Kawsar, S.M.A.; Rajia, S.; Fujii, Y.; Ogawa, Y.; Hirashima, K.; Kobayashi, N.; et al. Antiproliferative and antimicrobial potentials of a lectin from *Aplysia kurodai* (sea hare) eggs. *Mar. Drugs* **2021**, *19*, 394. [CrossRef] [PubMed]
10. Grinchenko, A.V.; von Kriegsheim, A.; Shved, N.A.; Egorova, A.E.; Ilyaskina, D.V.; Karp, T.D.; Goncharov, N.V.; Irina, Y.; Petrova, I.Y.; Kumeiko, V.V. A novel C1q domain-containing protein isolated from the mollusk *Modiolus kurilensis* recognizing glycans enriched with acidic galactans and mannans. *Mar. Drugs* **2021**, *19*, 668. [CrossRef] [PubMed]
11. Andreeva, A.; Budenkova, E.; Babich, O.; Sukhikh, S.; Dolganyuk, V.; Michaud, P.; Ivanova, S. Influence of carbohydrate additives on the growth rate of microalgae biomass with an increased carbohydrate content. *Mar. Drugs* **2021**, *19*, 381. [CrossRef] [PubMed]

marine drugs

MDPI

Article

N-Acetyl-D-Glucosamine-Binding Lectin in *Acropora tenuis* Attracts Specific Symbiodiniaceae Cell Culture Strains

Ryota Takeuchi [1,†], Mitsuru Jimbo [1,*], Fumika Tanimoto [1], Mariko Iijima [1,2], Hiroshi Yamashita [3], Go Suzuki [3], Saki Harii [4], Yoshikatsu Nakano [5], Ko Yasumoto [1] and Shugo Watabe [1]

1 School of Marine Biosciences, Kitasato University, 1-15-1 Kitasato, Minami, Sagamihara, Kanagawa 252-0373, Japan; r-takeuchi@aist.go.jp (R.T.); pooh.q-.-p.92@docomo.ne.jp (F.T.); m.iijima@aist.go.jp (M.I.); yasumoto@kitasato-u.ac.jp (K.Y.); swatabe@kitasato-u.ac.jp (S.W.)
2 Geological Survey of Japan, National Institute of Advanced Industrial Science and Technology (AIST), 1-1-1 Higashi, Tsukuba, Ibaraki 305-8567, Japan
3 Fisheries Technology Institute, Japan Fisheries Research and Education Agency, 148 Fukai-Ohta, Ishigaki, Okinawa 907-0415, Japan; hyamashita@fra.affrc.go.jp (H.Y.); gosuzu@fra.affrc.go.jp (G.S.)
4 Sesoko Station, Tropical Biosphere Research Center, University of the Ryukyus, 3422 Sesoko, Motobu, Okinawa 905-0227, Japan; sharii@lab.u-ryukyu.ac.jp
5 Research Support Division, Okinawa Institute of Science and Technology Graduate University, 1919-1 Tancha, Onna-son, Okinawa 904-0412, Japan; yoshikatsu.nakano2@oist.jp
* Correspondence: mjinbo@kitasato-u.ac.jp; Fax: +81-42-778-8007
† Present address: AIST-Osaka University Advanced Photonics and Biosensing Open Innovation Laboratory, AIST, Photonics Center, Osaka University P3, 2-1 Yamada-Oka, Suita, Osaka 565-0871, Japan.

Abstract: Many corals establish symbiosis with Symbiodiniaceae cells from surrounding environments, but very few Symbiodiniaceae cells exist in the water column. Given that the *N*-acetyl-D-glucosamine-binding lectin ActL attracts Symbiodiniaceae cells, we hypothesized that corals must attract Symbiodiniaceae cells using ActL to acquire them. Anti-ActL antibody inhibited acquisition of Symbiodiniaceae cells, and rearing seawater for juvenile *Acropora tenuis* contained ActL, suggesting that juvenile *A. tenuis* discharge ActL to attract these cells. Among eight Symbiodiniaceae cultured strains, ActL attracted NBRC102920 (*Symbiodinium tridacnidorum*) most strongly followed by CS-161 (*Symbiodinium tridacnidorum*), CCMP2556 (*Durusdinium trenchii*), and CCMP1633 (*Breviolum* sp.); however, it did not attract GTP-A6-Sy (*Symbiodinium natans*), CCMP421 (*Effrenium voratum*), FKM0207 (*Fugacium* sp.), and CS-156 (*Fugacium* sp.). Juvenile polyps of *A. tenuis* acquired limited Symbiodiniaceae cell strains, and the number of acquired Symbiodiniaceae cells in a polyp also differed from each other. The number of Symbiodiniaceae cells acquired by juvenile polyps of *A. tenuis* was correlated with the ActL chemotactic activity. Thus, ActL could be used to attract select Symbiodiniaceae cells and help Symbiodiniaceae cell acquisition in juvenile polyps of *A. tenuis*, facilitating establishment of symbiosis between *A. tenuis* and Symbiodiniaceae cells.

Keywords: *Acropora tenuis*; coral; chemoattraction; lectin

Citation: Takeuchi, R.; Jimbo, M.; Tanimoto, F.; Iijima, M.; Yamashita, H.; Suzuki, G.; Harii, S.; Nakano, Y.; Yasumoto, K.; Watabe, S. *N*-Acetyl-D-Glucosamine-Binding Lectin in *Acropora tenuis* Attracts Specific Symbiodiniaceae Cell Culture Strains. *Mar. Drugs* **2021**, *19*, 146. https://doi.org/10.3390/md19030146

Academic Editors: Yuki Fujii, Yasuhiro Ozeki and Marco Gerdol

Received: 12 February 2021
Accepted: 9 March 2021
Published: 11 March 2021

1. Introduction

Most reef-building corals establish symbiosis with symbiotic dinoflagellates, which belong to the family of Symbiodiniaceae cells. Symbiodiniaceae cells provide photosynthetic products, which enables corals to effloresce in sub-tropical and tropical regions. When corals lose symbiont cells or photosynthetic pigments of these cells, this event, bleaching, results in coral death [1,2]. Therefore, it is important for corals to symbiose with Symbiodiniaceae cells.

Symbiodiniaceae cells are divided into nine phylogenetically distinct genetic groups (clades A–I) [3,4], and recently, these clades have been proposed as different genera [5]. Each species of coral was reported to have maintained mainly specific genera of Symbiodiniaceae cells [6]. It has been reported that juvenile polyps of *Acropora* sp. acquired *Symbiodinium* and *Durusdinium*, although these genera were rare in the water column [7]. The number of Symbiodiniaceae cells acquired from juvenile polyps of *Acropora tenuis* differed according to the genus of Symbiodiniaceae cells [8,9], suggesting that some selection mechanism exists. Recently, Yamashita et al. reported that juvenile *A. tenuis* attracts some of the genera of Symbiodiniaceae cells during the initial stage of symbiosis [10]. Based on these results, they suggested that *Acropora* corals acquire a few specific genera of Symbiodiniaceae cells at the initial attraction step, followed by subsequent selective uptake by the coral.

Although the mechanism of selective acquisition of Symbiodiniaceae cells is not apparent, some factors for selection have been proposed. It has been reported that artificial green-fluorescent objects under blue LED light attracted free-living Symbiodiniaceae cells [11,12]. On the other hand, Fitt (1981) reported that cnidarians *Aiptasia* sp. and *Cassiopeia xamachana* attracted Symbiodiniaceae cells using ammonium derivatives [13]. In contrast, the planula larvae of *Lobactis scutaria* (referred to as *Fungia scutaria*) attracted these cells by using trehalose, which is also a low-molecular-weight compound [14]. Green fluorescent protein, ammonium derivatives, and trehalose are common substances for corals, and it seems difficult to explain the strict specificity of host–symbiont relationships by these compounds. Recently, Biquand et al. reported that smaller sizes of Symbiodiniaceae cells tend to be acquired by corals more readily [15], but this selection is the last step of Symbiodiniaceae cell acquisition. In the case of *A. tenuis*, it could not explain the acquisition of clades A and D in the larval stage, because they are rare genera in the water column [7].

Nowadays, sugar-binding proteins, lectins, of corals have attracted attention as candidates for the acquisition of Symbiodiniaceae cells. Lectins are involved in symbiosis as well as immunity, development, and differentiation [16–18]. In the case of the Nematode *Laxus oneistus*, a C-type lectin, Mermaid, maintained the symbiotic bacteria on their surface, and the subtle differences in their amino acid sequences resulted in the binding of different symbiotic bacteria [19]. In a solitary coral *Lobactis scutaria*, exogenous lectins and some carbohydrates inhibited acquisition of Symbiodiniaceae cells, suggesting that lectins/glycans are involved in the acquisition of Symbiodiniaceae cells in corals. Coral lectins are found in many corals [9,10,16,20], and some of them are involved in coral–Symbiodiniaceae symbiosis. A C-type lectin, Millectin, which was purified from a coral, *Acropora millepora*, bound to bacterial pathogens and Symbiodiniaceae cells [21], and was localized around symbiont cells in the host cells and in nematocysts [22], suggesting that Millectin is involved in immunity and symbiosis. Moreover, a lectin gene, *Pdc-Lectin*, found in the coral *Pocillopora damicornis*, was downregulated six days before a bleaching event [23]. These findings suggest the importance of lectins in symbiosis, but their function has not been examined.

Recently, we examined the function of lectins purified from the hard coral *A. tenuis*. *A. tenuis* is a model organism for examining the acquisition of Symbiodiniaceae cells because the planulae of this coral can survive for over a month in glass bottles [24], and they metamorphose easily by treatment with a hydra neuropeptide, Hym-248 [25]. Moreover, metamorphosed juvenile *A. tenuis* polyps acquired particular Symbiodiniaceae cell culture strains containing NBRC102920 (*Symbiodinium tridacnidorum*) [9,10]. We purified two lectins, AtTL-2 and ActL, from *A. tenuis* [10,26]. AtTL-2 is an *N*-acetyl galactosamine (GalNAc)/N-acetyl glucosamine (GlcNAc)-binding lectin and is similar to Tachylectin-2, which was isolated from a horseshoe crab, *Tachypleus tridentatus*. The lectin antibody inhibited the acquisition of the Symbiodiniaceae strain NBRC102920 [10]. Another lectin, ActL, was purified, and its N-terminal amino acid sequence was found to differ from that of AtTL-2 [26]. ActL binds to *N*-acetyl-D-glucosamine (GlcNAc) and shows chemotactic activity toward the Symbiodiniaceae strain NBRC102920. Moreover, GlcNAc inhibits the chemotactic activity of ActL, suggesting that ActL attracts NBRC102920 by binding to Symbiodiniaceae cell-surface carbohydrates. Since cell-surface carbohydrates differ among Symbiodiniaceae strains [27], it is feasible that ActL differentially binds to specific Symbiodiniaceae strains to attract them toward the corals. This may lead to the acquisition of specific Symbiodiniaceae cells by corals. In the present study, we examined whether GlcNAc-binding lectin ActL attracts specific Symbiodiniaceae strains for *A. tenuis* to acquire them.

2. Results

2.1. Carbohydrate Specificity of ActL

The sugar specificity of ActL was examined using rabbit erythrocytes. The hemagglutination activity of ActL was inhibited by GlcNAc and *N*-acetyl-D-neuramic acid (NANA), and the minimum inhibitory concentrations of GlcNAc and NANA were 1.56 mM and 3.13 mM, respectively.

2.2. Participation of ActL in Acquisition of Symbiodiniaceae Cells by Juvenile A. tenuis Polyps

Previously, we showed that the chemotaxis of Symbiodiniaceae cells to crude extract of *A. tenuis* was inhibited by GlcNAc [26]. To elucidate whether Symbiodiniaceae cell acquisition by polyps was also inhibited by GlcNAc, polyps were incubated with Symbiodiniaceae NBRC102920 cells in the presence of sugars, each at 20 mM concentration (Figure 1a). Symbiodiniaceae cell acquisition was significantly inhibited by GlcNAc (31.0 ± 13.8 cells/polyp, $p < 0.05$) and NANA (6.3 ± 2.2 cells/polyp, $p < 0.001$) compared with that in the absence of sugar (60.0 ± 22.1 cells/polyp) (Figure 1a), suggesting that ActL participates in the Symbiodiniaceae cell acquisition.

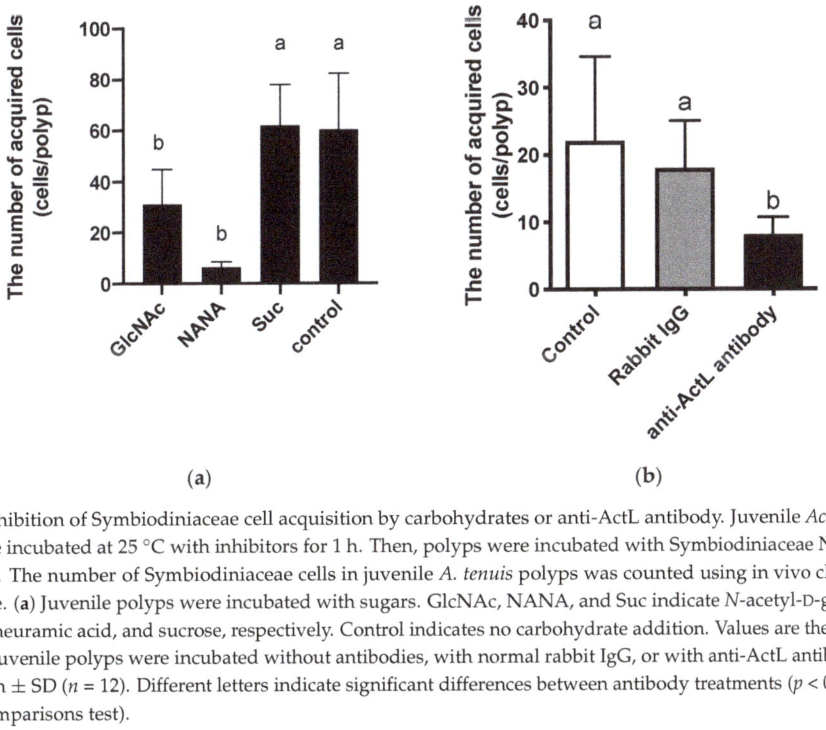

Figure 1. Inhibition of Symbiodiniaceae cell acquisition by carbohydrates or anti-ActL antibody. Juvenile *Acropora tenuis* polyps were incubated at 25 °C with inhibitors for 1 h. Then, polyps were incubated with Symbiodiniaceae NBRC102920 cells for 6 h. The number of Symbiodiniaceae cells in juvenile *A. tenuis* polyps was counted using in vivo chlorophyll a fluorescence. (**a**) Juvenile polyps were incubated with sugars. GlcNAc, NANA, and Suc indicate *N*-acetyl-D-glucosamine, *N*-acetyl-D-neuramic acid, and sucrose, respectively. Control indicates no carbohydrate addition. Values are the mean ± SD ($n = 4$). (**b**) Juvenile polyps were incubated without antibodies, with normal rabbit IgG, or with anti-ActL antibody. Values are the mean ± SD ($n = 12$). Different letters indicate significant differences between antibody treatments ($p < 0.05$, Tukey's multiple comparisons test).

Anti-ActL antibody treatment selectively abolished the hemagglutination activity of ActL against rabbit erythrocytes (data not shown). Using this antibody, we examined the participation of ActL on Symbiodiniaceae NBRC102920 cell acquisition. The number of cells acquired by polyps in the presence of the anti-ActL antibody (9.3 ± 5.0 cells/polyp, $p < 0.05$, Tukey's multiple comparisons test) was significantly reduced compared to that without the antibody (26.3 ± 17.3 cells/polyp) (Figure 1b). These results suggest that ActL participates in Symbiodiniaceae cell acquisition by polyps.

2.3. Detection of ActL from A. tenuis Rearing Seawater

Using the anti-ActL antibody, we examined by dot blotting, whether ActL was discharged from juvenile *A. tenuis* polyps. Anti-ActL antibody reacted with rearing seawater of juvenile *A. tenuis* polyps (Figure 2a). However, the antibody could also bind to antigens other than ActL, since it was purified using Protein A affinity chromatography. When the anti-ActL antibody was preincubated with ActL and this reaction mixture was used instead of the purified antibody in the experiment, it did not react with rearing seawater (Figure 2b). This indicates that the rearing seawater contained ActL, and juvenile *A. tenuis* polyps discharged ActL outside the body.

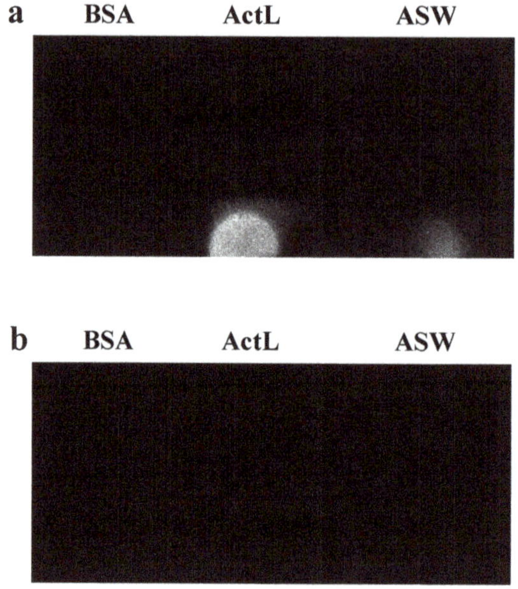

Figure 2. Examination of the occurrence of ActL in rearing seawater of juvenile *A. tenuis* by dot blotting assay. Samples were spotted on the PVDF membrane and detected using the anti-ActL antibody. Samples were 0.1 mg of bovine serum albumin (BSA), 0.1 mg of ActL, and artificial seawater (ASW), where 10 juvenile polyps were kept for 1 d. Binding of the anti-ActL antibody to purified ActL and rearing ASW of juvenile *A. tenuis* polyps was examined. (**a**) Each sample solution dot blotted onto a PVDF membrane. BSA was used as a negative control. Purified ActL was used as a positive control. (**b**) Anti-ActL antibody was preincubated with ActL, and the mixture was used for dot blotting assay instead of the purified anti-ActL antibody.

2.4. Chemotactic Activity of ActL to Each Symbiodiniaceae Strain

It is not known whether and how Symbiodiniaceae cells move toward ActL, although the chemotactic activity assay showed that ActL attracts Symbiodiniaceae NBRC102920 cells [26]. To clarify the chemotactic activity of ActL in detail, the movements of Symbiodiniaceae cells toward ActL were recorded (Figure 3a). The Symbiodiniaceae cells around the capillary moved toward the capillary, whereas the cells distant from the capillary moved linearly in different directions. Since Symbiodiniaceae NBRC102920 cells typically showed circular movement at the same point, this result indicated that ActL attracts Symbiodiniaceae cells.

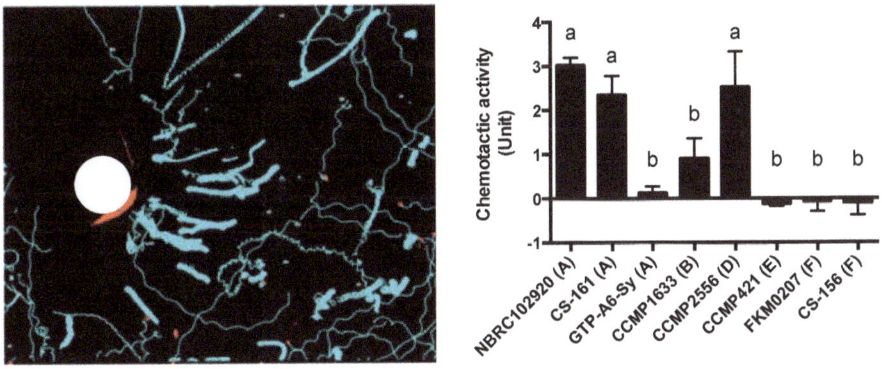

(a) (b)

Figure 3. The chemotactic activity of various Symbiodiniaceae strains. (**a**) A capillary containing ActL was placed in NBRC102920 culture in a 35 mm dish. The movements of the Symbiodiniaceae cells were video-recorded. White circle indicates the insertion point of the capillary, red indicates Symbiodiniaceae cells, and cyan lines indicate traces of Symbiodiniaceae cells. (**b**) The chemotactic assay was conducted for 1 h using Symbiodiniaceae cells at a concentration of 1×10^6 cells/mL. One unit (U) of chemotactic activity was defined as the activity at which 40 Symbiodiniaceae cells were attracted by sample solutions in 60 min. Values are the mean ± SD ($n = 3$). Different letters indicate a significant difference in chemotactic activity ($p < 0.05$, Tukey's multiple comparisons test). The letters in parentheses indicate genera of Symbiodiniaceae; A = genus *Symbiodinium*, B = genus *Breviolum*, D = genus *Durusdinium*, E = genus *Effrenium*, and F = genus *Fugacium*.

Next, we measured the chemotactic activity of ActL toward each Symbiodiniaceae strain using chemotactic activity assay [26]. The chemotactic activity was greatest for strain NBRC102920 (3.0 ± 0.2 U), CCMP2556 (2.5 ± 0.8 U), and CS-161 (2.3 ± 0.4 U), followed by CCMP1633 (0.9 ± 0.4 U), GTP-A6-Sy (0.1 ± 0.2 U), CCMP421 (−0.1 ± 0.1 U), FKM0207 (−0.1 ± 0.2 U), and CS-156 (−0.1 ± 0.2 U) (Figure 3). When the chemotactic activity of ActL decreased, the cell movement toward the capillary also decreased (data not shown). This result indicates that the chemotactic activity of ActL varies depending on the Symbiodiniaceae strains. Interestingly, the chemotactic activity is different even in the same genus, but different species.

2.5. The Number of Acquired Symbiodiniaceae Cells by Juvenile *A. tenuis* Polyps

The number of acquired Symbiodiniaceae cells was the largest for NBRC1029020 (22.8 ± 9.9 cells/polyp) and CCMP2556 (17.0 ± 11.9 cells/polyp), followed by CCMP1633 (8.9 ± 7.1 cells/polyp), CS-161 (4.7 ± 4.3 cells/polyp), GTP-A6-Sy (1.2 ± 2.9 cells/polyp), CS-156 (0.8 ± 2.5 cells/polyp), CCMP421 (0.2 ± 0.6 cells/polyp), and FKM0207 (0.0 ± 0.0 cells/polyp) (Figure 4).

Next, we explored the correlation between the number of acquired cells and chemotactic activity to examine whether chemoattraction by ActL participates in Symbiodiniaceae cell acquisition (Figure 5). The Pearson's correlation coefficient showed a positive correlation ($r = 0.87$). The point far from the regression line is a dataset of CS-161. When the point was removed, the Pearson's correlation coefficient decreased to 0.99.

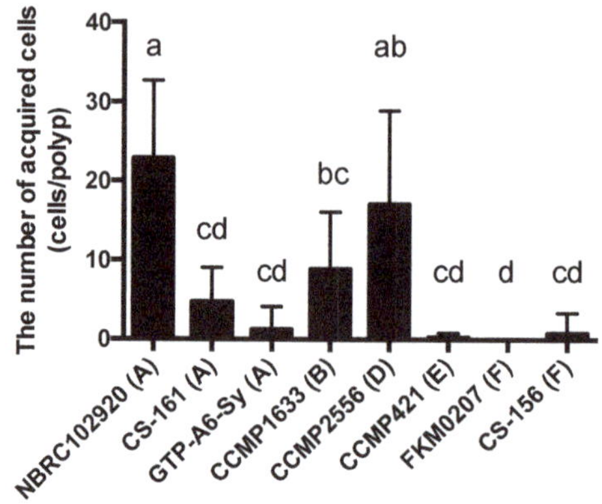

Figure 4. The number of acquired Symbiodiniaceae cells by juvenile *A. tenuis* polyps. Juvenile *Acropora tenuis* polyps were incubated with each Symbiodiniaceae cell strain for 24 h. Symbiodiniaceae cells acquired by juvenile *A. tenuis* polyps were counted using in vivo chlorophyll a fluorescence. Values are the mean ± SD (n = 10). Different letters indicate significant differences between Symbiodiniaceae strains ($p < 0.05$, Tukey's multiple comparisons test). The letters in parentheses indicate genera of Symbiodiniaceae; A = genus *Symbiodinium*, B = genus *Breviolum*, D = genus *Durusdinium*, E = genus *Effrenium*, and F = genus *Fugacium*.

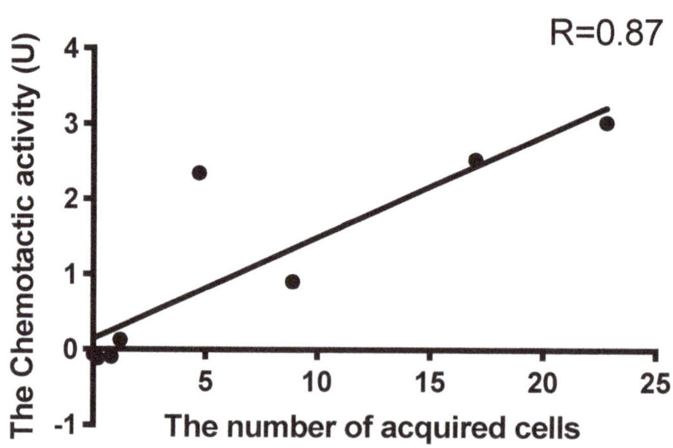

Figure 5. Correlation between the numbers of acquired cells and attracted cells. Pearson's correlation coefficient was calculated, resulting in a positive correlation (r = 0.87).

3. Discussion

Corals maintain particular Symbiodiniaceae according to the coral species, and many corals acquire Symbiodiniaceae cells from their surrounding environments. The acquisition process of Symbiodiniaceae cells could mainly be separated into the following two steps: (1) attraction of Symbiodiniaceae cells by corals and (2) phagocytosis of them. Many molecules should participate in this process, but few have been found. We previously purified a GlcNAc-binding lectin ActL from *A. tenuis* [26]. This lectin attracted the Symbiodiniaceae strain NBRC102920, but it was not obvious whether ActL was involved in the acquisition of Symbiodiniaceae cells by the *A. tenuis* polyps. In the present study, we found that the anti-ActL antibody inhibited the acquisition of the Symbiodiniaceae strain NBRC102920 (Figure 1). We also found that GlcNAc inhibited the chemoattraction of ActL [26]. Moreover, rearing artificial seawater (ASW) of *A. tenuis* contained ActL

(Figure 2). These results supported our hypothesis that juvenile polyps released ActL to acquire Symbiodiniaceae cells through chemotaxis.

In previous reports, some factors were found to participate in the attraction, including nitrogen-containing compounds such as nitrates and trehalose [13,14]. Many chemotactic compounds are low-molecular-weight substances, and they have the ability to diffuse rapidly and distantly. On the other hand, peptides and proteins, with their high molecular weights, cannot diffuse efficiently and seem not to be attractants. However, some proteins such as startrak have been shown to attract the sperm of starfish [27]. Thus, ActL could be an attractant for Symbiodiniaceae cells. It is possible that ActL could attract Symbiodiniaceae cells in regions close to polyps, since corals live in shallow coral reefs that are exposed to tidal currents and the attractant will be easily diluted. Recently, it has been shown that green fluorescence attracted Symbiodiniaceae cells [12]. Since light could reach a distant location and ActL could become attracted to within a very short range of the polyps, Symbiodiniaceae cells distant from polyps are likely to be attracted by light, and then attracted by ActL later in the process.

A few reports have described the selective attraction of Symbiodiniaceae cells. Blue light may be a factor of selective attraction (Yamashita et al., under review). In this study, Symbiodiniaceae strains were differentially attracted by ActL and *A. tenuis* juvenile polyps. When the number of Symbiodiniaceae strains attracted to ActL was plotted against the number of cells acquired by polyps (Figures 3 and 4), the number of attracted Symbiodiniaceae cells was positively correlated with the number of cells acquired by juvenile polyps (r = 0.87, Figure 5). These results indicate that ActL participates in the selection of Symbiodiniaceae strains during acquisition by *A. tenuis* through attraction, except for the Symbiodiniaceae strain CS-161.

CS-161 was highly attracted to ActL, but the number of cells acquired by juvenile polyps was low among the other strains tested (Figure 5). Symbiodiniaceae cell acquisition by juvenile polyps involves several steps, such as attraction and entry into the gastrodermal cells of polyps. Factors related to other steps also affect the number of Symbiodiniaceae cells acquired by juvenile polyps. Indeed, the cell size of Symbiodiniaceae affects the number of cells acquired by juvenile polyps [15]. The factors related to acquisition, except attraction, may inhibit the acquisition of CS-161 by polyps, and this may be a reason why the dataset of CS-161 is an outlier in Figure 5.

Since carbohydrates on the cell surface were different among each strain of Symbiodiniaceae cells [27], ActL could differentially bind to Symbiodiniaceae strains. It was reported that Symbiodiniaceae isolated from the soft coral *Plexaura kuna* had similar lectin binding specificity, although they belonged to different clades [28]. This means that the coral *P. kuna* selectively maintained specific Symbiodiniaceae according to the carbohydrate chains on its cell surface, regardless of clades. The fact that the corals maintained particular genera of Symbiodiniaceae leads to the hypothesis that the selection of Symbiodiniaceae cells by host corals is due to the binding of lectins with Symbiodiniaceae surface carbohydrates. Thus, ActL can attract particular Symbiodiniaceae genera to help polyps to acquire Symbiodiniaceae cells in *A. tenuis* and to maintain particular genera of Symbiodiniaceae. In addition, it is possible that coral colonies also select Symbiodiniaceae cells for maintenance.

On the other hand, Parkinson et al. reported that subtle differences in symbiont cell surface glycan did not explain species-specific colonization rate in the sea anemone, *Exaiptasia pallida* [29]. They showed that lectin masking did not inhibit Symbiodiniaceae cell acquisition by the host *E. pallida*. The paper mentioned that this resulted from the use of adult animals instead of larvae, and the different ages of the animals used. We used different cell densities of added Symbiodiniaceae, while the previously mentioned paper examined the acquisition assay at a very high density of Symbiodiniaceae cells (1×10^6 cells/mL), although Symbiodiniaceae cells at 2000 cells/mL were used in this study. Under high cell density conditions, *A. tenuis* larvae acquire even free-living-type Symbiodiniaceae; however, this was not achieved under low cell density conditions [10]. In our experience, when the density of Symbiodiniaceae cells is very high, Symbiodiniaceae cells will be acquired randomly by hosts (data not shown), and it seems difficult to examine whether Symbiodiniaceae cells will be selectively acquired by the host.

ActL attracted the strains that tend to be acquired; however, there are some conflicts. Although the Symbiodiniaceae strain CS-161 was attracted by ActL at the same level as CCMP2556, CS-161 was acquired less than CCMP2556 by polyps (Figure 5). It was reported that another lectin, *N*-acetyl-D-galactosamine (GalNAc), which is called AtTL-2, participated in *Symbiodinium* acquisition by *A. tenuis* polyps [9]. Since AtTL-2 did not show chemotactic activity in our capillary assay (data not shown), AtTL-2 might participate in other steps of Symbiodiniaceae cell acquisition except attraction, and the low number of acquisitions of CS-161 may be due to the lack of binding of CS-161 to AtTL-2. Light may also be another factor affecting attraction. This indicates that there are other variables that should be identified in the molecular mechanisms of Symbiodiniaceae cells in *A. tenuis*. In this study, we used purified ActL by GlcNAc-affinity chromatography, which is able to bind to AtTL-2. Purified ActL may contain a small amount of AtTL-2. In the future, we must examine whether AtTL-2 participates in chemoattraction and unveil the mechanism of selective chemoattraction by corals.

4. Materials and Methods

4.1. Materials

S. tridacnidorum-cultured strain NBRC102920 (clade A) [30], which was isolated from the giant clam *Tridacna maxima* in Palau, was obtained from the National Institute of Technology and Evaluation (Tokyo, Japan). Cultured strains GTP-A6-Sy (*Symbiodinium natans*) and FKM0207 (*Fugacium* sp.) were originally isolated by Yamashita and Koike (2013) [31]. Cultured strains of CCMP1633 (*Breviolum* sp.), CCMP2556 (*Durusdinium trenchii*), and CCMP421 (*Effrenium voratum*) were obtained from the National Center for Marine Algae and Microbiota (East Boothbay, ME, USA). Cultured strains of CS-161 (*S. tridacnidorum*) and CS-156 (*Fugacium* sp.) were obtained from the Commonwealth Scientific & Industrial Research Organization (Castray Esplanade, Tasmania, Australia). These strains were cultured in IMK medium for marine microalgae (FUJIJFILM Wako Pure Chemical Corporation, Osaka, Japan) at 25 °C with illumination at 80 µmol photons/m^2/s (12 h:12 h light/dark cycle; 08:00–20:00, light). Unless otherwise specified, all reagents were purchased from FUJIJFILM Wako Pure Chemical Corporation. *A. tenuis* specimens were collected from Sesoko Island, Okinawa, Japan. Collected *A. tenuis* were maintained in an aquarium for several days and then frozen at −70 °C in a freezer until use.

4.2. Preparation of ActL

ActL was prepared according to Takeuchi et al. [26]. Briefly, a piece of *A. tenuis* (typically 19 g) was extracted with three volumes of extraction solution (150 mM NaCl, 50 mM Tris–HCl (pH 8.5), and 10 mM $CaCl_2$). ActL was purified using a GlcNAc-binding Sepharose 6B column (1 mL) with 0.2 M GlcNAc in extraction solution. The ActL was dialyzed against 1000 volumes of IMK medium at 4 °C for more than 3 h. This process was repeated three times.

4.3. Hemagglutination Assay

Purified ActL (20 µL) was serially diluted two-fold using 50 mM Tris–HCl (pH 8.5) containing 150 mM NaCl and 10 mM $CaCl_2$ in a V-bottom 96-well microtiter plate. The well was supplemented with 4% rabbit erythrocyte suspension (20 µL) and incubated at 25 °C for 60 min. The titer of the maximum dilution showing positive agglutination was recorded, and the hemagglutination titer was defined as the reciprocal of the highest dilution. Hemagglutination units (HU) were calculated by multiplying the hemagglutination titer by the sample volume.

4.4. Carbohydrate Inhibition Test

Purified ActL was diluted to 16 HU using 50 mM Tris–HCl (pH 8.5) containing 150 mM NaCl and 10 mM $CaCl_2$. The diluted ActL (20 µL) was incubated for 1 h at 25 °C with diluted carbohydrates. Next, 4% rabbit erythrocyte suspension (20 µL) was added to the mixture, which was then incubated at 25 °C for 30 min, and hemagglutination was measured. The results are expressed as the minimum concentration of carbohydrates. The following carbohydrates were used: L-arabinose, L-fucose, D-ribose, deoxy-D-ribose, D-xylose, D-glucose, D-galactose, L-rhamnose, D-mannose, lactose, melibiose, maltose, D-glucosamine, D-galactosamine, D-mannosamine, *N*-acetyl-D-glucosamine, *N*-acetyl-D-galactosamine, and *N*-acetyl-D-neuramic acid.

4.5. Inhibition of Symbiodiniaceae Cell Acquisition by Carbohydrates and Anti-ActL Antibody

Juvenile polyps were prepared as described previously [9]. After 72 h, carbohydrates (GlcNAc, NANA, and sucrose) at 20 mM, rabbit IgG, or anti-ActL antibody (final concentration was 1 µg/mL) was added to ASW-containing juvenile *A. tenuis* polyps, which were then incubated at 25 °C for 1 h before the addition of 2000 NBRC102920 Symbiodiniaceae cells. After an additional incubation at 25 °C for 6 h, the number of Symbiodiniaceae cells within the juvenile *A. tenuis* polyps was counted based on the images obtained by confocal microscopy, as mentioned above.

4.6. Dot Blotting Using the Anti-ActL Antibody

Eight wells in a chamber cover grass were filled with 500 µL ASW, and 10 juvenile polyps were kept for 1 d in each of the eight wells. The ASW of 8 wells was then collected, and 30 mL of ice-cold acetone was added and the samples were kept overnight at −35 °C. The sample was centrifuged at 15,000× *g* for 15 min at 0 °C, the supernatant was removed, and the pellet was resuspended in 4 mL of ultrapure water. The solution was subjected to dialysis against 3000 mL ultrapure water at 4 °C for 3 h. This dialysis was repeated three times and the rearing ASW concentrate was obtained. 50 µL of the rearing ASW concentrate and ActL (0.1 mg/mL) were applied to a polyvinylidene difluoride membrane FluoroTrans® W (Pall, Port Washington, NY, USA) using a Bio-Dot apparatus (Bio-Rad Laboratories). After being washed with phosphate-buffered saline (PBS), the membrane was set to a SNAP i.d. 2.0 (Merck Millipore, Bedford, MA, USA), and dot blotting was carried out according to the manufacturer's protocol with some modifications. The membrane was then blocked with Blocking One (Nacalai Tesque, Kyoto, Japan). Next, it was soaked in 100 µg/mL of anti-ActL antibody diluted using Can Get Signal Solution 1 (TOYOBO, Osaka, Japan) for 10 min and washed with 0.1% Tween-20 in PBS (PBS-T). After that, the membrane was soaked in secondary antibody horseradish peroxidase (HRP)-labeled anti-rabbit IgG antibody (FUJIJFILM Wako Pure Chemical Corporation) in Can Get Signal Solution 2 (TOYOBO) for 10 min. After being washed with PBS-T, the membrane was then transferred onto a wrap and soaked in Luminata Forte Western HRP Substrate (MilliporeSigma, Brington, MA, USA) for 5 min, and the signal was detected using Ez-Capture MG (Atto, Tokyo, Japan). When performing the antibody adsorption test, 1 µg/mL of anti-ActL antibody was incubated with 100 µg/mL of ActL for 16 h. This mixture was then used instead of the anti-ActL antibody. Bovine serum albumin (0.1 mg/mL) was used as the negative control.

4.7. Chemotactic Activity Assay

The movements of Symbiodiniaceae cells toward ActL were examined as follows. IMK (2 mL) was added to a 35 mm dish, and a microcapillary containing 2 µL of ActL (100 µg/mL) in IMK medium was added to this dish. After addition of NBRC102920 (1×10^4 cells/mL), the motion around the capillary was video-recorded for 30 s. Symbiodiniaceae cell movements were traced using Particle Track, and analysis (https://github.com/arayoshipta/projectPTAj (accessed on 11 March 2021)).

The chemotactic assay was conducted according to Takeuchi et al. [26]. Briefly, 1.0×10^5 of Symbiodiniaceae cells containing the IMK medium were added to 1.5 mL

PROKEEP low protein binding tubes (Fukae Kasei, Kobe, Japan) and subjected to centrifugation at 860× *g* for 5 min at 25 °C. After removing the supernatant, 100 μL of the IMK medium was added to the pellet. The tubes containing pellets were allowed to stand for 24 h to hasten the recovery of Symbiodiniaceae motility, and then the capillaries (Capillary Calibrated Pipettes, Drummond Scientific Company, Broomall, PA, USA) containing 2 μL of sample solutions were inserted into the tubes containing the cells. After 60 min, the capillaries were removed from the tubes, and the solution around the capillary surface was carefully wiped off to remove attached Symbiodiniaceae cells. The solution in the capillaries was blown by mouth into a hemocytometer, and Symbiodiniaceae cells were counted under a microscope. The number of attracted cells was quantified by subtracting the number of Symbiodiniaceae cells in a capillary containing the IMK medium from that containing ActL. The chemotactic activity was calculated according to a standard curve of the number of attracted cells vs. protein concentrations of crude *A. tenuis* extract [26]. One unit (U) of chemotactic activity was defined as the activity at which 40 Symbiodiniaceae cells were attracted by sample solutions in 60 min. Each experiment was performed in triplicate.

4.8. Statistical Analysis

All data were analyzed using GraphPad Prism 6.0 for Macintosh (GraphPad Software, La Jolla, CA, USA). Results were analyzed by a one-way repeated-measures ANOVA followed by a Tukey's multiple comparisons test ($p < 0.05$).

Author Contributions: Conceptualization, R.T. and M.J.; methodology, R.T.; validation, R.T. and M.J.; formal analysis, R.T.; investigation, R.T.; resources, F.T., H.Y., G.S., S.H., Y.N., and M.I.; data curation, R.T.; writing—original draft preparation, R.T.; writing—review and editing, M.J., K.Y., and S.W.; visualization, R.T.; supervision, M.J.; project administration, M.J.; funding acquisition, M.J. and R.T. All authors have read and agreed to the published version of the manuscript.

Funding: This research was funded by a Grant-in-Aid for Scientific Research (C) from the Ministry of Education, Culture, Sports, Science and Technology of Japan to M.J. [22580228], [19K06240], and a grant-in-aid from The Mikimoto Fund for Marine Ecology, Japan, to R.T.

Institutional Review Board Statement: Not applicable.

Conflicts of Interest: The authors declare no conflict of interest.

References

1. Hoegh-Guldberg, O. Climate change, coral bleaching and the future of the world's coral reefs. *Mar. Freshw. Res.* **1999**, *50*, 839–866. [CrossRef]
2. Weis, V.M. Cellular mechanisms of cnidarian bleaching: Stress causes the collapse of symbiosis. *J. Exp. Biol.* **2008**, *211*, 3059–3066. [CrossRef] [PubMed]
3. Pochon, X.; Gates, R.D. A new *Symbiodinium* clade (Dinophyceae) from soritid foraminifera in Hawaii. *Mol. Phylogenet. Evol.* **2010**, *56*, 492–497. [CrossRef] [PubMed]
4. LaJeunesse, T.C. 'Species' radiations of symbiotic dinoflagellates in the Atlantic and Indo-Pacific since the Miocene–Pliocene transition. *Mol. Biol. Evol.* **2005**, *22*, 570–581. [CrossRef] [PubMed]
5. LaJeunesse, T.C.; Parkinson, J.E.; Gabrielson, P.W.; Jeong, H.J.; Reimer, J.D.; Voolstra, C.R.; Santos, S.R. Systematic revision of Symbiodiniaceae highlights the antiquity and diversity of coral endosymbionts. *Curr. Biol.* **2018**, *28*, 2570–2580.e6. [CrossRef]
6. Goulet, T.L. Most Corals May Not Change Their Symbionts. *Mar. Ecol. Prog. Ser.* **2006**, *321*, 1–7. [CrossRef]
7. Yamashita, H.; Suzuki, G.; Hayashibara, T.; Koike, K. *Acropora* recruits harbor "rare" *Symbiodinium* in the environmental pool. *Coral Reefs* **2013**, *32*, 355–366. [CrossRef]
8. Yuyama, I.; Hayakawa, H.; Endo, H.; Iwao, K.; Takeyama, H.; Maruyama, T.; Watanabe, T. Identification of symbiotically expressed coral mrnas using a model infection system. *Biochem. Biophys. Res. Commun.* **2005**, *336*, 793–798. [CrossRef]
9. Kuniya, N.; Jimbo, M.; Tanimoto, F.; Yamashita, H.; Koike, K.; Harii, S.; Nakano, Y.; Iwao, K.; Yasumoto, K.; Watabe, S. Possible involvement of tachylectin-2-like lectin from *Acropora tenuis* in the process of *Symbiodinium* acquisition. *Fish Sci.* **2015**, *81*, 473–483. [CrossRef]
10. Yamashita, H.; Suzuki, G.; Kai, S.; Hayashibara, T.; Koike, K. Establishment of coral–algal symbiosis requires attraction and selection. *PLoS ONE* **2014**, *9*, e97003. [CrossRef]
11. Hollingsworth, L.L.; Kinzie, R.A.; Lewis, T.D.; Krupp, D.A.; Leong, J.C. Phototaxis of motile zooxanthellae to green light may facilitate symbiont capture by coral larvae. *Coral Reefs* **2005**, *24*, 523. [CrossRef]

12. Aihara, Y.; Maruyama, S.; Baird, A.H.; Iguchi, A.; Takahashi, S.; Minagawa, J. Green fluorescence from cnidarian hosts attracts symbiotic algae. *Proc. Natl. Acad. Sci. USA* **2019**, *116*, 2118–2123. [CrossRef] [PubMed]
13. Fitt, W.K.; Chang, S.S.; Trench, R.K. Motility patterns of different strains of the symbiotic dinoflagellate *Symbiodinium* (=*Gymnodinium*) *microadriaticum* (Freudenthal) in culture. *Bull. Mar. Sci.* **1981**, *31*, 436–443.
14. Hagedorn, M.; Carter, V.; Zuchowicz, N.; Phillips, M.; Penfield, C.; Shamenek, B.; Vallen, E.A.; Kleinhans, F.W.; Peterson, K.; White, M.; et al. Trehalose is a chemical attractant in the establishment of coral symbiosis. *PLoS ONE* **2015**, *10*, e0117087. [CrossRef]
15. Biquand, E.; Okubo, N.; Aihara, Y.; Rolland, V.; Hayward, D.C.; Hatta, M.; Minagawa, J.; Maruyama, T.; Takahashi, S. Acceptable symbiont cell size differs among cnidarian species and may limit symbiont diversity. *ISME J.* **2017**, *11*, 1702–1712. [CrossRef]
16. Kawabata, S.; Iwanaga, S. Role of lectins in the innate immunity of horseshoe crab. *Dev. Comp. Immunol.* **1999**, *23*, 391–400. [CrossRef]
17. Sharon, N.; Lis, H. History of lectins: From hemagglutinins to biological recognition molecules. *Glycobiology* **2004**, *14*, 53R–62R. [CrossRef] [PubMed]
18. Gross, R.; Vavre, F.; Heddi, A.; Hurst, G.D.; Zchori-Fein, E.; Bourtzis, K. Immunity and symbiosis. *Mol. Microbiol.* **2009**, *73*, 751–759. [CrossRef]
19. Bulgheresi, S.; Gruber-Vodicka, H.R.; Heindl, N.R.; Dirks, U.; Kostadinova, M.; Breiteneder, H.; Ott, J.A. Sequence variability of the pattern recognition receptor Mermaid mediates specificity of marine nematode symbioses. *ISME J.* **2011**, *5*, 986–998. [CrossRef]
20. Kamiya, H.; Jimbo, M.; Koike, K.; Sakai, R. Roles of Coral Lectins in Morphological Change of Zooxanthellae. In *Animal Lectins*; Vasta, G.R., Ahmed, H., Eds.; CRC Press: Boca Raton, FL, USA, 2008; pp. 231–238.
21. Kvennefors, E.C.E.; Leggat, W.; Hoegh-Guldberg, O.; Degnan, B.M.; Barnes, A.C. An ancient and variable mannose-binding lectin from the coral *Acropora millepora* binds both pathogens and symbionts. *Dev. Comp. Immunol.* **2008**, *32*, 1582–1592. [CrossRef]
22. Kvennefors, E.C.E.; Leggat, W.; Kerr, C.C.; Ainsworth, T.D.; Hoegh-Guldberg, O.; Barnes, A.C. Analysis of evolutionarily conserved innate immune components in coral links immunity and symbiosis. *Dev. Comp. Immunol.* **2010**, *34*, 1219–1229. [CrossRef] [PubMed]
23. Vidal-Dupiol, J.; Adjeroud, M.; Roger, E.; Foure, L.; Duval, D.; Mone, Y.; Ferrier-Pages, C.; Tambutte, E.; Tambutte, S.; Zoccola, D.; et al. Coral bleaching under thermal stress: Putative involvement of host/symbiont recognition mechanisms. *BMC Physiol.* **2009**, *9*, 14. [CrossRef]
24. Nishikawa, A.; Katoh, M.; Sakai, K. Larval settlement rates and gene flow of broadcast-spawning (*Acropora tenuis*) and planula-brooding (*Stylophora pistillata*) Corals. *Mar. Ecol. Prog. Ser.* **2003**, *256*, 87–97. [CrossRef]
25. Iwao, K.; Fujisawa, T.; Hatta, M. A Cnidarian neuropeptide of the GLW amide family induces metamorphosis of reef-building corals in the genus *Acropora*. *Coral Reefs* **2002**, *21*, 127–129. [CrossRef]
26. Takeuchi, R.; Jimbo, M.; Tanimoto, F.; Tanaka, C.; Harii, S.; Nakano, Y.; Yasumoto, K.; Watabe, S. Establishment of a model for chemoattraction of *Symbiodinium* and characterization of chemotactic compounds in *Acropora tenuis*. *Fish. Sci.* **2017**, *83*, 479–487. [CrossRef]
27. Logan, D.D.K.; LaFlamme, A.C.; Weis, V.M.; Davy, S.K. Flow-cytometric characterization of the cell-surface glycans of symbiotic dinoflagellates (*Symbiodinium* spp.). *J. Phycol.* **2010**, *46*, 525–533. [CrossRef]
28. Miller, R.L.; Vogt, R. An N-Terminal Partial Sequence of the 13 kDa *Pycnopodia helianthoides* sperm chemoattractant "startrak" possesses sperm-attracting activity. *J. Exp. Biol.* **1996**, *199*, 311–318. [PubMed]
29. Parkinson, J.E.; Tivey, T.R.; Mandelare, P.E.; Adpressa, D.A.; Loesgen, S.; Weis, V.M. Subtle differences in symbiont cell surface glycan profiles do not explain species-specific colonization rates in a model cnidarian-algal symbiosis. *Front. Microbiol.* **2018**, *9*, 842. [CrossRef] [PubMed]
30. Lee, S.Y.; Jeong, H.J.; Kang, N.S.; Jang, T.Y.; Jang, S.H.; LaJeunesse, T.C. *Symbiodinium tridacnidorum* sp. nov., a dinoflagellate common to Indo-Pacific giant clams, and a revised morphological description of *Symbiodinium microadriaticum* Freudenthal, emended Trench & Blaank. *Eur. J. Phycol.* **2015**, *50*, 155–172.
31. Yamashita, H.; Koike, K. Genetic identity of free-living *Symbiodinium* obtained over a broad latitudinal range in the Japanese coast. *Phycol. Res.* **2013**, *61*, 68–80. [CrossRef]

marine drugs

Article

Oncolytic Vaccinia Virus Expressing White-Spotted Charr Lectin Regulates Antiviral Response in Tumor Cells and Inhibits Tumor Growth In Vitro and In Vivo

Xue Wang, Ningning Zhou, Tingting Liu, Xiaoyuan Jia, Ting Ye, Kan Chen and Gongchu Li *

College of Life Sciences and Medicine, Zhejiang Sci-Tech University, Hangzhou 310018, China; wx108@mails.zstu.edu.cn (X.W.); znnzjlg@mails.zstu.edu.cn (N.Z.); liutt@mails.zstu.edu.cn (T.L.); xyjia@zstu.edu.cn (X.J.); yeting@zstu.edu.cn (T.Y.); chenkan@zstu.edu.cn (K.C.)
* Correspondence: lgc@zstu.edu.cn

Abstract: Oncolytic vaccina virus (oncoVV) used for cancer therapy has progressed in recent years. Here, a gene encoding white-spotted charr lectin (WCL) was inserted into an oncoVV vector to form an oncoVV-WCL recombinant virus. OncoVV-WCL induced higher levels of apoptosis and cytotoxicity, and replicated faster than control virus in cancer cells. OncoVV-WCL promoted IRF-3 transcriptional activity to induce higher levels of type I interferons (IFNs) and blocked the IFN-induced antiviral response by inhibiting the activity of IFN-stimulated responsive element (ISRE) and the expression of interferon-stimulated genes (ISGs). The higher levels of viral replication and antitumor activity of oncoVV-WCL were further demonstrated in a mouse xenograft tumor model. Therefore, the engineered oncoVV expressing WCL might provide a new avenue for anticancer gene therapy.

Keywords: white-spotted charr lectin; oncolytic vaccinia virus; interferon; antiviral response

Citation: Wang, X.; Zhou, N.; Liu, T.; Jia, X.; Ye, T.; Chen, K.; Li, G. Oncolytic Vaccinia Virus Expressing White-Spotted Charr Lectin Regulates Antiviral Response in Tumor Cells and Inhibits Tumor Growth In Vitro and In Vivo. *Mar. Drugs* **2021**, *19*, 292. https://doi.org/10.3390/md19060292

Academic Editors: Yuki Fujii, Marco Gerdol and Yasuhiro Ozeki

Received: 29 April 2021
Accepted: 20 May 2021
Published: 21 May 2021

1. Introduction

Lectins are a group of proteins that bind carbohydrates reversibly and specifically, and play important roles in various biological processes such as cell division, fertilization, congenital immunity, and cell recognition [1]. Lectins are widely distributed in marine bioresources such as marine cyanobacteria, algae, invertebrate animals and fish. In addition, lectins have shown remarkable abilities in inducing apoptosis and inhibiting angiogenesis, suggesting the future use in treating cancers [2,3]. For example, *Eucheuma serra* agglutinin in red macroalga induced apoptosis in various colon adenocarcinoma, breast cancer and osteosarcoma cells [4–6]. *Crenomytilus grayanus* lectin extracted from homonymous bivalve showed globotriaosylceramide (Gb3)-dependent cytotoxicity in breast cancer and Burkitt's lymphoma cells [7,8]. N-acetyl sugar-binding lectin isolated from *Ibacus novemdentatus* elicited acytotoxic effects on various cancer cells through glycoconjugate interaction [9].

Vaccinia viruses can be used as replicating vectors harboring therapeutic genes to directly lyse tumor cells, or as cancer vaccines to stimulate antitumor immunity [10–12]. JX-594, the most famous oncolytic vaccinia virus carrying a human granulocyte-macrophage colony-stimulating factor (GM-CSF) gene with the thymidine kinase (TK) gene deletion, has been advanced to clinical phase III for the treatment of advanced hepatocellular carcinoma (HCC) and clinical phase II trial for renal cell carcinoma [13–17]. In our previous work, genes encoding marine lectins *Tachypleus tridentatus* lectin (TTL) and *Aphrocallistes vastus* lectin (AVL) were inserted into the oncolytic vaccinia virus (oncoVV) vector, forming oncoVV-TTL and oncoVV-AVL recombinant viruses, respectively [18,19]. Compared with the control virus, both oncoVV-TTL and oncoVV-AVL showed faster replication and significant antitumor activities in vitro and in vivo.

Previously, some marine fish lectins have been shown to elicit antitumor effects [2]. In the presented studies, we constructed a new recombinant vaccinia virus oncoVV-WCL, which was generated by inserting a gene encoding white-spotted charr lectin (WCL) into the oncoVV vector. WCL is a type of rhamnose-binding lectin from white-spotted Charr (*Salvelinus leucomaenis*) eggs [20], which has not been previously investigated in oncoVV vectors. We then investigated the antitumor effect of oncoVV-WCL mainly on hepatocellular carcinoma in vitro and in vivo. Lung and cervical cancer cell lines were also used for verification. Simultaneously, the underlying mechanisms of oncoVV-WCL involved in interferon (IFN) production and the IFN-induced antiviral response were elucidated.

2. Results

2.1. Cytotoxicity of oncoVV-WCL in Cancer Cells

Hepatocellular carcinoma cells PLC/PRF/5 and Huh-7, as well as lung carcinoma cell H460, were infected with oncoVV or oncoVV-WCL at five multiplicities of infection (MOI) or 15MOI for 48 h and 72 h. An assay of 3-(4,5-dimethylthiazol-2-yl)-5-(3-carboxymethoxyphenyl)-2-(4-sulfophenyl)-2H-tetrazolium (MTT) was performed to determine cell viability. PBS served as the negative control. The cell viability of the oncoVV-WCL group was decreased compared to the oncoVV group (Figure 1a). The results indicate that exogenous WCL expression enhanced the cytotoxicity of oncoVV in H460, PLC/PRF/5 and Huh-7 cell lines.

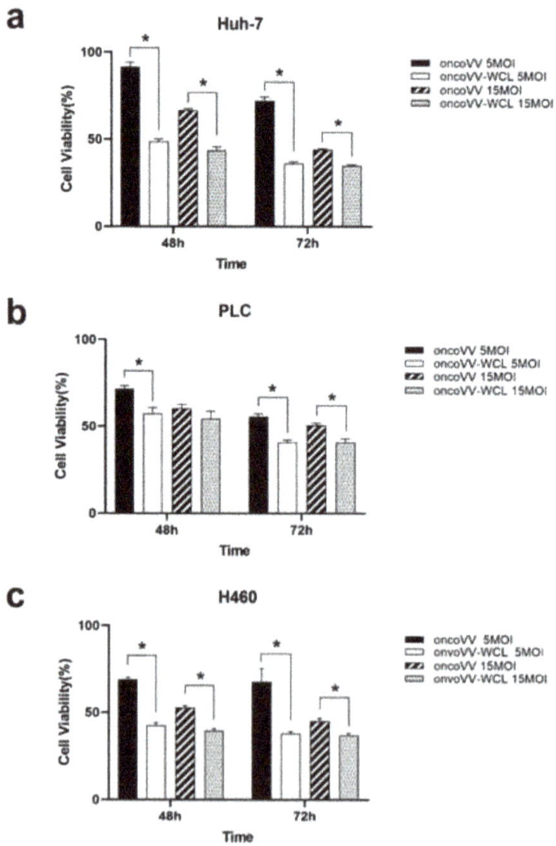

Figure 1. The antiproliferative effect of oncoVV-WCL in tumor cells. Cell viability was measured by MTT assay in Huh-7 cells (**a**), PLC/PRF/5 cells (**b**), H460 cells (**c**). Data are expressed as the mean ± SEM from at least three separate experiments. (* $p < 0.05$)

2.2. Apoptotic Effect of oncoVV-WCL in Huh-7 Cells

To investigate whether the oncoVV-WCL can induce the apoptosis in cancer cells, Huh-7 cells were treated with oncoVV, oncoVV-WCL and PBS respectively, followed by apoptotic analysis through flow cytometry analysis and Western blot. As shown in Figure 2a,b, compared with PBS and oncoVV, 5 MOI oncoVV-WCL triggered higher apoptosis rates. The infection of oncoVV-WCL increased the expression of caspase-3 and cleaved caspase-9 (Figure 2c), suggesting that the intrinsic apoptotic pathway may play a role in oncoVV-WCL induced cell death.

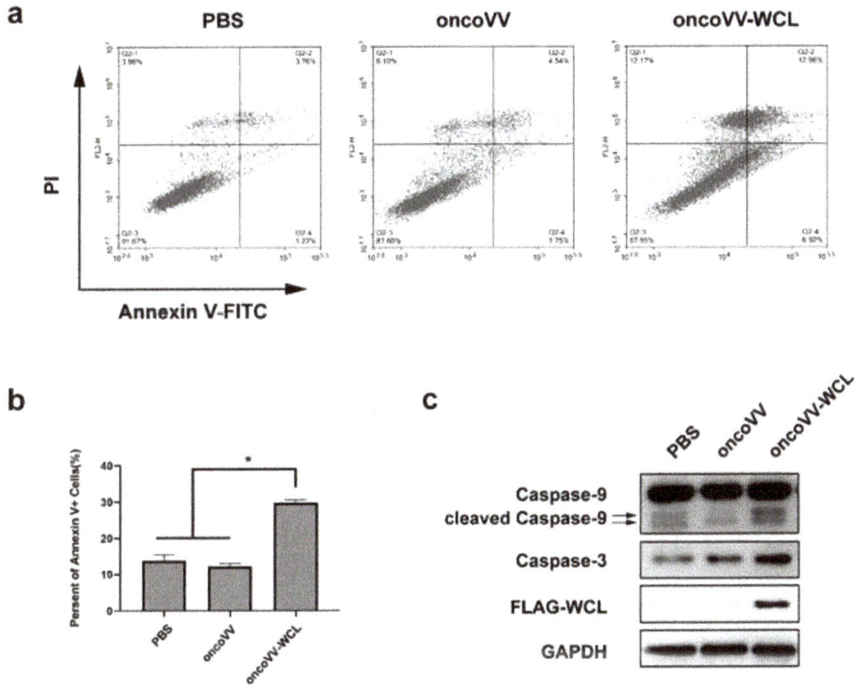

Figure 2. VV-WCL inducing apoptosis in Huh-7 cells. (**a**) Flow cytometric examination. Huh-7 cells were treated with oncoVV or oncoVV-WCL at 5MOI as well as PBS control for 72 h. Cells were stained with Annexin V-FITC and PI followed by analysis under a flow cytometer; (**b**) the percentage of apoptotic cells. Three repeats are represented as mean ± SEM (* $p < 0.05$). (**c**) The expression of Caspase-3, Caspase-9 and FLAG. The expression level was detected by Western blot. GAPDH served as a loading control.

2.3. Replication of Oncolytic Vaccinia Virus Improved by WCL

To clarify the underlying mechanism of the antiproliferative effect of oncoVV-WCL, the replication abilities of oncoVV and oncoVV-WCL were tested through the $TCID_{50}$ method in different cancer cells. The results demonstrated that the reproductive number of oncoVV-WCL was much higher than that of oncoVV in Huh-7 cells (Figure 3a), H460 cells (Figure 3b) and Hela-S3 cells (Figure 3c). The expression level of protein A27L, which is located on the surface of the intracellular mature virus [21], was also detected in hepatocellular carcinoma cells Huh-7 and PLC/PRF/5 (Figure 3d). We found the expression level of A27L in oncoVV-WCL treatment group was significantly higher than that in oncoVV treatment group, suggesting that WCL harboring dramatically promotes the replication of vaccinia virus.

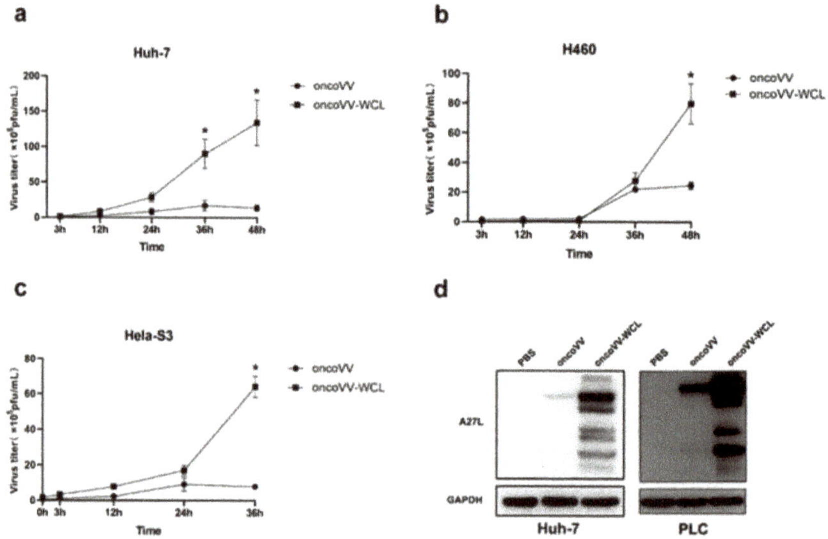

Figure 3. oncoVV-WCL replication in multiple tumor cell lines. The replication of oncoVV-WCL and oncoVV control in Huh-7 cells (**a**), H460 cells (**b**), Hela-S3 cells (**c**). Viral replication was determined by $TCID_{50}$ assay. Data are expressed as the mean ± SEM from at least three separate experiments. (* $p < 0.05$). (**d**) The expression level of A27L was detected by Western blot. PBS served as negative treatment control, GAPDH served as a loading control.

2.4. Onco VV-WCL Stimulated the Production of Type I IFNs

It is well-recognized that IFNs are secreted glycoproteins and are produced from cells in response to virus infection [22]. Vaccinia virus infection activates IFN regulatory factor 3 (IRF3) and IRF7 pathways, and further causes translocation of transcription factors NF-κB and AP-1 into the nucleus, and finally IRF3/7, NF-κB and AP-1 form a complex known as the enhanceosome, driving the transcription of type I IFNs [22]. To investigate the production of interferons by oncoVV-WCL, RNA was extracted from infected Huh-7 cells, and the mRNA levels of IFN-α and IFN-β were evaluated by qRT-PCR. The results demonstrated that the transcription of IFN-α and IFN-β in oncoVV-WCL-treated Huh-7 cells increased dramatically as compared to oncoVV-treated cells (Figure 4a), indicating that WCL enhanced the oncoVV induced production of type I IFNs.

We next preformed the dual-luciferase reporter assay to access the effect of oncoVV-WCL on IRF-3, IRF-7, NF-κB and AP-1 transcription factors. As shown in Figure 4b, the activity of IRF-3 promoters in the oncoVV-WCL treatment group significantly increased compared to that of oncoVV treatment group, while there was no significant difference between oncoVV-WCL and oncoVV treatment groups in IRF-7, AP-1 and NF-κB reporter assays (Figure 4c–e). Therefore, the results suggest that WCL harboring may promote IRF-3 transcription activity to induce the expression of type I IFNs.

2.5. OncoVV-WCL Regulated the IFN-Induced Signaling and the Expression of IFN-Stimulated Genes

Because IFNs can induce the transcription of hundreds of interferon-stimulated genes (ISGs) [22], it is reasonable to further profile the effect of oncoVV-WCL on hepatocellular carcinoma cells. Therefore, the gene expressions of Huh-7 cells treated with oncoVV-WCL, oncoVV and PBS were examined through transcriptomic analysis, then a heatmap was created to graphical visualization of differential gene expression. The results showed that 34 genes related to response/defense response to virus were differentially expressed, and among them 25 genes expressions were downregulated by oncoVV-WCL as compared to oncoVV control (Figure 5a). Results suggested that oncoVV-WCL downregulated ISGs expression.

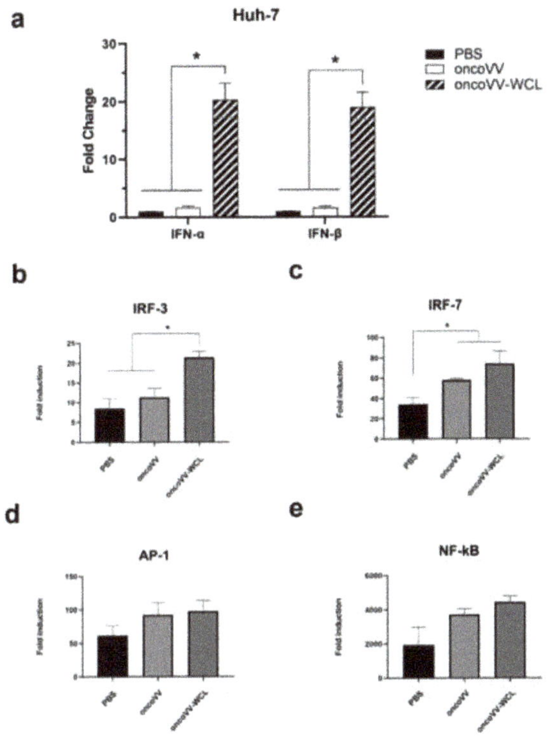

Figure 4. oncoVV-WCL regulates interferon transcription. (**a**) Quantitative RT-PCR evaluations of IFNα/β transcription 48 h after infection of Huh-7 cells with oncoVV or oncoVV-WCL. PBS served as the negative control. The comparative Ct (cycle threshold) method was used to determine RNA expression. Statistically significant differences are presented as * $p < 0.05$. (**b**) oncoVV-WCL activated the transcription activity of IRF-3. Transfection efficiencies were normalized to the pTK-RL luciferase plasmid. Data are expressed as the mean ± SEM from at least three separate experiments. (* $p < 0.05$). Reporter assays of IRF-7 (**c**), AP-1 (**d**), NF-κB (**e**) were detected as described above.

The IFNs induced by virus infection can engage with their cognate receptors and induce Janus-associated kinase (JAK) and the signal transducer and activator of the transcription (STAT) pathway [22,23]. In IFN-α and IFN-β signaling, STAT1 and STAT2 form heterodimers, then associate with IFN regulatory factor 9 (IRF9) to form a trimeric complex, which further translocate into the nucleus to drive the transcription of ISGs associated with the IFN-stimulated responsive element (ISRE) [22,23]. Given that ISGs such as ISG15 and OASL were downregulated (Figure 5a), we postulated that the JAK-STAT signaling cascade pathway might be influenced by the infection of oncoVV-WCL. Therefore, STAT1 expression was tested by Western blot, and the ISRE activity was monitored with the dual-luciferase reporter assay. The results showed that oncoVV-WCL downregulated the activity of ISRE and led to STAT1 cleavage (Figure 5b,c), suggesting that oncoVV-WCL regulated the antiviral response in cancer cells through STAT1 cleavage.

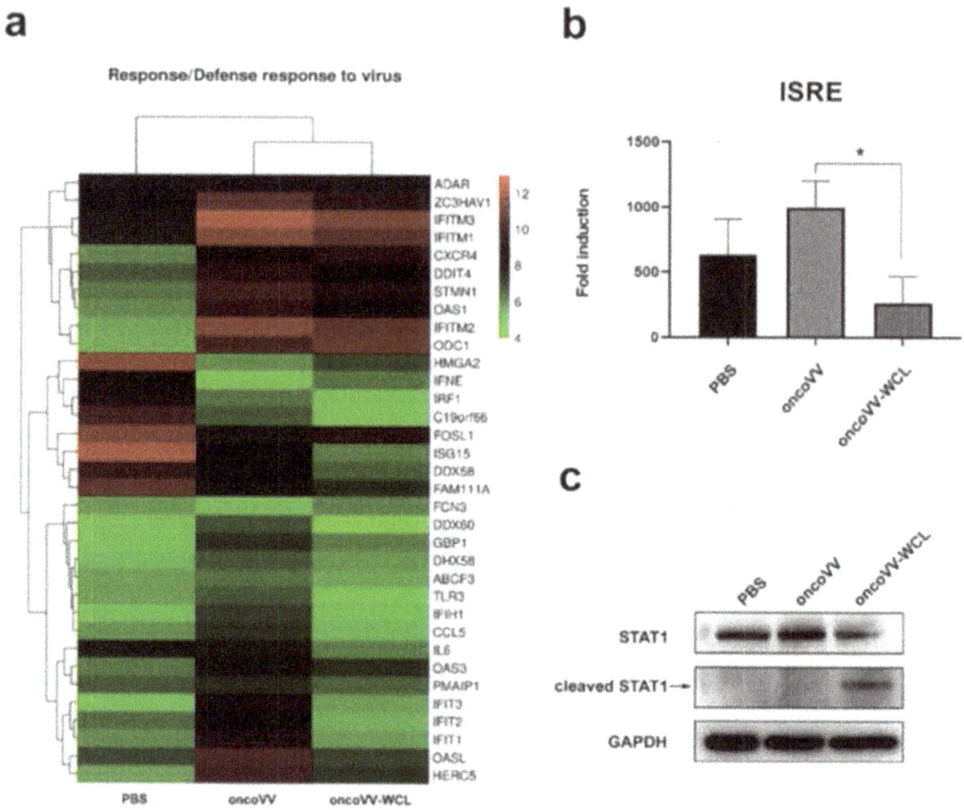

Figure 5. Transcriptomic validation of oncoVV-WCL. (**a**) Heatmap of genes. Huh-7 cells were infected with oncoVV and oncoVV-WCL at 5 MOI for 24 h followed by transcriptomic analysis. The key includes a histogram of the distribution of $\log_2$ fold change values for all the included genes. (**b**) Activity of ISRE promoter was analyzed through a dual-luciferase reporter assay kit. The transcription activity of ISRE was evaluated by the dual-luciferase reporter assay. Data are represented as mean ± SEM from at least three independent experiments (* $p < 0.05$). (**c**) The expression level of STAT1 was detected by Western blot. GAPDH was used as a loading control.

2.6. OncoVV-WCL Suppressed Hepatocellular Carcinoma Cell Growth In Vivo

To investigate the antitumor efficacy of oncoVV-WCL in vivo, subcutaneous tumors were established in Balb/c nude mice with Huh-7 cells and then administrated with saline, oncoVV and oncoVV-WCL, respectively. As shown in Figure 6a, the oncoVV-WCL treatment group demonstrated a significant antitumor effect compared with the saline and oncoVV treatment groups. Subsequently, RNA and proteins were extracted from the mice tumors, and semiquantitative PCR and Western blot were performed to test the expression of A27L of virus (Figure 6b,c). Furthermore, the A27L level of oncoVV-WCL and control treatment in mice tumors were also confirmed by immunohistochemistry assay (Figure 6d). The results suggest that the remarkable antitumor effect of oncoVV-WCL may be due to the increasing replication of viruses harboring WCL.

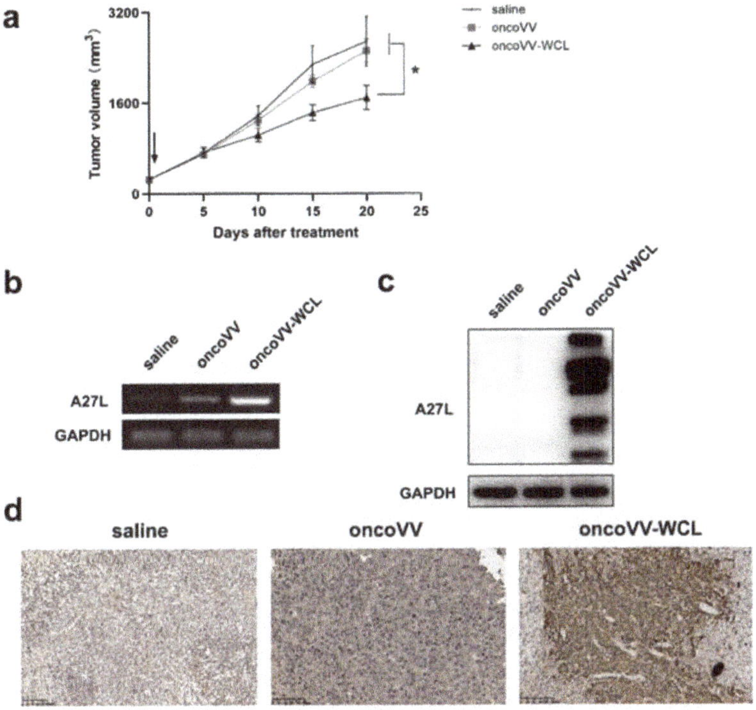

Figure 6. The antitumor effect of oncoVV-WCL in vivo. (**a**) Huh-7 cells were inoculated into the Balb/c nude mice on the back. Tumors were then injected with saline, oncoVV, or oncoVV-WCL, respectively. Data are expressed as the mean ± SEM. (* $p < 0.05$). (**b**) Semiquantitative PCR validated A27L gene expression at the level of RNA. GAPDH served as a loading control. The extracted RNA was reverse transcribed into cDNA and performed this cDNA to semiquantitative PCR followed by agarose gel electrophoresis. GAPDH served as a loading control. (**c**) The expression level of A27L was detected by Western blot. GAPDH served as a loading control. (**d**) The immunohistochemical results of A27L on Huh-7 tumors. Scale bars show 100 μm.

3. Discussion

WCL, a family member of lectin, was originally isolated from white-spotted charr eggs in Salmonidae [20]. Since the first report of WCL, there have been few reports studying the function of this protein. Based on our previous study, utilizing an adenovirus or a vaccinia virus carrying a specific lectin manifested a strong antitumor effect. Therefore, investigating the function of an unknown WCL by means of the vaccinia virus intrigued us. To our knowledge, this is the first demonstration of combining a WCL gene with vaccinia virus. This study reports oncoVV harboring WCL can effectively inhibit tumor cell growth both in vitro and in vivo. We further revealed that the recombinant oncoVV-WCL demonstrates high replication in tumor cells and induces type I IFNs production, while inhibiting the expression of ISGs via downregulating the JAK-STAT signaling pathway.

VV used for antitumor therapy has progressed in the past decades [10,24–27]. It is well recognized that immediately after VV invasion, the innate immunity of the host is activated and then a cascade of signaling pathways involved. To restrict VV replication and spread, the host cells secrete IFNs which launch an immediate large response to attack the virus. Consequently, the viruses express many proteins and evolve multiple strategies to hinder the effect that IFNs produce, including preventing production of IFN, blocking binding of IFNs to IFN receptors, blocking IFN-induced signaling pathways and blocking the antiviral action of ISGs [22]. In our previous reports, oncoVV carrying lectin TTL suppressed the production of IFN-β [18], while in the present study, oncoVV carrying lectin WCL promoted the expression of IFN-α and IFN-β in liver cancer cells. Given that IRF3 NF-κB and AP-1 forming the enhanceosome and binding to IFN-β gene promoter are necessary for the expression of the antiviral IFN-β gene [22,28], our further investigation showed the induction of IFNs might be due to the activation of the key transcription factor IRF-3. Since the VV can impede the effect of IFNs by interfering with the IFN-induced JAK-STAT signaling pathway and the production of ISGs, testing the changes of the JAK-STAT signaling pathway and ISGs is necessary. As shown in the transcriptomic analysis, some ISGs were downregulated by oncoVV-WCL. ISG15 is a ubiquitin-like protein which is induced by viral infection, and IFN-α and -β, and can conjugate to target proteins such as IRF3 [22,29]. IFIT1/2/3 are strongly induced ISGs and function as both sensor and effector molecules of the cellular innate immune system [30,31]. OASL is also a member of ISGs and induced by virus infection rapidly via IFN signaling [32]. The IFIH1 gene encodes a RIG-I-like receptor involved in the sensing of viral RNA [33]. The downregulation of these ISGs might help oncoVV-WCL escape the intracellular antiviral response, thereby achieving a significantly higher level of viral replication in cancer cells both in vitro and in vivo.

We present a novel WCL expressing oncoVV in this study. OncoVV-WCL not only induced type I IFN production, which could elicit antitumor activity but also inhibit IFN-induced ISG production, possibly due to STAT1 cleavage, which helps oncoVV escape elimination. However, oncoVV-WCL only elicited a weak antitumor effect in vivo, as shown. Further investigations into signaling pathways regulating oncoVV-WCL could help to enhance its antitumor activity.

4. Materials and Methods

4.1. Cell Culture

The human embryonic kidney cell line 293A, hepatocellular carcinoma cell lines Huh-7 and PLC/PRF/5, lung carcinoma cell line H460 and cervical carcinoma cell lines HELA-S3 were provided by the Chinese Academy of Sciences. Cells were cultured in Dulbecco's Modified Eagle Medium (DMEM)/high glucose medium (Gibco, Thermo Fisher Scientific, Waltham, MA, USA) supplemented with 10% fetal bovine serum (HyClone Laboratories) and 1% Penicillin-Streptomycin on the condition of 37 °C and 5% CO_2.

4.2. Recombination of oncoVV-WCL

The *WCL* gene (GenBank: AB077045.1) was synthesized in Shanghai Generay Biotech Co., Ltd. (Shanghai, China). The oncoVV-WCL recombinant virus was constructed following the steps described previously [18,19]. Briefly, the pCB-Flag-WCL plasmid was constructed by inserting the *WCL* gene into a pCB vector with a thymidine kinase (TK) gene deletion. The 293A cells were infected by wild type vaccinia virus (Western Reverse) for about 3 h and then transfected with pCB-Flag-WCL. The recombinant oncolytic vaccinia viruses were formed 48 h later. Subsequently, the selection of mycophenolic acid, hypoxanthine and dioxopurine were used to select for effective oncoVV-WCL. After three repetitions, purified viruses were obtained. The titers of oncoVV-WCL were determined by $TCID_{50}$ (median tissue culture infective dose).

4.3. Western Blot

After cell extraction, equal amounts of proteins were electrophoresed by SDS-PAGE and transferred onto polyvinylidene fluoride membranes (Millipore, Bedford, MA, USA). Then, the membranes were blocked with 5% bovine serum albumin solution at room temperature for 2 h, followed by incubating with primary antibodies at 4 °C overnight. The membranes were washed and incubated with secondary antibodies for 1 h at room temperature. After washing, the membranes were detected using a Tanon 5500 chemiluminescence image system (Tanon Inc., Shanghai, China). The following primary and secondary antibodies were used: Caspase-9 (Abcam, Cambridge, UK, ab202068,1:2000), caspase-3 (Cell Signaling Technology, Danvers, MA, USA, 9662, 1:1000), STAT1 (Santa Cruz, Dallas, TX, USA, sc-346,1:1000), A27L (Abcam, ab35219,1:2000), GAPDH (Cell Signaling Technology, 2118 s, 1:1000), HRP conjugated Goat antiRabbit IgG(H + L) (ABclonal, AS014, 1:5000), HRP conjugated Goat antiMouse IgG(H + L) (ABclonal, Woburn, MA, USA, AS003, 1:5000), and AntiDDDDK-tag pAb (Medical & Biological Laboratories, PM020, 1:1000).

4.4. Animal Experiments

Hepatocellular carcinoma tumor-bearing mouse models were built using 6-week-old female Balb/c nude mice (Shanghai Slack Animal Laboratory, Shanghai, China). Mice were housed and raised in accordance with the standards of Animal Care and Use Committee of Zhejiang Sci-Tech University. Huh-7 cells (5×10^6) were subcutaneously inoculated into the back region of mice. When the tumor had grown to reach approximately 250 mm^3 in size, 1×10^7 plaque-forming units (PFU) oncoVV or oncoVV-WCL, as well as saline, were injected in situ, respectively. Then, tumor volume was recorded and measured every five days. Tumor volume (V) was calculated using the formula: $V(mm^3) = 1/2$ (length $\times$ width $\times$ width) [34,35]. Width was considered as the shorter diameter and length as the longer diameter of the tumor in mm.

4.5. Cell Viability Detection and Flow Cytometry Assay

Seeding of 5×10^3/well cells was done in 96-well plates. Cells were infected with oncoVV or oncoVV-WCL at MOIs of 5 and 15. Cell viability was detected by MTT assay. Apoptosis analysis was carried out using the Annexin V-FITC Apoptosis Detection Kit (BD Biosciences, San Jose, CA, USA) according to the manufacturer's instructions, and then analyzed by flow cytometer (Accuri C6; BD Biosciences).

4.6. Virus Replication Assay

To test the replication of the virus in different cell lines, cells were plated on 24-well plates at a density of 6×10^4/well. Different cells were infected with oncoVV, oncoVV-WCL at and MOI of 5 for 0 h, 3 h, 12 h, 24 h, 36 h and 48 h, respectively. The viral titers were determined through $TCID_{50}$ assay on 293A cells.

4.7. Reporter Assay

The dual-luciferase assay kit (GeneCopoeia, Inc., Rockville, MD, USA) was used to perform the reporter assay according to the manufacturer's instructions. Huh-7 cells were cotransfected with 0.005 μg of pTK-RL plasmid, an internal control encoding Renilla luciferase, and 0.25 μg luciferase reporter plasmids for IRF3, IRF7, NF-κB, AP-1 and ISRE, and then treated with PBS at 5 MOI of oncoVV or oncoVV-WCL for 48 h. After the cells were lysed, the luciferase activities were assessed. The transfection efficiency was normalized to the internal control pTK-RL. All experiments were carried out at least three times.

4.8. RNA Extraction and Semi-Quantitative PCR

Total RNA was extracted using Trizol (Invitrogen) reagent according to the manufacturer's protocol. Then, 2 μg of total RNA was reverse transcribed into complementary DNA (cDNA) using the Sigma Enhanced Avian HS RT–PCR Kit. The products of reverse transcription were used for semiquantitative PCR and detected the expression level of mRNA.

4.9. qRT-PCR

IFN-α and IFN-β expression in Huh-7 cells was evaluated by qRT-PCR after infection with oncoVV and oncoVV-WCL for 48 h. After RNA extraction, cDNA was obtained. The qRT-PCR reaction was carried out using a MYiQ single-color real-time PCR detection system. The reaction volume of 20μL contained 1 uL cDNA,320 nM each primer and 10 uL iQSYBR green supermix. The reaction performed at 95 °C for 10 min, then 40 PCR cycles of denaturation at 95 °C for 10 s and annealing at 60 °C for 1 min. Targeted gene expression was normalized to GAPDH messenger RNA. Experiments were repeated three times.

4.10. Transcriptomic Analysis

Huh-7 cells were infected with oncoVV and oncoVV-WCL at five MOI for 24 h. PBS treatment served as negative control. The cells were harvested in TRIzol reagent (Invitrogen, Waltham, MA, USA) and submitted to Shanghai Baygene Biotechnologies Co., Ltd. (Shanghai, China) for analysis and profiling the differentially expressed gene. Briefly, the total RNA was extracted using Trizol and a gene chip assay was performed using Clariom D human Affymetrix gene chip (Affymetrix, Shanghai, China), GeneChip WT PLUS Reagent Kit (Affymetrix, 902923), and GeneChip Hybridization, Wash and Stain Kit (Affymetrix, 900720).

4.11. Immunohistochemistry

On day 10 post-injection of viruses, the tumors were harvested for immunohistochemistry (IHC) analysis. The IHC was performed by HaoKe Biotechnology Co. Ltd. (Hangzhou, China). Briefly, after fixation, the tumors were embedded in paraffin, deparaffinized with xylene, and rehydrated with graded alcohol washes and H_2O_2. Antigen retrieval was performed with EDTA retrieval buffer. The endogenous peroxidase of samples was blocked followed by protein block with 3% BSA. Subsequently, slides were incubated with primary AntiA27L rabbit antibody at 4 °C overnight, washed with PBS and incubated with secondary HRP conjugated goat antiRabbit IgG (Abcam, ab97051, 1:200) for 50 min at room temperature then stained with DAB solution, counterstained with hematoxylin, dehydrated and mounted for examination. The PBS treatment group served as negative control.

4.12. Statistical Analysis

Statistical analysis was determined by student's *t*-test. Significance was considered as $p < 0.05$.

5. Conclusions

Our study showed the newly constructed oncoVV-WCL induced significant apoptotic cell death in hepatocellular carcinoma cells and lung carcinoma cells. The animal experiments verified the in vivo antitumor activity of oncoVV-WCL. WCL harboring promoted vaccinia virus replication and production of type I IFNs via upregulation of IRF-3 activity in hepatocellular carcinoma cells. The underlying mechanisms of antitumor activity of oncoVV-WCL might be due to the inhibition of the transcriptional activity of ISRE and the expression of ISGs thereby resulted in the inhibition of the antiviral response and the promotion of viral replication in cancer cells. Our findings may provide insights into oncolytic viral therapies armed with WCL.

Author Contributions: Investigation, X.W., N.Z., T.L., X.J., T.Y., K.C. and G.L.; data curation, T.Y. and G.L.; writing—original draft preparation, X.W. and T.Y.; writing—review and editing, X.J. and G.L.; supervision, G.L.; project administration, G.L.; funding acquisition, G.L. and K.C. All authors have read and agreed to the published version of the manuscript.

Funding: This research was funded by National Natural Science Foundation of China grant number 81972281 and 81572986.

Institutional Review Board Statement: Animal studies were approved by the Institutional Animal Care and Use Committee (IACUC) of Zhejiang Sci-Tech University (Identification code: 20210302-1, date of approval: 2 March 2021).

Informed Consent Statement: Not applicable.

Data Availability Statement: Data are contained within the article.

Conflicts of Interest: The authors declare no conflict of interest.

References

1. Sharon, N. Lectins: Carbohydrate-specific Reagents and Biological Recognition Molecules. *J. Biol. Chem.* **2007**, *282*, 2753–2764. [CrossRef]
2. Catanzaro, E.; Calcabrini, C.; Bishayee, A.; Fimognari, C. Antitumor Potential of Marine and Freshwater Lectins. *Mar. Drugs* **2019**, *18*, 11. [CrossRef]
3. Yau, T.; Dan, X.; Ng, C.C.; Ng, T.B. Lectins with potential for anti-cancer therapy. *Molecules* **2015**, *20*, 3791–3810. [CrossRef]
4. Hayashi, K.; Walde, P.; Miyazaki, T.; Sakayama, K.; Nakamura, A.; Kameda, K.; Masuda, S.; Umakoshi, H.; Kato, K. Active Targeting to Osteosarcoma Cells and Apoptotic Cell Death Induction by the Novel Lectin *Eucheuma serra* Agglutinin Isolated from a Marine Red Alga. *J. Drug Deliv.* **2012**, *2012*, 842785. [CrossRef]
5. Sugahara, T.; Ohama, Y.; Fukuda, A.; Hayashi, M.; Kawakubo, A.; Kato, K. The cytotoxic effect of *Eucheuma serra* agglutinin (ESA) on cancer cells and its application to molecular probe for drug delivery system using lipid vesicles. *Cytotechnology* **2001**, *36*, 93–99. [CrossRef]
6. Omokawa, Y.; Miyazaki, T.; Walde, P.; Akiyama, K.; Sugahara, T.; Masuda, S.; Inada, A.; Ohnishi, Y.; Saeki, T.; Kato, K. In vitro and in vivo anti-tumor effects of novel Span 80 vesicles containing immobilized *Eucheuma serra* agglutinin. *Int. J. Pharm.* **2010**, *389*, 157–167. [CrossRef]
7. Chernikov, O.; Kuzmich, A.; Chikalovets, I.; Molchanova, V.; Hua, K.-F. Lectin CGL from the sea mussel *Crenomytilus grayanus* induces Burkitt's lymphoma cells death via interaction with surface glycan. *Int. J. Biol. Macromol.* **2017**. [CrossRef]
8. Liao, J.-H.; Chien, C.-T.H.; Wu, H.-Y.; Wu, K.-F.; Wang, I.; Ho, M.-R.; Tu, I.-F.; Lee, I.-M.; Li, W.; Shih, Y.-L. A Multivalent Marine Lectin from *Crenomytilus grayanus* Possesses Anti-cancer Activity through Recognizing Globotriose Gb3. *J. Am. Chem. Soc.* **2016**, *138*, 4787–4795. [CrossRef]
9. Fujii, Y.; Fujiwara, T.; Koide, Y.; Hasan, I.; Sugawara, S.; Rajia, S.; Kawsar, S.M.A.; Yamamoto, D.; Araki, D.; Kanaly, R.A. Internalization of a novel, huge lectin from *Ibacus novemdentatus* (slipper lobster) induces apoptosis of mammalian cancer cells. *Glycoconj. J.* **2017**. [CrossRef]
10. Shen, Y.; Nemunaitis, J. Fighting cancer with vaccinia virus: teaching new tricks to an old dog. *Mol. Ther. J. Am. Soc. Gene Ther.* **2005**, *11*, 180–195. [CrossRef] [PubMed]
11. Kirn, D.H.; Thorne, S.H. Targeted and armed oncolytic poxviruses: A novel multi-mechanistic therapeutic class for cancer. *Nat. Rev. Cancer* **2009**, *9*, 64–71. [CrossRef] [PubMed]
12. Ekeke, C.N.; Russell, K.L.; Joubert, K.; Bartlett, D.L.; Luketich, J.D.; Soloff, A.C.; Guo, Z.S.; Lotze, M.T.; Dhupar, R. Fighting Fire with Fire: Oncolytic Virotherapy for Thoracic Malignancies. *Ann. Surg. Oncol.* **2021**. [CrossRef]
13. Heo, J.; Reid, T.; Ruo, L.; Breitbach, C.J.; Rose, S.; Bloomston, M.; Cho, M.; Lim, H.Y.; Chung, H.C.; Kim, C.W.; et al. Randomized dose-finding clinical trial of oncolytic immunotherapeutic vaccinia JX-594 in liver cancer. *Nat. Med.* **2013**, *19*, 329–336. [CrossRef] [PubMed]
14. Aitcheson, G.; Pillai, A.; Dahman, B.; John, B.V. Recent Advances in Systemic Therapies for Advanced Hepatocellular Carcinoma. *Curr. Hepatol. Rep.* **2021**, 1–11. [CrossRef]
15. Breitbach, C.J.; Thorne, S.H.; Bell, J.C.; Kirn, D.H. Targeted and armed oncolytic poxviruses for cancer: The lead example of JX-594. *Curr. Pharm. Biotechnol.* **2012**, *13*, 1768–1772. [CrossRef]
16. Kim, S.-G.; Ha, H.K.; Lim, S.-n.; Silva, N.S.D.; Pelusio, A.; Mun, J.H.; Patt, R.H.; Breitbach, C.J.; Burke, J.M. Phase II trial of pexa-vec (pexastimogene devacirepvec; JX-594), an oncolytic and immunotherapeutic vaccinia virus, in patients with metastatic, refractory renal cell carcinoma (RCC). *J. Clin. Oncol.* **2018**, *36* (Suppl. 6), 671–671. [CrossRef]
17. Breitbach, C.J.; Moon, A.; Burke, J.; Hwang, T.H.; Kirn, D.H. A Phase 2, Open-Label, Randomized Study of Pexa-Vec (JX-594) Administered by Intratumoral Injection in Patients with Unresectable Primary Hepatocellular Carcinoma. *Methods Mol. Biol.* **2015**, *1317*, 343–357.
18. Li, G.; Cheng, J.; Mei, S.; Wu, T.; Ye, T. Tachypleus tridentatus Lectin Enhances Oncolytic Vaccinia Virus Replication to Suppress In Vivo Hepatocellular Carcinoma Growth. *Mar. Drugs* **2018**, *16*, 200. [CrossRef]
19. Wu, T.; Xiang, Y.; Liu, T.; Wang, X.; Ren, X.; Ye, T.; Li, G. Oncolytic Vaccinia Virus Expressing Aphrocallistes vastus Lectin as a Cancer Therapeutic Agent. *Mar. Drugs* **2019**, *17*, 363. [CrossRef]
20. Tateno, H.; Ogawa, T.; Muramoto, K.; Kamiya, H.; Saneyoshi, M. Distribution and molecular evolution of rhamnose-binding lectins in *Salmonidae*: isolation and characterization of two lectins from white-spotted Charr (*Salvelinus leucomaenis*) eggs. *Biosci. Biotechnol. Biochem.* **2002**, *66*, 1356–1365. [CrossRef]
21. Vázquez, M.I.; Rivas, G.; Cregut, D.; Serrano, L.; Esteban, M. The vaccinia virus 14-kilodalton (A27L) fusion protein forms a triple coiled-coil structure and interacts with the 21-kilodalton (A17L) virus membrane protein through a C-terminal alpha-helix. *J. Virol.* **1998**, *72*, 10126–10137. [CrossRef] [PubMed]
22. Smith, G.L.; Talbot-Cooper, C.; Lu, Y. How Does Vaccinia Virus Interfere with Interferon? *Adv. Virus Res.* **2018**, *100*, 355–378. [PubMed]
23. Aaronson, D.S.; Horvath, C.M. A road map for those who don't know JAK-STAT. *Science* **2002**, *296*, 1653–1655. [CrossRef]
24. Puhlmann, M.; Gnant, M.; Brown, C.K.; Alexander, H.R.; Bartlett, D.L. Thymidine kinase-deleted vaccinia virus expressing purine nucleoside phosphorylase as a vector for tumor-directed gene therapy. *Hum. Gene Ther.* **1999**, *10*, 649–657. [CrossRef]
25. Kim, M. Replicating poxviruses for human cancer therapy. *J. Microbiol.* **2015**, *53*, 209–218. [CrossRef]

26. Guo, Z.S.; Lu, B.; Guo, Z.; Giehl, E.; Feist, M.; Dai, E.; Liu, W.; Storkus, W.J.; He, Y.; Liu, Z.; et al. Vaccinia virus-mediated cancer immunotherapy: cancer vaccines and oncolytics. *J. Immunother. Cancer* **2019**, *7*, 6. [CrossRef]
27. Jia, X.; Chen, Y.; Zhao, X.; Lv, C.; Yan, J. Oncolytic vaccinia virus inhibits human hepatocellular carcinoma MHCC97-H cell proliferation via endoplasmic reticulum stress, autophagy and Wnt pathways. *J. Gene Med.* **2016**, *18*, 211–219. [CrossRef]
28. Maniatis, T.; Falvo, J.V.; Kim, T.H.; Kim, T.K.; Lin, C.H.; Parekh, B.S.; Wathelet, M.G. Structure and function of the interferon-beta enhanceosome. *Cold Spring Harb. Symp. Quant. Biol.* **1998**, *63*, 609–620. [CrossRef]
29. Villarroya-Beltri, C.; Guerra, S.; Sánchez-Madrid, F. ISGylation—A key to lock the cell gates for preventing the spread of threats. *J. Cell Sci.* **2017**, *130*, 2961–2969. [CrossRef] [PubMed]
30. Diamond, M.S. IFIT1: A dual sensor and effector molecule that detects non-2′-O methylated viral RNA and inhibits its translation. *Cytokine Growth Factor Rev.* **2014**, *25*, 543–550. [CrossRef]
31. Pidugu, V.K.; Pidugu, H.B.; Wu, M.M.; Liu, C.J.; Lee, T.C. Emerging Functions of Human IFIT Proteins in Cancer. *Front. Mol. Biosci.* **2019**, *6*, 148. [CrossRef] [PubMed]
32. Zhu, J.; Ghosh, A.; Sarkar, S.N. OASL-a new player in controlling antiviral innate immunity. *Curr. Opin. Virol.* **2015**, *12*, 15–19. [CrossRef]
33. Asgari, S.; Schlapbach, L.J.; Anchisi, S.; Hammer, C.; Bartha, I.; Junier, T.; Mottet-Osman, G.; Posfay-Barbe, K.M.; Longchamp, D.; Stocker, M.; et al. Severe viral respiratory infections in children with IFIH1 loss-of-function mutations. *Proc. Natl. Acad. Sci. USA* **2017**, *114*, 8342–8347. [CrossRef] [PubMed]
34. Euhus, D.M.; Hudd, C.; LaRegina, M.C.; Johnson, F.E. Tumor measurement in the nude mouse. *J. Surg. Oncol.* **1986**, *31*, 229–234. [CrossRef]
35. Tomayko, M.M.; Reynolds, C.P. Determination of subcutaneous tumor size in athymic (nude) mice. *Cancer Chemother. Pharmacol.* **1989**, *24*, 148–154. [CrossRef]

marine drugs

MDPI

Article

Influence of Carbohydrate Additives on the Growth Rate of Microalgae Biomass with an Increased Carbohydrate Content

Anna Andreeva [1], Ekaterina Budenkova [1], Olga Babich [1], Stanislav Sukhikh [1], Vyacheslav Dolganyuk [1,2], Philippe Michaud [3,*] and Svetlana Ivanova [4,5,*]

[1] Institute of Living Systems, Immanuel Kant Baltic Federal University, A. Nevskogo Street 14, 236016 Kaliningrad, Russia; AnnaAndreeva@kantiana.ru (A.A.); abudenkova@kantiana.ru (E.B.); olich.43@mail.ru (O.B.); stas-asp@mail.ru (S.S.); dolganuk_vf@mail.ru (V.D.)

[2] Department of Bionanotechnology, Kemerovo State University, Krasnaya Street 6, 650043 Kemerovo, Russia

[3] CNRS, SIGMA Clermont, Institut Pascal, Université Clermont Auvergne, F-63000 Clermont-Ferrand, France

[4] Natural Nutraceutical Biotesting Laboratory, Kemerovo State University, Krasnaya Street 6, 650043 Kemerovo, Russia

[5] Department of General Mathematics and Informatics, Kemerovo State University, Krasnaya Street 6, 650043 Kemerovo, Russia

* Correspondence: Philippe.michaud@uca.fr (P.M.); pavvm2000@mail.ru (S.I.); Tel.: +33-473407425 (P.M.); +7-384-239-6832 (S.I.)

Citation: Andreeva, A.; Budenkova, E.; Babich, O.; Sukhikh, S.; Dolganyuk, V.; Michaud, P.; Ivanova, S. Influence of Carbohydrate Additives on the Growth Rate of Microalgae Biomass with an Increased Carbohydrate Content. *Mar. Drugs* **2021**, *19*, 381. https://doi.org/10.3390/md19070381

Academic Editors: Marialuisa Menna, Yuki Fujii, Marco Gerdol, Yasuhiro Ozeki and Hitoshi Sashiwa

Received: 22 May 2021
Accepted: 29 June 2021
Published: 1 July 2021

Abstract: Our study focused on investigating the possibilities of controlling the accumulation of carbohydrates in certain microalgae species (*Arthrospira platensis* Gomont, *Chlorella vulgaris* Beijer, and Dunaliella *salina* Teod) to determine their potential in biofuel production (biohydrogen). It was found that after the introduction of carbohydrates (0.05 $g \cdot L^{-1}$) into the nutrient medium, the growth rate of the microalgae biomass increased, and the accumulation of carbohydrates reached 41.1%, 47.9%, and 31.7% for *Arthrospira platensis, Chlorella vulgaris*, and *Dunaliella salina*, respectively. *Chlorella vulgaris* had the highest total carbohydrate content (a mixture of glucose, fructose, sucrose, and maltose, 16.97%) among the studied microalgae, while for *Arthrospira platensis* and *Dunaliella salina*, the accumulation of total carbohydrates was 9.59% and 8.68%, respectively. Thus, the introduction of carbohydrates into the nutrient medium can stimulate their accumulation in the microalgae biomass, an application of biofuel production (biohydrogen).

Keywords: *Chlorella vulgaris*; *Dunaliella salina*; *Arthrospira platensis*; growth rate; accumulation of carbohydrates; biohydrogen

1. Introduction

Recently, microalgae have attracted attention as a new raw material for biofuel production [1]. Algae biomass has several advantages over land-based energy crops in biofuel production. Microalgae are considered to be the most efficient organisms in converting solar energy. Additionally, microalgae do not require cultivated land, so they do not compete with food crops on arable land [2,3].

Today, microalgae cultivation for lipid production is receiving a large amount of attention [4,5]. After esterification, lipids are used to produce biofuels [6]. Current microalgae research focuses on culture technology to optimize the lipid content of microalgae biomass [7]. However, as an alternative, other algal biomass components can also be used to produce biofuels using biotechnology and thermochemical conversion technology [8,9].

Algae biomass contains different amounts of the most important organic compounds: carbohydrates, proteins, and lipids. Although carbohydrates have a lower energy value compared to other microalgae compounds, they are the best choice or primary raw material for the production of biofuels (such as bioethanol, biobutanol, and biohydrogen) through biotechnological conversion [10–12].

Microalgae produce carbohydrates for two main purposes: they serve as a structural component in the cell wall and as a component for intracellular storage [13,14]. As a storage compound, carbohydrates provide the energy needed for the metabolism of the microalgae and allow them to survive temporarily in the dark when needed [15,16]. Microalgae components (such as proteins, lipids, and carbohydrates) allow them to adapt to changing environmental conditions for their growth [17,18]. Carbohydrates are a broad category that includes sugars (monosaccharides) and their polymers (disaccharides, oligosaccharides, and polysaccharides) [19,20].

The content of carbohydrates in biomass depends on the type of microalgae and growth conditions [21,22]. Certain species of microalgae, for example, *Porphyridium cruentum* (40–57%), and *Spirogyra* sp. (33–64%), contain many carbohydrates [23,24]. However, to maximize biofuel production, it is necessary to combine a high carbohydrate content with the ability of microalgae to produce biomass in significant quantities. Therefore, if the microalgae have a high growth rate or other advantages, the cultivation conditions can be controlled to obtain a higher carbohydrate value. Microalgae carbohydrates are used for the production of biofuels, especially biohydrogen [25,26].

The main factors affecting the carbohydrate content of microalgae are nutrient content, salt stress, light intensity, temperature, and metabolism (autotrophic, heterotrophic, and polytrophic). Besides solar energy and carbon dioxide, microalgae also need nutrients, such as nitrogen, phosphorus, and potassium, to grow. The strategy of decreasing or increasing the amount of nutrients is considered an affordable way to produce carbohydrate-rich microalgae [27]. This is possible as it is relatively easy to control the nutrient content of the medium [27,28].

The biomass accumulation level and carbohydrate productivity are important effectiveness indicators of a microalgae strain intended for use in third-generation biofuel production processes. These parameters are influenced by many conditions, including the concentration and composition of nutrients, temperature, pH, and light. A change in these parameters leads to a change in the biomass composition; often the content and composition of various substances (including carbohydrates) can vary significantly in microalgae, which must be considered when scaling the biomass production process. The optimum growth temperature for the most commonly used microalgae is in the 15–35 °C range, depending on the strain. Certain microalgae strains are highly stress-resistant at high temperatures and high concentrations of CO_2 and NO [29].

In general, the pH values for microalgae cultivation are in the range of pH 4.4–7.9, depending on the strain. The pH of the medium influences not only the microalgae, but also the solubility of CO_2 required for their growth. In certain cases, a steady increase in the microalgae biomass is observed at extremely low pH values. The illumination intensity influences the photosynthesis in chloroplasts of microalgae. The microalgae growth stops with photoinhibition, which occurs after the saturation of their photosystem with photons and the formation of active oxygen forms [30]. In addition to the illumination intensity, the spectral characteristics of light and the light cycle also influence microalgae cultivation. Both a light flux with a wide spectrum (white light) and various LEDs with different spectral characteristics are used, since microalgae absorb light in a narrow range corresponding to the photosynthesis process (up to 700 nm). Several studies have shown that the intensity and spectral characteristics of light affect the accumulation of biomass and carbohydrates [31].

Microalgae absorb organic substances through various mechanisms, such as phosphorylation (for sugars), simple diffusion into cells (for glycerol) or through membrane proteins (for organic acids) [32]. With a mixed [32] or heterotrophic diet [33], Chlorella can be high in carbohydrates [34,35]. However, there is no research into the effect of culture methods on carbohydrate accumulation.

This study aimed to investigate the possibilities of controlling the accumulation of carbohydrates in certain microalgae species (*Arthrospira platensis* Gomont, *Chlorella vulgaris* Beijer, and *Dunaliella salina* Teod) to determine their potential in biofuel production (biohydrogen). To achieve this goal, we planned to sample microalgae, determine their morphology, study the introduction of carbohydrate additives into the nutrient medium, extract chlorophyll, isolate the carbohydrate fraction, determine the carbohydrate content and conduct a statistical analysis of the results.

2. Results

2.1. Microalgae Identification Results

A comparative analysis of the sequences of 16S rRNA and 18S rRNA genes of microalgae is presented in Appendix A, Appendix B, Appendix C. It was found that *Chlorella vulgaris Beijer, Dunaliella salina Teod,* and *Arthrospira platensis Gomont* were isolated from natural sources (soil, water, and sand).

2.2. The Results of the Introduction of Carbohydrates into Nutrient Media for Microalgae Cultivation

Fructose, sucrose, maltose, and glucose were studied as carbohydrates accumulated in microalgae. The results of studies of *Chlorella vulgaris*, *Arthrospira platensis*, and *Dunaliella salina* cultivation processes with the introduction of carbohydrates to the nutrient medium are presented in Figures 1–4.

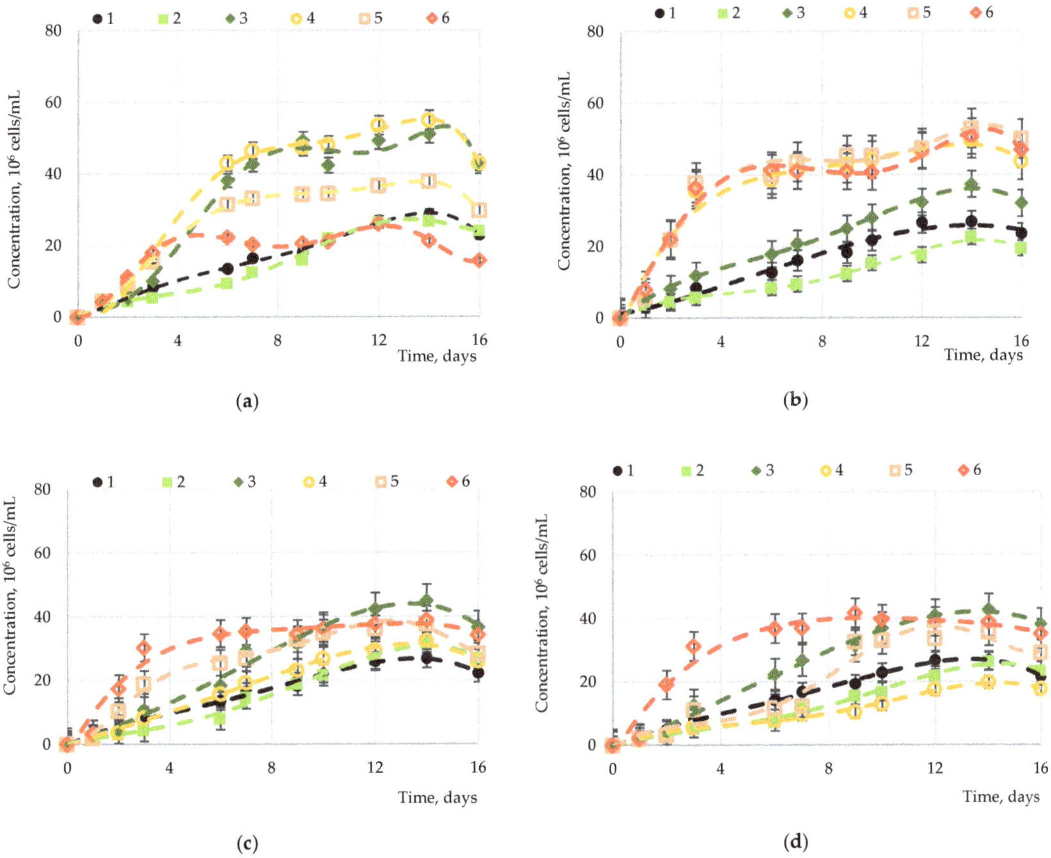

Figure 1. Dependence of the *Chlorella vulgaris* biomass on various sugars: (**a**) fructose, (**b**) glucose, (**c**) maltose, and (**d**) sucrose at various concentrations: 1—control (0.00 g·L^{-1}); 2—0.05 g·L^{-1}; 3—0.10 g·L^{-1}; 4—1.00 g·L^{-1}; 5—2.00 g·L^{-1}; 6—5.00 g·L^{-1}.

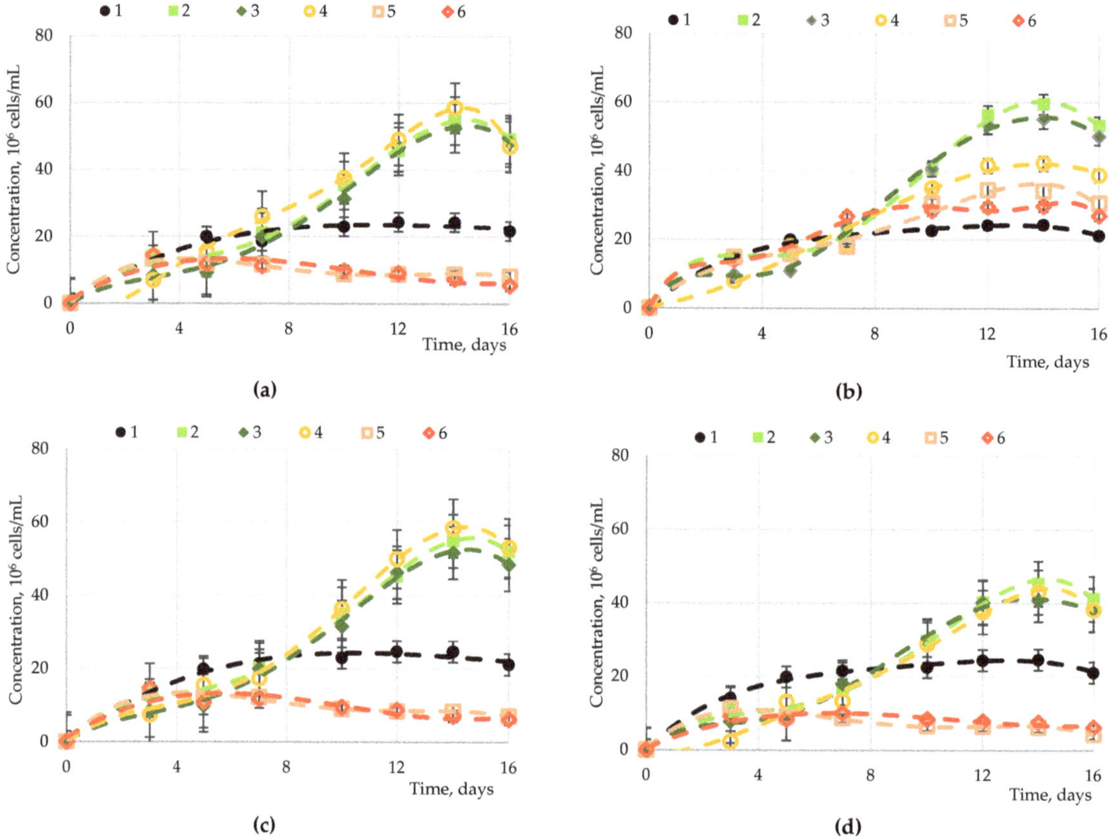

Figure 2. Dependence of the *Arthrospira platensis* biomass on various sugars: (**a**) fructose, (**b**) glucose, (**c**) maltose, and (**d**) sucrose at various concentrations1—control (0.00 g·L^{-1}); 2—0.05 g·L^{-1}; 3—0.10 g·L^{-1}; 4—1.00 g·L^{-1}; 5—2.00 g·L^{-1}; 6—5.00 g·L^{-1}.

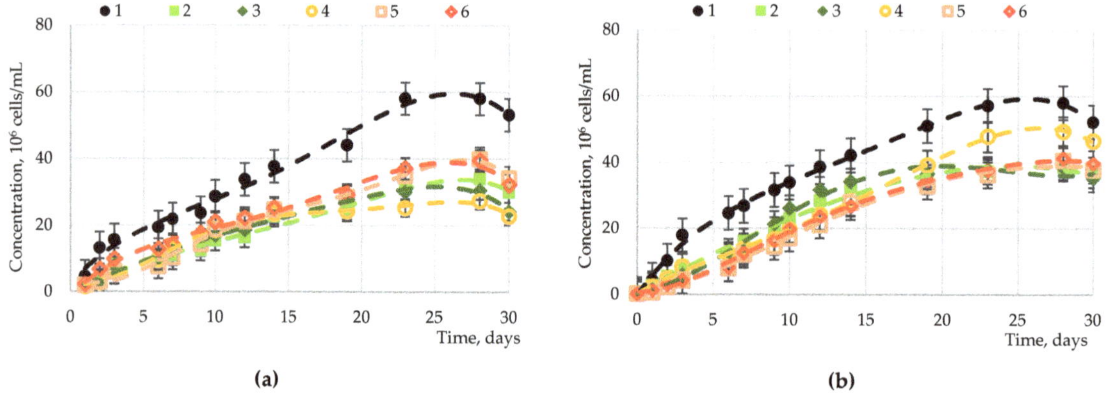

Figure 3. *Cont.*

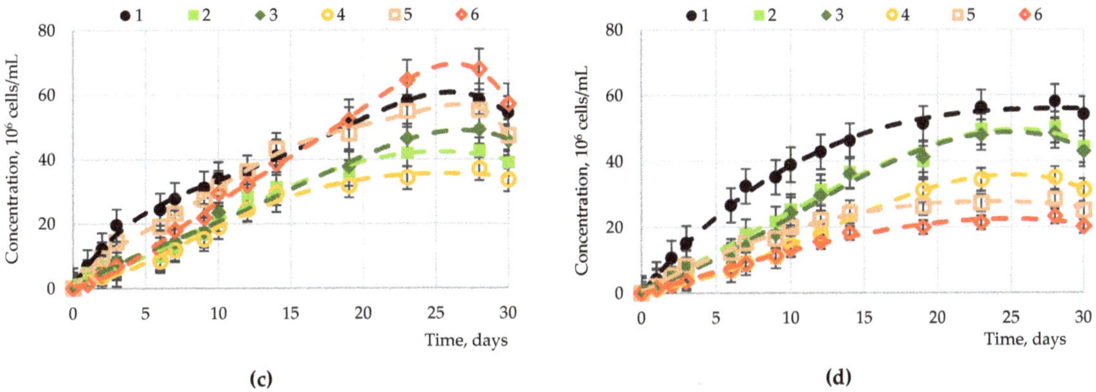

Figure 3. Dependence of the *Dunaliella salina* biomass on various sugars: (**a**) fructose, (**b**) glucose, (**c**) maltose, and (**d**) sucrose at various concentrations: 1—control (0.00 g·L^{-1}); 2—0.05 g·L^{-1}; 3—0.10 g·L^{-1}; 4—1.00 g·L^{-1}; 5—2.00 g·L^{-1}; 6—5.00 g·L^{-1}.

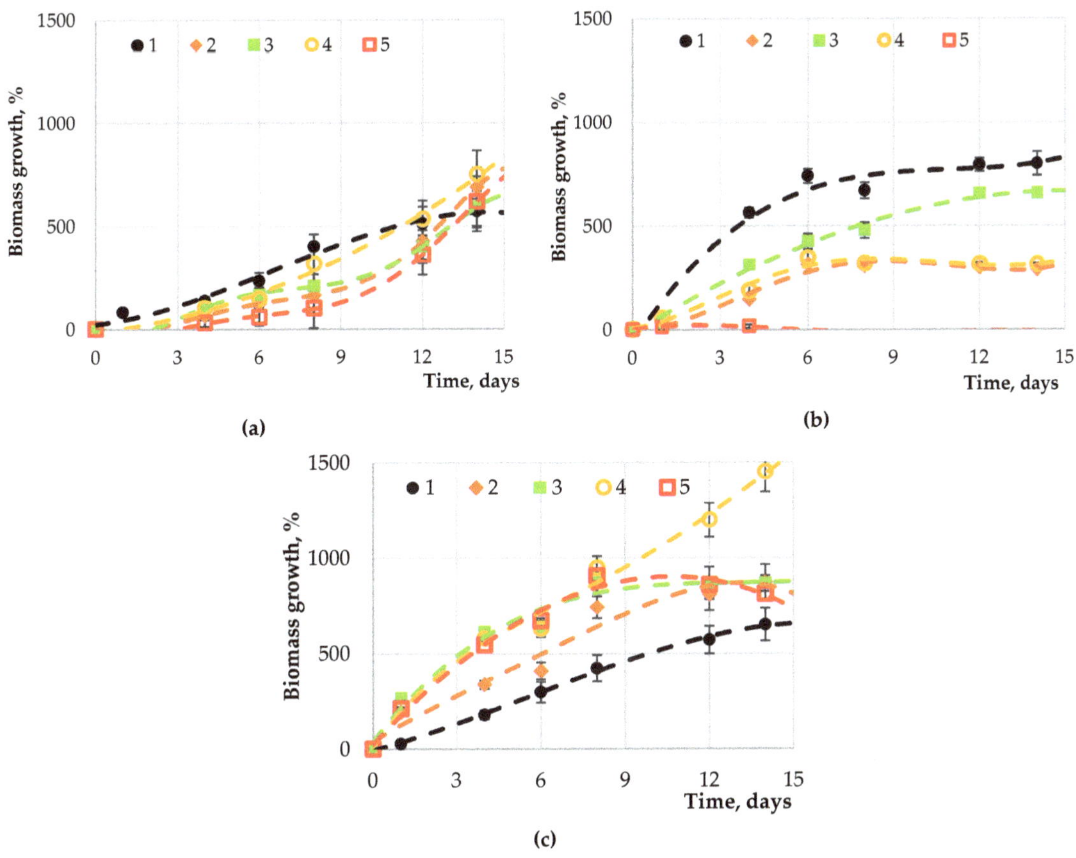

Figure 4. Dependence of the (**a**) *Chlorella vulgaris*, (**b**) *Arthrospira platensis*, and (**c**) *Dunaliella salina* biomass with carbohydrate concentration of 0.05 g·L^{-1} from the concentration medium: 1—control (without carbohydrates); 2—with added fructose; 3—with added glucose; 4—with added maltose; 5—with added sucrose.

Statistically significant differences (p-value < 0.05, Duncan's test) in biomass accumulation were observed for *Chlorella vulgaris* (Figure 1) upon application of fructose of different concentrations (from 0 to 5 $g \cdot L^{-1}$). No differences were observed at concentrations of 0.00, 0.05, or 5.00 $g \cdot L^{-1}$, as well as at 0.10 or 1.00 $g \cdot L^{-1}$. The concentrations of 0.00 and 0.05 are not sufficient to significantly affect the biomass growth, while 5.00 $g \cdot L^{-1}$, despite a significant increase at the initial stage, did not lead to significant biomass accumulation, apparently being the concentration of saturation of the medium with this carbohydrate. This is confirmed by the data corresponding to the concentration of 2.00 $g \cdot L^{-1}$. The introduction of glucose at a concentration from 1.00 to 5.00 $g \cdot L^{-1}$ caused significant biomass accumulation, while the biomass growth indicators did not differ significantly. Despite a significant increase in the initial period (up to 3 days) at a concentration of 5.00 $g \cdot L^{-1}$, the concentration of maltose introduced into the cultivation medium did not significantly affect the biomass growth. However, at a concentration of 0.01 $g \cdot L^{-1}$ with a shorter cultivation period (10 days), a slight increase was observed compared to other cultivation modes. The introduction of sucrose in an amount of 5.00 $g \cdot L^{-1}$ led to a significant increase in biomass at the initial stage (3 days), the values of which were not exceeded either at lower concentrations of sucrose or with an increase in the cultivation duration to 14 days. The highest biomass growth rates were achieved with the introduction of fructose and glucose with approximately equal effects.

For *Arthrospira platensis* (Figure 2), statistically significant differences in the accumulation of biomass were observed upon the introduction of fructose at various concentrations (from 0 to 5 $g \cdot L^{-1}$). The introduction of fructose at a concentration of up to 1.0 $g \cdot L^{-1}$ led to an increase in microalgae biomass. At the same time, an increase in the concentration of fructose of 2.00 $g \cdot L^{-1}$ and above led to a significant decrease in biomass accumulation compared to the control. There were no significant differences at concentrations from 0.05 to 1.00 $g \cdot L^{-1}$, or from 2.00 to 5.00 $g \cdot L^{-1}$. The dependence of the biomass accumulation upon the addition of glucose at a concentration from 0.05 to 5.00 $g \cdot L^{-1}$ had the opposite effect; with an increase in the glucose concentration, the biomass accumulated more slowly. The greatest increase in biomass was observed at a concentration of 0.05 and 0.10 $g \cdot L^{-1}$ (p-value < 0.05, Duncan's test). At the initial stage of concentration (up to 3 days), no significant differences in biomass accumulation dependent on the maltose concentration were observed. With further cultivation (more than 7 days), the process of biomass accumulation at a maltose concentration in the cultivation medium from 0.05 to 1.00 $g \cdot L^{-1}$ significantly differed from the control and from the cultivation process at high maltose concentrations (p-value = 0.918–0.922 and 0.894–0.942, respectively). The introduction of sucrose in an amount of more than 2.00 $g \cdot L^{-1}$ led to the inhibition of overall biomass growth throughout the entire time interval of cultivation. When cultured for more than 7 days, the biomass accumulation was significantly different (p-value > 0.05, Duncan's test) at a carbohydrate concentration from 0.05 to 2.00 $g \cdot L^{-1}$. The greatest biomass growth indicators were achieved with the introduction of fructose, glucose, or maltose, with approximately equal effects.

The addition of maltose at a concentration of 5.00 $g \cdot L^{-1}$ had a statistically significant effect on *Dunaliella salina* biomass accumulation (Figure 3) compared to the control cultivation. A lower maltose concentration led to a decrease in the growth rate of the biomass of this microalga. The introduction of sucrose led to a significant decrease in the accumulation of biomass in comparison with the control.

The highest indicators of biomass growth were achieved with the introduction of maltose; the rest of the carbohydrates led to a negative impact on the biomass during cultivation.

Thus, for *Chlorella vulgaris*, the maximum biomass growth rate was observed during cultivation with intermittent stirring and the introduction of glucose (1–5 $g \cdot L^{-1}$), fructose (0.1–1.0 $g \cdot L^{-1}$), maltose (2–5 $g \cdot L^{-1}$), and sucrose (5 $g \cdot L^{-1}$). When a higher concentration of carbohydrates (at least 1 $g \cdot L^{-1}$) was introduced into the medium, strong adhesion/aggregation of cells was noted, which complicated mixing and sampling. The contamination risk also increased. The optimal results were achieved by introducing glucose and fructose additives at a concentration of up to 1 g·L^{-1}. For *Dunaliella salina*, the maximum biomass growth rate was observed during cultivation under constant stirring and the introduction of maltose to the medium at a rate of 0.05 g·L^{-1} two weeks after the start of growth. The introduction of maltose with a higher concentration and the introduction of fructose, sucrose, and maltose did not significantly affect the growth rate. For *Arthrospira platensis* Gomont, the maximum biomass growth rate was observed during cultivation with constant stirring and the introduction of glucose to the medium at a rate of 0.05–1.00 g·L^{-1} two weeks after the start of growth. The introduction of glucose in a higher concentration and the introduction of fructose, sucrose, and maltose led to the inhibition of culture growth, aggregation of algal cells, and death.

No significant differences (p-value > 0.05, Duncan's test) were observed when introducing carbohydrates (fructose, glucose, maltose, or sucrose) of the same concentration (0.05 g·L^{-1}) to the culture medium (Figure 4) of *Chlorella vulgaris*. The introduction of carbohydrates led to a significantly negative effect on the *Arthrospira platensis* biomass growth compared to the control, while maltose had a positive effect on *Dunaliella salina*.

2.3. Results of Determining the Content of Carbohydrates in Microalgae

The results of determining the content of carbohydrates in microalgae are presented in Table 1.

Table 1. Microalgae carbohydrate content before extraction.

Carbohydrates	Concentration, % of Dry Matter					
	I		II		III	
	exp.	Control	exp.	Control	exp.	Control
1	0.12 ± 0.02 ^{a}	0.10 ± 0.01	0.39 ± 0.08 ^{a}	0.42 ± 0.05	0.38 ± 0.01 ^{a}	0.33 ± 0.01
2	11.26 ± 0.72 ^{b}	9.78 ± 0.22	3.41 ± 0.62 ^{b}	2.32 ± 0.30	1.78 ± 0.29 ^{b}	1.60 ± 0.22
3	0.12 ± 0.03 ^{a}	0.11 ± 0.02	0.21 ± 0.02 ^{a}	0.20 ± 0.02	0.21 ± 0.01 ^{a}	0.22 ± 0.01
4	0.33 ± 0.06 ^{a}	0.27 ± 0.03	0.31 ± 0.09 ^{a}	0.32 ± 0.07	0.11 ± 0.06 ^{a}	0.10 ± 0.02
5	16.97 ± 0.64 ^{c}	15.12 ± 0.48	9.59 ± 0.20 ^{c}	7.81 ± 0.76	8.68 ± 0.74 ^{c}	7.54 ± 0.30

1—fructose; 2—lucose; 3—maltose; 4—sucrose; 5—total glucose, fructose, sucrose, and maltose; I—*Chlorella vulgaris*; II—*Arthrospira platensis*; III—*Dunaliella salina*; exp.—experimental samples; control—no added carbohydrates. Data presented as a mean ± SD (n = 3). Values in a columns followed by the same letter do not differ significantly (p-value > 0.05), as assessed by the post hoc test (Duncan's test).

No significant difference (p-value < 0.05, Duncan's test) was found in the accumulation of fructose, maltose, and sucrose in all microalgae under the chosen culture regime. At the same time, the accumulation of glucose significantly differed both between the samples of different microalgae and compared to the control. The same was observed for the complex of carbohydrates; apparently, the greatest contribution of its quantitative formation also relates to glucose.

After the introduction of a carbohydrate additive to the nutrient medium, the accumulation of carbohydrates in the microalgae biomass was 46.4%, 47.9%, 42.3%, and 38.4% for *Chlorella vulgaris*; 35.9%, 41.1%, 37.1%, and 30.2% for *Arthrospira platensis*; 28.6%, 31.7%, 27.6%, and 25.3% for *Dunaliella salina*, respectively, of fructose, glucose, maltose, sucrose, and total carbohydrates (Table 1). *Chlorella vulgaris* was distinguished by the highest content of total carbohydrates (a mixture of glucose, fructose, sucrose, and maltose, 16.97%) among the studied microalgae. While for *Arthrospira platensis* and *Dunaliella salina*, this figure was only 9.59% and 8.68%, respectively.

Chlorella vulgaris had the greatest glucose content (11.26%); the lowest content was observed in *Dunaliella salina* (1.78%). Fructose content prevailed in *Arthrospira platensis* (0.39%); the lowest content was observed in *Chlorella vulgaris* (0.12%). *Chlorella vulgaris* had the greatest sucrose content (0.33%). The highest maltose content was found in *Arthrospira platensis* (0.31%), and the same amount (0.21%) in *Arthrospira platensis* and *Dunaliella salina*.

The carbohydrate content of microalgae after extraction is shown in Table 2.

Table 2. Amount of carbohydrate in microalgae after extraction.

Carbohydrates	Concentration, % of Dry Matter					
	I		II		III	
	exp.	Control	exp.	Control	exp.	Control
1	0.11 ± 0.02 [a]	0.10 ± 0.01	0.37 ± 0.08 [a]	0.42 ± 0.05	0.33 ± 0.01 [a]	0.33 ± 0.01
2	10.48 ± 0.72 [b]	9.78 ± 0.22	3.40 ± 0.62 [b]	2.32 ± 0.30	1.76 ± 0.29 [b]	1.60 ± 0.22
3	0.11 ± 0.03 [a]	0.11 ± 0.02	0.19 ± 0.02 [a]	0.20 ± 0.02	0.20 ± 0.01 [a]	0.22 ± 0.01
4	0.29 ± 0.06 [a]	0.27 ± 0.03	0.30 ± 0.09 [a]	0.32 ± 0.07	0.10 ± 0.06 [a]	0.10 ± 0.02
5	15.99 ± 0.64 [c]	15.12 ± 0.48	8.98 ± 0.20 [c]	7.81 ± 0.76	8.66 ± 0.74 [c]	7.54 ± 0.30

1—fructose; 2—lucose; 3—maltose; 4—sucrose; 5—total glucose, fructose, sucrose, and maltose; I—*Chlorella vulgaris;* II—*Arthrospira platensis;* III—*Dunaliella salina;* exp.—experimental samples; control—no added carbohydrates. Data presented as a mean ± SD (n = 3). Values in a columns followed by the same letter do not differ significantly ($p > 0.05$), as assessed by the post hoc test (Duncan's test).

The concentration of carbohydrates (Tables 1 and 2) remained practically unchanged after extraction, which indicates that almost all carbohydrates were transferred to the extract.

The amount of residual carbohydrates in nutrient media after microalgae cultivation was determined spectrophotometrically after acid hydrolysis. The results of determining the residual amount of carbohydrates in the nutrient medium are presented in Table 3. The dynamics of the total consumption of sugars over time by microalgae during cultivation is presented in Table 4.

Table 3. Residual amounts of carbohydrates in nutrient media.

Carbohydrates	Concentration, % of Dry Matter		
	I	II	III
1	1.0 ± 0.2 [a]	3.0 ± 0.6 [a]	2.3 ± 0.3 [a]
2	1.3 ± 0.2 [a]	4.1 ± 0.5 [b]	9.1 ± 0.6 [b]
3	2.1 ± 0.3 [b]	1.2 ± 0.2 [c]	2.0 ± 0.1 [a]
4	2.0 ± 0.3 [b]	2.0 ± 0.3 [c]	1.1 ± 0.2 [c]
5	1.9 ± 0.4 [b]	1.1 ± 0.2 [c]	0.9 ± 0.4 [c]

1—fructose; 2—glucose; 3—maltose; 4—sucrose; 5—total glucose, fructose, sucrose, and maltose; I—*Chlorella vulgaris* culture medium; II—*Arthrospira platensis* culture medium; III—*Dunaliella salina* culture medium; exp. —experimental samples; control—no added carbohydrates. Data presented as a mean ± SD (n = 3). Values in a columns followed by the same letter a, b or c do not differ significantly (p-value > 0.05), as assessed by the post hoc test (Duncan's test).

Table 4. Total consumption of sugars by microalgae over time.

Cultivation Time, Days	Concentration, % of Dry Matter		
	I	II	III
0	0.08 ± 0.02 [a]	0.09 ± 0.03 [a]	0.02 ±0.06 [a]
4	2.43 ± 0.31 [b]	1.59 ± 0.05 [b]	0.45 ± 0.14 [b]
8	9.88 ± 0.46 [c]	5.29 ± 0.18 [c]	0.78 ± 0.23 [b]
12	16.97 ± 0.64 [d]	9.59 ± 0.20 [d]	1.13 ± 0.51 [c]
16	4.58 ± 0.34 [e]	2.62 ± 0.12 [e]	1.27 ± 0.62 [c]
20	-	-	1.69 ± 0.66 [c]
24	-	-	2.37 ±0.71 [d]
28	-	-	8.68 ± 0.74 [e]
30	-	-	0.97 ± 0.47 [b]

I—*Chlorella vulgaris;* II—*Arthrospira platensis;* III—*Dunaliella salina;* exp.—experimental samples; control—no added carbohydrates. Data presented as a mean ± SD (n = 3). Values in a columns followed by the same letter a, b, c, d or e do not differ significantly (p-value > 0.05), as assessed by the post hoc test (Duncan's test).

A significant decrease in the content of residual carbohydrates in nutrient media after microalgae cultivation (Table 3) and the dynamics of total consumption of sugars by microalgae (Table 4) lead to the conclusion that most carbohydrates transfer from the nutrient medium into microalgae.

No significant effect (p-value < 0.05, Duncan's test) of the fructose, maltose or sucrose introduction into the cultivation medium on the carbohydrate content in the culture medium or on the biomass of microalgae (Tables 2 and 3) was found. Glucose and a mixture of carbohydrates had a statistically significant effect. The residual amount of carbohydrates in nutrient media after growing microalgae decreased by approximately 10 times (Table 2). Such a sharp decrease in the amount of carbohydrates indicates that most of the carbohydrates converted into microalgae biomass during cultivation. After extraction, the amount of carbohydrates was almost identical to that in the algae biomass before extraction.

The total carbohydrate content significantly differed in specialized media (with carbohydrate content) from the control after 8 h of cultivation for all studied algae species. The accumulation of the total amount of carbohydrates during microalgae cultivation period occurred gradually.

3. Discussion

Microalgae attract the interest of researchers as a potential source of useful components, including carbohydrates. Ho et al. [36] studied three microalgae isolates for their ability to produce carbohydrates. Among them, *Chlorella vulgaris* FSP-E demonstrated a relatively high rate of cell growth and carbohydrate accumulation. The ability of *C. vulgaris* FSP-E to produce carbohydrates has been further improved through engineering strategies. The results show that using a suitable light intensity and a reasonable inoculum volume was accompanied by cell growth and increased carbohydrate productivity. As a result of nitrogen starvation for 4 days, the carbohydrate content in microalgae reached 51.3%. In our case, the accumulation of carbohydrates ranged from 30.2% in *Arthrospira platensis* to 47.9% in *Chlorella vulgaris*. Under optimal conditions, the maximum productivity of biomass and carbohydrates was 1.437 and 0.631 $g \cdot L^{-1}$ per day^{-1}, respectively. Since glucose accounts for almost 93% of the carbohydrates accumulated in *C. vulgaris* FSP-E, microalgae are a promising raw material for bioethanol fermentation.

Markou et al. [18] reported that during the cultivation of microalgae, the carbohydrate content increased with the development of cultivation, and the protein content increased with the later cultivation of diatoms *Rhodomonas sp.*, and decreased in *I. galbana*, *P. lutheri*, and *T. suecica*. *Rhodomonas* sp., and *C. calcitrans* showed lower daily productivity in semi-continuous cultivation. Carbohydrates increased with culture growth, reaching 53.10% and 48.35%, respectively. When comparing the data, we concluded that the results of studying the accumulation of carbohydrates during the cultivation of microalgae, obtained by us and Markou et al. [18], agree, since the studied microalgae and those in our study are multicellular organisms. Microalgae were cultivated at changing operating characteristics (concentration and composition of nutrients, temperature, pH, and light). Changes in these parameters had a significant effect on microalgae biomass composition, as in our studies.

For *Rhodomonas* sp., the highest carbohydrate level peaked in the late resting stage and was 40.24%. In the case of *T. suecica*, carbohydrates accumulate during growth, accounting for 43.23%. A continuous increase in carbohydrate content was observed in the diatoms *P. tricornutum* and *C. calcitruns*, reaching 25.0% and 11.32%, respectively. Except for hybrids, carbohydrate levels (expressed as a percentage) in all tested species increased as the reproductive process intensified.

Chou et al. [37] confirm this trend for *T. suecicu*. In a semi-continuous culture, the production of all carbohydrates is limited, except for *H. akashiwo*.

Liu et al. [38] proposed a method for obtaining carbohydrates from *Arthrospira platensis*. In open industrial reservoirs, microalgae grew in conditions of nitrogen deficiency. The maximum yield of biomass and carbohydrates was 27.5 and 26.2 g/m^2 per day, respectively. By homogenization under pressure, the carbohydrates were extracted with hot water and purified by flocculation. In our study, *Arthrospira platensis* accumulated slightly more carbohydrates—30.2%, as carbohydrates were introduced not independently, but in a complex (glucose, fructose, and maltose).

El-Ahmady El-Naggar et al. [39] extracted and identified water-soluble polysaccharides from the microalga *Chlorella* to use as a plant growth stimulant.

Gaignard et al. [40] studied 166 species of marine microalgae and cyanobacteria to identify strains producing original extracellular polysaccharides. Furthermore, 45 strains with the required characteristics were isolated. Eight new genera of microalgae have been discovered that produce extracellular polysaccharides, including polymers with an original composition.

Dolganyuk et al. [41] found during their study that, in comparison with other cells of microalgae, cultures of microalgae, *Chlorella vulgaris* Beijer, *Arthrospira platensis* Gomont, and *Dunaliella salina* Teod, have characteristics of increased carbohydrate content: 27.36% ± 0.76% and 21.95% ± 0.74%, respectively. The mass fraction of carbohydrates in the biomass of *Cellulopsis obliquus* reached 13.69 ± 0.34 g/m^2 per day. The content of carbohydrates in the biomass of the microalga *Nannochloropsis gaditana* was 15.34 ± 0.51 g/m^2 per day. The amount of synthetic carbohydrates in *Chlamydomonas reinhardtii* and *Neochloris cohaerens* was 12.48% ± 0.34% and 12.58% ± 0.34% g/m^2 per day, respectively [41]. In our study, the microalgae biomass accumulated more carbohydrates (*Chlorella vulgaris* by 13–14%; *Arthrospira platensis* by 3–4%; *Dunaliella salina* by 20–22%), apparently due to the well-chosen dosages of carbohydrates added to the cultivation medium.

In all the studies [36–41], stirring during the cultivation of microalgae was a factor that reduces their ability to produce carbohydrates, as in our study.

4. Materials and Methods

4.1. Microalgae Sampling

To determine the research objects, samples of microalgae were taken from natural sources (water, sand, and soil) in the period from October 2020 to December 2020 in various regions of the Kaliningrad Oblast Lake Vištytis (54°25′37″ N 22°43′30″ E), Lake Chaika (56°03′49″ N 29°04′50″ E), Lake Yantarnoye (56°01′44″ N 30°44′03″ E), Curonian Lagoon (55°07′00″ N 21°01′00″ E), Strait of Baltiysk (59°43′ N 28°24′ E), Baltic Sea coast (54°42.4′0″ N 20° 30.4′0″ E), and Lake Krasnoye (54°25′59″ N 22°30′27″ E)).

Microalgae were sampled with a box-shaped bottom sampler [42], developed at the Institute for Biology of Inland Waters of the Russian Academy of Sciences (IBIW) (Borok, Russia), covering a square area of the bottom 160 × 160 mm in size with a maximum immersion depth of 440 mm in bottom sediments; a 400 mm^2 sample was taken. Immediately after transportation to the shore, test cores were taken using plastic tubes with an inner diameter of 45 mm. The tubes were sealed at both ends and stored in an upright position at +4 °C. In the laboratory, the core was cut lengthwise and halved using two thin stainless steel plates inserted into the cut. The core halves were then divided into transverse samples (slices) with a step of 5–10 mm. All samples were stored at –20 °C in the dark, in air-tight plastic bags, from which samples of microalgae were taken for research [42].

Pure microalgae cultures were isolated, and strains capable of actively accumulating biomass and target products (lipids, proteins, and carbohydrate–mineral complex) and suitable for cultivation in laboratory conditions from enrichment cultures in which their growth was observed, were identified. The studied samples of natural sources (water, sand, and soil) were introduced into a standard BBM nutrient medium (Stylab, Moscow, Russia) to obtain enrichment cultures to obtain enrichment cultures, purchased from Stylab, Moscow, Russia. For this research, 128 samples of natural sources, taken in various regions of the Kaliningrad region, were used, of which 27 samples showed the growth and development of microalgae at the initial stage of obtaining enrichment cultures.

To identify isolated from the enrichment culture strains of microorganisms (microalgae), partial sequences of the 18S and/or 16S rRNA gene were determined, after which a comparative analysis was performed with the known sequences from the Genbank database. The results of the comparative analysis (Appendix A, Appendix B, Appendix C) of the 18S rRNA gene sequence indicate that the following microalgae were isolated from natural sources (soil, water, and sand): *Chlorella vulgaris* Beijer, *Dunaliella salina* Teod, and *Arthrospira platensis* Gomont.

4.2. Microalgae Biomass Cultivation

Microalgae cultivation and the biomass production were carried out on a standard nutrient medium recommended by IPPAS (cellreg.org) (UNIQUE SCIENTIFIC INSTALLATION COLLECTION OF MICROALGAE AND CYANOBACTERIA IPPAS IPP RAS, Moscow, Russia) under conditions of red-white light (~80 ± 10 μE m^{-2} s^{-1}), at a temperature of 25 ± 1 °C, with constant and intermittent stirring for 16 days in the case of *Arthrospira platensis* Gomot and *Chlorella vulgaris* Beijer and 30 days in the case of *Dunaliella salina* Teod. Further cultivation is futile and leads to a decrease in the accumulation of microalgae biomass. Shihira-Ishikawa medium (LLC "MicroTech", Moscow, Russia) was used to culture *Chlorella vulgaris* Beijer; Zarrouk medium (LLC "MicroTech", Moscow, Russia) was used to culture *Arthrospira platensis* Gomont and produce its biomass; Omarov's medium (LLC "MicroTech", Moscow, Russia) was used to produce *Dunaliella salina* Teod biomass. The culture media were sterilized by autoclaving. The microelements of the Zarruk medium were sterilized by filtration through a filter with a pore diameter of 0.22 μm and added after autoclaving into culture media cooled to room temperature [43]. The biomass growth was assessed through the absorption levels of algae samples in the culture medium at 750 nm (Shimadzu, Kyoto, Japan).

Microalgae were cultivated until the required amount of biomass of the studied samples was obtained.

The cell concentration was counted under an AxioScope A1 microscope (Zeiss, Oberkochen, Germany) applying Goryaev's cell counting chamber (MiniMed LLC, Bryansk region, Russia) by direct counting of the total number of cells in 1 mL of suspension (OFS.1.7.2.0008.15 determination of concentration microbial cells, Ministry of Health of the Russian Federation). The number of cells in 5 horizontal and 15 diagonal large squares was counted, and the number of cells (X) in 1 mL of the suspension under study was determined by the following equation:

$$X = a \cdot 12499 \cdot b, \quad (1)$$

where a—number of cells in 20 squares; b—dilution of the initial microorganism suspension.

The relative increase in the microalgae biomass was determined by the following equation [44]:

$$X = \frac{(m1 - m2)}{m1} \cdot 100\%, \quad (2)$$

where m1—microalgae mass through the entire growing period, mg; m2—microalgae mass each cultivation day.

4.3. Microalgae Morphology Determination

The morphology of microalgae was determined at 40× magnification using a binocular microscope MC-300 (Micros, Vienna, Austria) [45].

4.4. Introduction of Carbohydrate Additives to the Nutrient Medium

Fructose, sucrose, maltose, and glucose (LLC "Moskhimtorg", Moscow, Russia) were used as carbohydrate additives. Carbohydrates were added to the nutrient medium in amounts of 0.05, 0.10, 1.00, 2.00, and 5.00 g·L^{-1}. Control—cultivation medium without added carbohydrates.

4.5. Chlorophyll Extraction

A 1 mL microalgae suspension sample was centrifuged for 20 min at 3400 rpm, and the supernatant was removed. After that, 10 mL of ethanol (96%) was added. The mixture was incubated for 10 min at 25 °C in a water bath and centrifuged for 20 min at 3400 rpm (the sediment is the discolored algae biomass; the liquid phase is the pigment extract). Then, 0.3 g of activated carbon was added to the liquid phase, gently mixed for several seconds, and filtered (to prevent partial ingress of pigments in the liquid phase into the precipitate) [46]. All reagents were purchased from LLC "Moskhimtorg", Moscow, Russia.

4.6. Carbohydrate Fraction Isolation

The carbohydrate fraction was isolated as follows. An amount of 1 mL of sulfuric acid (72%) was added to a weighed portion of dry algae (100 mg) in a glass flask, incubated at 30 °C for 1 h, and 28 mL of distilled water was added, then it was autoclaved for 1 h at 120 °C. Furthermore, it was quickly cooled, and 1 mL of the sample was taken and centrifuged at 8000 rpm for 5 min. Hydrochloric acid was used to isolate fructose. All reagents were purchased from LLC "Moskhimtorg", Moscow, Russia. Additionally, the method of high-performance liquid chromatography (HPLC) was used applying an HPLC NGC chromatograph (Bio-rad, Berkeley, CA, USA) on a Uno-Q1 column (Bio-rad, Berkeley, CA, USA) in the mode of gradient pH 2.5–8.9. The eluent comprised the following buffer solutions: phase A—citrate-phosphate buffer with pH 2.5; B—tris-glycine buffer pH 8.5; gradient phase B 0—100% for 15 column volumes (1 column volume—1 mL). Chromatography parameters were as follows: run: 04; trace type: λ 3 (280 nm); best fit: 8; slope: 10; sensitivity: medium; size: N/A.

4.7. Determination of Carbohydrate Content

Experimental samples were cultured under conditions similar to those previously described. The carbohydrate additive was a mixture of glucose, fructose, and maltose; the concentration of each carbohydrate was 0.1 $g{\cdot}L^{-1}$. The concentrated carbohydrate solution was sterilized and added to the experimental flasks with the medium (20 μL each). Samples for analysis were taken at regular intervals once a day under sterile conditions.

The content of carbohydrates in the sample was determined by the phenol-sulfuric acid method; a weighed portion of dry algae (10 mg) was dissolved in distilled water (10 mL). Then, 1 mL of the sample was introduced into a glass flask, and 3 mL of sulfuric acid (72%) and 1 mL of phenol (5%) were added. To determine the fructose content, 1 mL of resorcinol (0.1%) and 3 mL of concentrated hydrochloric acid were added to the sample and kept in a water bath for 5 min at 90 °C. Next, calibration solutions were prepared with a known concentration of carbohydrates (glucose, fructose, sucrose, and maltose mixture), and the absorption was measured on a spectrophotometer at 490 nm relative to glucose and fructose, and at 440 nm for sucrose and at 545 nm for maltose (the absorption maximum was checked using standard solutions in fivefold repetition). The carbohydrate content determination was carried out on cultures in the exponential growth phase (determined spectrophotometrically, 750 nm). The spectrophotometer was calibrated using the dry weight method. The dry residue of the biomass of multicellular microalgae was dissolved in carbon tetrachloride (to obtain a solution with a microalgae content of 200 mg/mL), and the light absorption of the solution was investigated at 5.0–9.0 nm with respect to carbon tetrachloride. The applied method guaranteed the accuracy of the data obtained [43]. All reagents were purchased from LLC "Moskhimtorg", Moscow, Russia.

4.8. Determination of the Residual Amount of Sugars in Nutrient Media

Acid hydrolysis was used to extract residual sugar from the culture medium. A fixed amount of the collected culture medium (15 g/L, resuspended in distilled H_2O) was used as a substrate for analyses. Various concentrations of sulfuric acid (47, 94, 188, 281, and 563 mM) were tested to determine the most effective concentration. Hydrolysis analyses were performed in Erlenmeyer flasks. The reaction proceeded at 100 °C for 30 min (Waiser Lab. Products NC EST–011). Samples were cooled at room temperature and then centrifuged at 3200× *g* at 20 °C for 8 min (Excelsa ®II model 206 BL). The supernatant containing residual sugars was collected, and its pH was adjusted to 5.5 using 1 M NaOH.

The residual sugar concentration was analyzed by the DNS (dinitrosalicylic acid) method with glucose as the standard for the calibration curves. After mixing 0.75 mL of glucose with 0.5 mL of DNS reagent, the samples were heated at 100 °C for 5 min. The samples were cooled to room temperature, and then 3 mL of water was added. Sugar concentrations were determined spectrophotometrically (Varian, Inc. Cary ®50 UV-Vis) at 540 nm.

4.9. Statistical Analysis

One-way analysis of variance (ANOVA) was performed to minimize the risk of misjudgment of a type 1 error in the case of multiple comparisons. The correspondence of the samples used to the normal distribution was checked by the *t*-test (mathematical expectations) for independent samples and Fisher's test (variance). Post hoc analysis (Duncan's test) was undertaken to identify samples that were significantly different from each other. The equality of the variances of the extracted samples was checked using the Levene test [47]. Significant differences were considered as a *p*-value < 0.05.

5. Conclusions

The accumulation of carbohydrates in certain species of microalgae (*Arthrospira platensis* Gomont, *Chlorella vulgaris* Beijer, and *Dunaliella salina* Teod) was studied to determine their potential for biofuel production (biohydrogen). It was found that after the introduction of carbohydrates ($0.05\ g \cdot L^{-1}$) into the nutrient medium, the growth rate of the microalgae biomass increased, and the accumulation of carbohydrates reached 41.1%, 47.9%, and 31.7% for *Arthrospira platensis, Chlorella vulgaris*, and *Dunaliella salina*, respectively. *Chlorella vulgaris* had the highest total carbohydrate content (the sum of glucose, fructose, sucrose, and maltose was 16.97%).

The introduction of carbohydrates to culture media can be used to produce microalgal biomass enriched with these biopolymers. For certain technologies, carbohydrates play a key role in biomass conversion.

However, the accumulation of carbohydrates in biomass is significantly influenced by the species of microalgae and the initial growth conditions; therefore, further research focused on optimizing, and, where possible, standardizing the conditions for cultivating microalgae is needed. The high carbohydrate content in microalgae, especially monosaccharides, such as glucose, contributes to the biofuel production process. Carbohydrate-rich microalgae can also be used as raw materials to produce ethylene glycol and 1,2-propanediol using environmentally friendly chemical reactions. Therefore, an urgent and important task is to obtain the biomass of microalgae enriched with carbohydrates or starch, which can be used as a raw material for subsequent chemical or biochemical transformations [13].

Author Contributions: O.B. and S.I. conceived and designed the research; S.S. and V.D. analyzed and interpreted the data; A.A. and E.B. contributed reagents, materials, and analysis tools or data; writing—review and editing, O.B., P.M., and S.I. All authors have read and agreed to the published version of the manuscript.

Funding: This research was funded by the Russian Foundation for Basic Research, grant number 19-316-60001.

Institutional Review Board Statement: Not applicable.

Informed Consent Statement: Not applicable.

Data Availability Statement: The data are included in the manuscript.

Conflicts of Interest: The authors declare no conflict of interest.

Appendix A. Sequences of the 18S Ribosomal RNA Gene of the *Chlorella vulgaris* Beijer

TGTAGTCATATGCTTGTCTCAAA GATTAAGCCATGCATGTCTAAGTATAAACT GCTTTATACTGTGAAACTGCGAATGGCTCAT TAAATCAGTTATAGTTTATTTGATG-GTACCTAC TACTCGGATACCCGTAGTAAATCTAGAGCTA ATACGTGCGTAAATC-

CCGACTTCTGGA AGGGACGTATTTATTAGATAAAAG GCCGACCGGGCTCTGCCC-GACTCGCGGT GAATCATGATAACTTCACGAATCGCATGGC CTTGTGCCGGCGAT-GTTTCATTCAAAT TTCTGCCCTATCAACTTTCGATGGT AGGATAGAGGCCTACCATG-GTGGTA ACGGGTGACGGAGGATTAGGGTTCGAT TCCGGAGAGGGAGCCTGAGAA ACG GCTACCACATCCAAGGAAGGC AGCAGGCGCGCAAATT ACCCAATCCTGA-CACAG GGAGGTAGTGACAATAAAT AACAATACTGGGCCTTT TCAGGTCTGGTAAT TGGA ATGAGTACAATCTAAA CCCCTTAACGAGGATCAA TTGGAGGGCAAGTCTG-GTGC CAGCAGCCGCGGTAAT TCCAGCTCCAATAGCGTA TATTTAAGTTGCTGCA GTTAAA AAGCTCGTAGT TGGATTTCGGGTGGGGCC TGCCGGTCCGCCGTTTCGG TGTGCACTGGCAGGGCCC ACCTTGTTGCCGGGGACGG GCTCCTGGGCTTCACT-GTC CGGGACTCGGAGTCGGCGC TGTTACTTTGAGTAAATTAG AGTGTTCAAAGCAG G CCTACGCTCTGAATACA TTAGCATGGAATAACAC GATAGGACTCTGGCCTAT CCT-GTTGGTCTGTAGGACC GGAGTAATGATTAAGAGGGA CAGTCGGGGGCATTCGTAT TTCA TTGTCAGAGGTGAAATTCTTGGATT TATGAAAGACGAACTACTGCGAAAG-CAT TTGCCAAGGATGTTTTCATTAATCAAGAACGAA AGTTGGGGGCTCGAAGAC-GATTAGATACCGTCCTAGTCTCAA CCATAAACGATGCCGACTAG GGATCG GCG-GATGTTTCTTCGATGA CTCCGCCGGCACCTTATGAGAAATCAAAGTTTTTGGGT TC-CGGGGGGAGTATGGTCGCA AGGCTGAAACTTAAAGGA ATTGACGGAAGGGCAC-CACCAG GCGTGGAGCCTGCGGCTTAATTTGACTCA ACACGGGAAAACTTACCAGG TCCAGACATAGTGAG GATTGACAGATTGAGAGCTCT TTCTTGATTCTATGGGTG-GTGGTGCATG GCCGTTCTTAGTTGGTGGGTTGCC TGTCAGGTTGATTCCGGTAAC-GAACGAG ACCTCAGCCTGCTAAATAGTC ACGGTTGGCTCGCCAGCCGGCGGACT TCTT AGAGGGACTATTGGCGACTAGCC AATGAAGCATGAGGCAATAA CAGGTCT-GTGATGCCCTTAGATGTT CTGGG CCGCACGCGCGCTACACTGATGC ATTCAAC-GAGCTTAGCCTTGGCCGAGAG GCCCGGGTAATCTTTGAAACTGCATC GTGATGG GGATAGATTATTGCAA TTATTAATCTTCAACGAGGAATGCCT AGTAAGCGCAAGT-CATCAGCTTGCGTTG ATTACGTCCCTGCCCTTTGTACACACCGCCC GTCGCTCC-TACCGATTGGGTGTGC TGGTGAAGTGTTCGGATTGGC GACCGGGGGCGGTCTC-CGCTCTCG GCCGCCGAGAAGTTCATTAAACC CTCCCACCTAGAGGAAGGAG AAG TCGTAACAAG GTTTCCGTAGGTGAACCTGCGGAAGGATCA

Appendix B. Sequences of the 18S Ribosomal RNA Gene of the *Arthrospira platensis* Gomont

AGAGTTTGATCCTGGCTCAGGATGAACGCT GGCGGT CTGCTTAACACATG-CAAGTCGAACGGGCT CTTCGGAGCTAGTGGC GGACGGGTGAGTAACACGTGA-GAATCTGGCT CCCGGTCGGGGACAACAGAGGGAAACT TCTGCTAATCCCGGAT-GAG CCGA AAGGTAAAAGATTTATCG CCGGGAGATGAGCTCG CGTCTGATTAGC-TAGTTGGT GAGGTAAAGGCTCACCAAG GCGACGATCAGTAGCTGGT CTGAGAG-GATGATCAGCCACA CTGGGACTGAG ACACGGCC CAG ACTCC TACGGG AGGCA GCAGTGGAGAATTTTCCGC AATGGGCG CAA GCCTG ACGG AGCAAGACCGCG TGGGGGAGGAAGGCTCTTG GGTTGTAAACCCCTTTTCTCAAGGAAGA ACACAAT-GACGGTACTTGAGGAATAAGCC TCGGCTAACTCCGTGCCAGCAGCCGCG GTAAT-ACGGAGGAGGCAAGCGTTATCC GGAATGATTGGGCGTAAAGCGTCCGTAGG TG-GCAGTTCAAGTCTGCTGT CAA AG ACAGTAG CTCAACTACTGAAAGGCAGTG-GAAAC TGAACAGCTAGAGTACG GTAGGGGCAGAGGGAA TTCCCGGTGTAGCG-GTGAAATGCGTAGATATCGGGAAGAACAC CGGTGGCGAAAGCGCTCTGCT GGGC-CGTAACTGACACTGAGGGACGAAA GCTAGGGGAGCGAATGGGATTAGATACCC CAGTAG TCCTAGCCGTAAACGATGGAAACTAGGTGTAGCCTGTAT CGACCCGGGCT-GTGCCGAAGCTAACGCGTTA AGTTTCCCGCCTGGGGAGTACGCACGCAAG TGT-GAAACTCAAAGGAATTGACGGGGGCCC GCACAAGCGGTGGAGTATGTGG TTTAAT TCGATGCAACGCGAAGAACCTT ACCAGGGCTTGACATGTCCGGAATCTTGGTG AAA GCCGAGAGTGCCTTCGGGAGCC GGAACACAGGTGGTGCATGGCTG TCGTCAGCTC GTGTCGTGAGGTGTT GGGTTAAGTCCCGCAACGAGCGCAAC CCTCGTCCTTAGTTG CCATCATTCA GTTGGGCACTTTAGGGAGACTGCC GGTGACAAACCGGAGGAAG-

GTGGGGA TGACGTCAAGTCATCATGCCCCTTAC GTCCTGGGCTACACACGTACTA-CAATGG GGGGGACAAAGGGTAGCCAAGACGCGA GTCTGAGCCAATCCCGTAAA CCTCT CCTCAGTTCAGATTGCAGGCTGC AACTCGCCTGCATGAAGGAGGAATC GC-TAGTAATCGCAGGTCAGCATACTGCG GTGAATCCGTTCCCGGGCCTTGTACACAC-CGC CCGTCACACCATGGAAGTTAGCCACG CCCGAAGTCGTTACTCTAACCGTTCGC GG AG GAGGATGCCGAAGGCAGGGCTGATGAC TGGGGTG AAGTCGTAACAAG-GTAGCCGT AC CGGAAGGTGTGGCTGGATCACCTCCTT TTTAGGGAG ACCTAC TTCGAGATATCGC GCCTTAACAACTATAGCCGTGTCTTGA GGTCATCCTTAGGTCG-GATGGGG CGGTCAGAGAGCTTTCAA ACTTTAGGGTTCGTGTTATGG GCTATTAGCT CAGGTGGTTAGA GCGCACCCCTGATAAGG GTGAGGTC CCTGGTTCAAGTCCAGGA TG GCCCACA TCCACCCCAAACTGGGGGTATAGCTCAGT TGGTAGAGCGCTGC-CTTTGCACGGCAG AAGTCAGCGGTTCGAGTCCGCTT ACCTCCACTCTCCTTTGT-GATGGTGCT AGTTGGGGTGAGATGAGATGAGATGAC CTCTGATAGATAATTTAT-CACTGTAC AGCTCCTAAATCTTTAGATGTTAGT CTGAGATTGGATAGCTGGACATCTG TT CCAG TCAGAACCTTGAAAACTGCAT AGAGAAAAGCATAATGGTGTAGGAAAA CGTCGTAAAGACAATTCCAATG TAGGTCAAGCTACAAAGGGCTAACGG TGGAAC-CTAGGCAC ACAGAGCGGCCGCAAA

Appendix C. Sequences of the 18S Ribosomal RNA Gene of the *Dunaliella salina* Teod

CTCGTTGATCCTGCCAGTAGTCATAT GCTTGTCTCAAAGATTAAGCC ATGCAT-GTCTAAGTATAA ACTGCTTATACTGTGAAA CTGCGAATGGCTCATTAAA TCAGT-TATAGTTTATTTGAT GGTACCTTTACTCGGATAAC CGTAGTAATTCTAGAGCTAAT-ACGT GCGTAAATCCAGACTTCTGGAAGGG ACGTATTTATTAGATAAAAGGCCAG CCGGGCTTGCCCGACTCTTGGCG AATCATGATAACTTCACGAATCG CACGGCTT-TATGCCGGCGATG TTTCATTCAAATTTCTGCCCT ATCAACTTTCGATGGTAGGATA GAGGCCTACCATGGTGGTAAC GGGTGACGGAGGATTAGGG TTCGATTCCGGAGAG GGAG CCTGAGAAACGGCTACCAC ATCCAAGGAAGGCAGCAGG CGCGCAAAT-TACCCAATCC CAACACGGGGAGGTAGTG ACAATAAATAACAATACCG GGCATTTT TGTCTGGTAAT T GGAATGAGTACAATCTAAAT CCCTTAACGAGTATCCATTG GAG GGCAAGTCTGGTGCCA GCAGCCGCGGTAATTCCAGCT CCAATAGCGTATATTTAAG TTGT TGCAGTTAAAAAGCTCGTAGT TGGATTTCGGGTGGGTTGT AGCGGTCAGC-CTTTGGTTA GTACTGCTACGGCCTACCTT TCTGCCGGGGACGAG CTCCTGGGCT-TAAC TGTCCGGGACT CGG AATCGGCGAGGTTA CTTTGAGTAAATTAG AGTGTTCAA A GCAAGC CTACGCTCTGAATAC ATTAGCATGGAATA ACACGATAGGACTC TGGCT-TATCTTGTTGGT CTGTAAGACCGGAGTAA TGATTAAGAGGGACAGT CGGGGGCATT CGTATTTCA TTGTCAGAGGTGAAATTC TTGGATTTATGAAAG ACGAACTTCTGCG AAAGCATTTGCCAAGG ATGTTTTCATTAATCAA GAACGAAAGTTGGGGG CTCGAA-GACGATTAGATACCGTCGTAGTCTCAA CCATAAACGATGCCG ACTAGGGATTGCCA G GTGTTTCGTTGATGA CCCTGCCAGCACCTTA TGAGAAATCAAAGTTTT TGGGTTC-CGGGGGGAGT ATGGTCGCAAGGCTGAAA CTTAAAGGAATTGACGGA AGGGCAC-CACCAGGCGT TAACTTAGCAGCAAGCT CAGCGCCTCAAAGTCG AAGGGAAAC-CTTTGG CTAGTATCTGGGTGT AGATTTCACCTAAGT GCAACACTGTTCAAA TTGCGG GAAAGCC CTAAAGCTTTGCTAACC AAGCTGTCCTAGAAATG GGATGGTGGCCAGG TG AAAGACCTTGGGTACG GTAAAATCAGCAAAGA TGCAACAATGGGCAAT CCGCA GCCAAGCTCCT ACGGGCTGTCAAAGC CTATGGAGAAGGTTCA GAGACTAAATG-GCAGT GGGCAAGCATGGC AATG CTTGCTTAAGATATAGT CCGTCCCAGCTGAGAA GCTGCCTATGAGAGGAAT GCCGTAAGGCAGGAGAGCT AATAGGAAGTAAGTGTC TTTAATCAACTTACTTGG ATTCCACGGGAGCCTGCG GCTTAATTTGACTCAACA CGG GAAAACTTACCAGGT CCAGACACGGGGAGGATT GACAGATTGAGAGCTCTT TCTT GATTCTGTGGGTGG TGGTGCATGGCCGTTCTTAG TTGGTGGGTTGCCTTGTCAG GTT GATTCCGGTAACGAACG AGACCTCAGCCTGCTAAATA GTCACGTCTACCTCGGTA GG CGCCTGACTTCTTAGAGGGA CTATTGGCGTTTAGCCAATG GAAGTGTGAGGCA ATAACAG GTCTGTGATGCCCTTAGATGT TCTGGGCCGCACGCGCGCT ACACTGAT-

GCATTCAACGAGCCTATCCTTGGCCGAGAGGTCCG GGTAATCTTTGAAACTGCATC GTGATGGGGATAGATTATTGCA ATTATTAGTCTTCAACGAG GAATGCCTAGTAAGCG CGA GTCAT CAG CT CGC GTT GA TTACGTCCCTGCC CTTTGTACACACCGCCCG TCGCTCCTACCGATT GGGTGTGCTGGTGAAGT GTTTGGATCGGTACCAATGG GGGG AAACCTCTGTTGGT ACTGAGAAGAACATTAAACCCT CC CAC CTAGAGGAAGGA-GAAGTCGTAACAAGGTTTCCGTAGGTGAACCT GCAGAAGGATCA

References

1. Zhou, X.; Lin, W.; Tong, L.; Liu, X.; Zhong, K.; Liu, L.; Wang, L.; Zhou, S. Hypolipidaemic effects of oat flakes and beta-glucans derived from four Chinese naked oat (*Avena nuda*) cultivars in Wistar-Lewis rats. *J. Sci. Food Agric.* **2016**, *96*, 644–649. [CrossRef]
2. Kharwar, S.; Bhattacharjee, S.; Mishra, A.K. disentangling the impact of sulfur limitation on exopolysaccharide and functionality of Alr2882 by in silico approaches in *Anabaena* sp. PCC 7120. *Appl. Biochem. Biotechnol.* **2021**, *193*, 1447–1468. [CrossRef]
3. Vieira, M.V.; Pastrana, L.M.; Fuciños, P. Microalgae encapsulation systems for food, pharmaceutical and cosmetics applications. *Mar. Drugs* **2020**, *18*, 644. [CrossRef]
4. Ma, R.; Wang, B.; Chua, E.T.; Zhao, X.; Lu, K.; Ho, S.H.; Shi, X.; Liu, L.; Xie, Y.; Lu, Y.; et al. Comprehensive utilization of marine microalgae for enhanced co-production of multiple compounds. *Mar. Drugs* **2020**, *18*, 467. [CrossRef] [PubMed]
5. Rumin, J.; Nicolau, E.; Junior, R.G.O.; Fuentes-Grünewald, C.; Picot, L. Analysis of scientific research driving microalgae market opportunities in Europe. *Mar. Drugs* **2020**, *18*, 264. [CrossRef] [PubMed]
6. Bodachivskyi, I.; Kuzhiumparambil, U.; Bradley, G.; Williams, D. High yielding acid-catalysed hydrolysis of cellulosic polysaccharides and native biomass into low molecular weight sugars in mixed ionic liquid systems. *Chem. Open* **2019**, *8*, 1316–1324. [CrossRef] [PubMed]
7. Koyande, A.K.; Show, P.L.; Guo, R.; Tang, B.; Ogino, C.; Chang, J.S. Bio-processing of algal bio-refinery: A review on current advances and future perspectives. *Bioengineered* **2019**, *10*, 574–592. [CrossRef]
8. Lam, G.P.; Giraldo, J.B.; Vermue, M.H.; Olivieri, G.; Eppink, M.H.; Wijffels, R.H. Understanding the salinity effect on cationic polymers in inducing flocculation of the microalga *Neochloris oleoabundans*. *J. Biotechnol.* **2016**, *225*, 10–17. [CrossRef]
9. Lakatos, G.E.; Ranglová, K.; Manoel, J.C.; Grivalský, T.; Kopecký, J.; Masojídek, J. Bioethanol production from microalgae polysaccharides. *Folia. Microbiol.* **2019**, *64*, 627–644. [CrossRef]
10. Andreeva, A.; Budenkova, E.; Babich, O.; Sukhikh, S.; Ulrikh, E.; Ivanova, S.; Prosekov, A.; Dolganyuk, V. Production, purification, and study of the amino acid composition of microalgae proteins. *Molecules* **2021**, *26*, 2767. [CrossRef]
11. Klanchui, A.; Dulsawat, S.; Chaloemngam, K.; Cheevadhanarak, S.; Prommeenate, P.; Meechai, A. An improved genome-scale metabolic model of *Arthrospira platensis* C1 (iAK888) and its application in glycogen overproduction. *Metabolites* **2018**, *8*, 84. [CrossRef]
12. Aytenfisu, A.H.; Yang, M.; MacKerell, A.D. Drude polarizable force field for glycosidic linkages involving pyranoses and furanoses. *J. Chem. Theory Comput.* **2018**, *14*, 3132–3143. [CrossRef]
13. Cheng, D.; Li, D.; Yuan, Y.; Zhou, L.; Li, X.; Wu, T.; Wang, L.; Zhao, Q.; Wei, W.; Sun, Y. Improving carbohydrate and starch accumulation in *Chlorella* sp. AE10 by a novel two-stage process with cell dilution. *Biotechnol. Biofuels* **2017**, *10*, 75. [CrossRef] [PubMed]
14. Schulze, C.; Wetzel, M.; Reinhardt, J.; Schmidt, M.; Felten, L.; Mundt, S. Screening of microalgae for primary metabolites including β-glucans and the influence of nitrate starvation and irradiance on β-glucan production. *J. Appl. Phycol.* **2016**, *28*, 2719–2725. [CrossRef]
15. Khan, M.I.; Lee, M.G.; Seo, H.J.; Shin, J.H.; Shin, T.S.; Yoon, Y.H.; Kim, M.Y.; Choi, J.I.; Kim, J.D. Enhancing the feasibility of *Microcystis aeruginosa* as a feedstock for bioethanol production under the influence of various factors. *Biomed. Res. Int.* **2016**, *2016*, 4540826. [CrossRef]
16. Fimbres-Olivarría, D.; López-Elías, J.A.; Carvajal-Millán, E.; Márquez-Escalante, J.A.; Martínez-Córdova, L.R.; Miranda-Baeza, A.; Enríquez-Ocaña, F.; Valdéz-Holguín, J.E.; Brown-Bojórquez, F. *Navicula* sp. sulfated polysaccharide gels induced by Fe(III): Rheology and microstructure. *Int. J. Mol. Sci.* **2016**, *17*, 1238. [CrossRef]
17. Mooij, P.R.; de Jongh, L.D.; van Loosdrecht, M.C.; Kleerebezem, R. Influence of silicate on enrichment of highly productive microalgae from a mixed culture. *J. Appl. Phycol.* **2016**, *28*, 1453–1457. [CrossRef] [PubMed]
18. Markou, G.; Depraetere, O.; Vandamme, D.; Muylaert, K. Cultivation of *Chlorella vulgaris* and *Arthrospira platensis* with recovered phosphorus from wastewater by means of zeolite sorption. *Int. J. Mol. Sci.* **2015**, *16*, 4250–4264. [CrossRef] [PubMed]
19. Bhatt, N.C.; Panwar, A.; Bisht, T.S.; Tamta, S. Coupling of algal biofuel production with wastewater. *Sci. World J.* **2014**, *2014*, 210504. [CrossRef]
20. Patel, D.S.; He, X.; MacKerell, A.D., Jr. Polarizable empirical force field for hexopyranose monosaccharides based on the classical Drude oscillator. *J. Phys. Chem. B* **2015**, *119*, 637–652. [CrossRef]
21. Patel, D.S.; Pendrill, R.; Mallajosyula, S.S.; Widmalm, G.; MacKerell, A.D., Jr. Conformational properties of α- or β-(1→6)-linked oligosaccharides: Hamiltonian replica exchange MD simulations and NMR experiments. *J. Phys. Chem. B* **2014**, *118*, 2851–2871. [CrossRef]
22. Park, D.; Jagtap, S.; Nair, S.K. Structure of a PL17 family alginate lyase demonstrates functional similarities among exotype depolymerases. *J. Biol. Chem.* **2014**, *289*, 8645–8655. [CrossRef] [PubMed]

23. Mühlroth, A.; Li, K.; Røkke, G.; Winge, P.; Olsen, Y.; Hohmann-Marriott, M.F.; Vadstein, O.; Bones, A.M. Pathways of lipid metabolism in marine algae, co-expression network, bottlenecks and candidate genes for enhanced production of EPA and DHA in species of Chromista. *Mar. Drugs* **2013**, *11*, 4662–4697. [CrossRef] [PubMed]
24. Dai, L.; Tan, L.; Jin, X.; Wu, H.; Houbo, W.; Tao, L.; Wenzhou, X. Evaluating the potential of carbohydrate-rich microalga *Rhodosorus* sp. SCSIO-45730 as a feedstock for biofuel and β-glucans using strategies of phosphate optimization and low-cost harvest. *J. Appl. Phycol.* **2020**, *32*, 3051–3061. [CrossRef]
25. Blockx, J.; Verfaillie, A.; Thielemans, W.; Muylaert, K. Unravelling the mechanism of chitosan-driven flocculation of microalgae in seawater as a function of pH. *ACS Sustain. Chem. Eng.* **2018**, *6*, 11273–11279. [CrossRef]
26. Chng, L.M.; Lee, K.T.; Chan, D.J.C. Synergistic effect of pretreatment and fermentation process on carbohydrate-rich *Scenedesmus dimorphus* for bioethanol production. *Energy Convers. Manag.* **2017**, *141*, 410–419. [CrossRef]
27. Corrêa, D.O.; Duarte, M.E.R.; Noseda, M.D. Biomass production and harvesting of *Desmodesmus subspicatus* cultivated in flat plate photobioreactor using chitosan as flocculant agent. *J. Appl. Phycol.* **2018**, *31*, 857–866. [CrossRef]
28. Gerchman, Y.; Vasker, B.; Tavasi, M.; Mishael, Y.; Kinel-Tahan, Y.; Yehoshua, Y. Effective harvesting of microalgae: Comparison of different polymeric flocculants. *Bioresour. Technol.* **2017**, *228*, 141–146. [CrossRef]
29. Varshney, P.; Beardall, J.; Bhattacharya, S.; Wangikar, P.P. Isolation and biochemical characterisation of two thermophilic green algal species—*Asterarcys quadricellulare* and *Chlorella sorokiniana*, which are tolerant to high levels of carbon dioxide and nitric oxide. *Algal Res.* **2018**, *30*, 28–37. [CrossRef]
30. Berteotti, S.; Ballottari, M.; Bassi, R. Increased biomass productivity in green algae by tuning nonphotochemical quenching. *Sci. Rep.* **2016**, *6*, 21339. [CrossRef]
31. Schulze, P.S.C.; Pereira, H.G.C.; Santos, T.F.C.; Schueler, L.; Guerra, R.; Barreira, L.A.; Perales, J.A.; Varela, J.C.S. Effect of light quality supplied by light emitting diodes (LEDs) on growth and biochemical profiles of *Nannochloropsis oculat* and *Tetraselmis chuii*. *Algal Res.* **2016**, *16*, 387–398. [CrossRef]
32. Gupta, S.K.; Kumar, N.M.; Guldhe, A.; Ansari, F.A.; Rawat, I.; Nasr, M.; Bux, F. Wastewater to biofuels: Comprehensive evaluation of various flocculants on biochemical composition and yield of microalgae. *Ecol. Eng.* **2018**, *117*, 62–68. [CrossRef]
33. Krayesky-Self, S.; Phung, D.; Schmidt, W.; Sauvage, T.; Butler, L.; Fredericq, S. First report of endolithic members of *Rhodosorus marinus (Stylonematales, Rhodophyta)* growing inside rhodoliths offshore Louisiana, Northwestern Gulf of Mexico. *Front. Mar. Sci.* **2020**, *7*, 1–8. [CrossRef]
34. Li, T.; Xu, J.; Wu, H.; Jiang, P.; Chen, Z.; Xiang, W. Growth and biochemical composition of *Porphyridium purpureum* SCS-02 under different nitrogen concentrations. *Mar. Drugs* **2019**, *17*, 124. [CrossRef] [PubMed]
35. Dolganyuk, V.; Andreeva, A.; Budenkova, E.; Sukhikh, S.; Babich, O.; Ivanova, S.; Prosekov, A.; Ulrikh, E. Study of morphological features and determination of the fatty acid composition of the microalgae lipid complex. *Biomolecules* **2020**, *10*, 1571. [CrossRef] [PubMed]
36. Ho, S.-H.; Huang, S.-W.; Chen, C.-Y.; Hasunuma, T.; Kondo, A.; Chang, J.-S. Characterization and optimization of carbohydrate production from an indigenous microalga *Chlorella vulgaris* FSP-E. *Bioresour. Techno.* **2013**, *135*, 157–165. [CrossRef]
37. Chou, N.T.; Cheng, C.F.; Wu, H.C.; Lai, C.P.; Lin, L.T.; Pan, I.H.; Ko, C.H. *Chlorella sorokiniana*-induced activation and maturation of human monocyte-derived dendritic cells through NF-κB and PI3K/MAPK pathways. *Evid. Based Complement. Altern. Med.* **2012**, *2012*, 735396. [CrossRef]
38. Liu, Q.; Yao, C.; Sun, Y.; Chen, W.; Tan, H.; Cao, X.; Xue, S.; Yin, H. Production and structural characterization of a new type of polysaccharide from nitrogen-limited *Arthrospira platensis* cultivated in outdoor industrial-scale open raceway ponds. *Biotechnol. Biofuels* **2019**, *12*, 131. [CrossRef] [PubMed]
39. El-Naggar, N.E.; Hussein, M.H.; Shaaban-Dessuuki, S.A.; Dalal, S.R. Production, extraction and characterization of *Chlorella vulgaris* soluble polysaccharides and their applications in AgNPs biosynthesis and biostimulation of plant growth. *Sci. Rep.* **2020**, *10*, 3011. [CrossRef] [PubMed]
40. Gaignard, C.; Laroche, C.; Pierre, G.; Dubessay, P.; Delattre, C.; Gardarin, C.; Gourvil, P.; Probert, I.; Dubuffet, A.; Michaud, P. Screening of marine microalgae: Investigation of new exopolysaccharide producers. *Algal Res.* **2019**, *44*, 101711. [CrossRef]
41. Dolganyuk, V.; Belova, D.; Babich, O.; Prosekov, A.; Ivanova, S.; Katserov, D.; Patyukov, N.; Sukhikh, S. Microalgae: A promising source of valuable bioproducts. *Biomolecules* **2020**, *10*, 1153. [CrossRef] [PubMed]
42. Sero, E.T.; Siziba, N.; Bunhu, T.; Shoko, R.; Jonathan, E. Biophotonics for improving algal photobioreactor performance: A review. *Int. J. Energy Res.* **2020**, *44*, 5071–5092. [CrossRef]
43. Toor, M.; Kumar, S.S.; Malyan, S.K.; Bishnoi, N.R.; Mathimani, T.; Rajendran, K.; Pugazhendhi, A. An overview on bioethanol production from lignocellulosic feedstocks. *Chemosphere* **2019**, *242*, 125080. [CrossRef] [PubMed]
44. Yamin, W.A.; Shaleh, S.R.M.; Ching, F.F.; Othman, R.; Manjaji-Matsumoto, M.; Mustafa, S.; Shigeharu, S.; Kandasamy, G. Harvesting Chaetoceros gracilis by flocculation using chitosan. *IOP Conf. Ser. Earth Environ. Sci.* **2019**, *236*, 012123. [CrossRef]
45. Zhu, Q.; Wu, S. Water-soluble beta-1,3-glucan prepared by degradation of curdlan with hydrogen peroxide. *Food Chem.* **2019**, *283*, 302–304. [CrossRef] [PubMed]

46. Rojo-Cebreros, A.H.; Ibarra-Castro, L.; Martínez-Brown, J.M.; Velasco-Blanco, G.; Martínez-Téllez, M.A.; Medina-Jasso, M.A.; Nieves-Soto, M.; Quintana-Zavala, D. Potential of Nannochloropsis in beta glucan production. In *Nannochloropsis: Biology, Biotechnological, Potential and Challenges*; Jan, M., Kazik, P., Eds.; Nova Sciences Publishers Inc.: New York, NY, USA, 2017; pp. 181–225.
47. Welham, S.J.; Gezan, S.A.; Clark, S.J.; Mead, A. *Statistical Methods in Biology: Design and Analysis of Experiments and Regression*; CRC Press (Taylor & Francis Group): Boca Raton, FL, USA, 2014.

marine drugs

MDPI

Article

Antiproliferative and Antimicrobial Potentials of a Lectin from *Aplysia kurodai* (Sea Hare) Eggs

Rubaiya Rafique Swarna [1], A. K. M. Asaduzzaman [1], Syed Rashel Kabir [1], Nawshin Arfin [1], Sarkar M. A. Kawsar [2], Sultana Rajia [3,4], Yuki Fujii [5], Yukiko Ogawa [5], Keisuke Hirashima [4], Nanae Kobayashi [4,6], Masao Yamada [4], Yasuhiro Ozeki [4] and Imtiaj Hasan [1,*]

1 Department of Biochemistry and Molecular Biology, University of Rajshahi, Rajshahi 6205, Bangladesh; rubaiyarafique5252@gmail.com (R.R.S.); jonyasad2005@yahoo.com (A.K.M.A.); rashelkabir@ru.ac.bd (S.R.K.); arfin.nawshin@gmail.com (N.A.)
2 Department of Chemistry, University of Chittagong, Chittagong 4331, Bangladesh; akawsarabe@yahoo.com
3 Center for Interdisciplinary Research, Varendra University, Rajshahi 6204, Bangladesh; rajia_bio@yahoo.com
4 School of Science, Yokohama City University, 22-2 Seto, Kanazawa-ku, Yokohama 236-0027, Japan; n195274c@yokohama-cu.ac.jp (K.H.); kobayashi.nanae444@gmail.com (N.K.); yamada.mas.ug@yokohama-cu.ac.jp (M.Y.); ozeki@yokohama-cu.ac.jp (Y.O.)
5 Graduate School of Pharmaceutical Sciences, Nagasaki International University, 2825-7, Huis Ten Bosch-cho, Sasebo, Nagasaki 859-3298, Japan; yfujii@niu.ac.jp (Y.F.); yogawa@niu.ac.jp (Y.O.)
6 School of Medicine, Yokohama City University, 3-9, Fukuura, Kanazawa-ku, Yokohama 236-0004, Japan
* Correspondence: hasanimtiaj@yahoo.co.uk or imtiajbio@ru.ac.bd; Tel.: +880-721-71-1109

Abstract: In recent years, there has been considerable interest in lectins from marine invertebrates. In this study, the biological activities of a lectin protein isolated from the eggs of Sea hare (*Aplysia kurodai*) were evaluated. The 40 kDa *Aplysia kurodai* egg lectin (or AKL-40) binds to D-galacturonic acid and D-galactose sugars similar to previously purified isotypes with various molecular weights (32/30 and 16 kDa). The N-terminal sequence of AKL-40 was similar to other sea hare egg lectins. The lectin was shown to be moderately toxic to brine shrimp nauplii, with an LC_{50} value of 63.63 μg/mL. It agglutinated Ehrlich ascites carcinoma cells and reduced their growth, up to 58.3% in vivo when injected into Swiss albino mice at a rate of 2 mg/kg/day. The morphology of these cells apparently changed due to AKL-40, while the expression of apoptosis-related genes (p53, Bax, and Bcl-XL) suggested a possible apoptotic pathway of cell death. AKL-40 also inhibited the growth of human erythroleukemia cells, probably via activating the MAPK/ERK pathway, but did not affect human B-lymphoma cells (Raji) or rat basophilic leukemia cells (RBL-1). In vitro, lectin suppressed the growth of Ehrlich ascites carcinoma and U937 cells by 37.9% and 31.8%, respectively. Along with strong antifungal activity against *Talaromyces verruculosus*, AKL showed antibacterial activity against *Staphylococcus aureus*, *Shigella sonnei*, and *Bacillus cereus* whereas the growth of *Escherichia coli* was not affected by the lectin. This study explores the antiproliferative and antimicrobial potentials of AKL as well as its involvement in embryo defense of sea hare.

Keywords: antibacterial activity; anticancer activity; antifungal activity; *Aplysia kurodai*; apoptosis; Ehrlich ascites carcinoma; lectin

Citation: Swarna, R.R.; Asaduzzaman, A.K.M.; Kabir, S.R.; Arfin, N.; Kawsar, S.M.A.; Rajia, S.; Fujii, Y.; Ogawa, Y.; Hirashima, K.; Kobayashi, N.; et al. Antiproliferative and Antimicrobial Potentials of a Lectin from *Aplysia kurodai* (Sea Hare) Eggs. *Mar. Drugs* **2021**, *19*, 394. https://doi.org/10.3390/md19070394

Academic Editor: Hitoshi Sashiwa

Received: 7 May 2021
Accepted: 12 July 2021
Published: 14 July 2021

1. Introduction

Lectins are carbohydrate-binding proteins, omnipresent in almost all life forms, which specially recognize carbohydrate structures of glycoproteins and glycolipids present on the cell surface [1]. Marine invertebrates possess diverse classes of lectins with various protein foldings and different carbohydrate-binding specificities [1–5]. Many of these lectins show affinity towards galactose-related carbohydrates and perform biological functions including antitumor and antimicrobial activities.

Endogenous lectins interact with cell-surface glycans to mediate numerous biological functions in cells [6]. Compared to normal cells, glycosylation pattern of cancer cells are altered, which affects multiple cellular mechanisms performed by lectins through their association with corresponding glycans [7]. As a result, this alteration supports their neoplastic progression [8,9]. For example, galactose-binding lectins or galectins aid the binding of tumor cells through the interaction with galactose-containing carbohydrate ligands on tumor cells [9]. Many marine invertebrate lectins have been reported to bind with glycans on the tumor cell surface and killed those inducing apoptosis [3,10–12]. On the other hand, lectins can also modulate the entry and subcellular targeting of drugs into cancer cells [13,14]. Both these phenomena exhibit considerable influence of lectin–glycan interaction on the proliferation and regulation of tumor cells [5,10–12,14].

As a component of innate immunity, lectins present in marine invertebrates agglutinate both Gram-negative and Gram-positive bacteria through the interaction with lipopolysaccharides and peptidoglycans on their cell walls [4,5,15,16]. Galactose residues present as terminal sugars in lipopolysaccharides (LPS) of bacteria impact the intracellular composition of bacteria and maintain the synthesis of UDP-galactose for LPS [17,18]. Bacterial endotoxins also possess galactose as a constituent [19]. On the contrary, antifungal activity of lectins is not well reported yet.

Sea hares are marine gastropod mollusks found in common coastal areas. Various lectins have been reported to be present in eggs and gonads of sea hares due to their possible roles in early development and body defense. D-galacturonic acid and D-galactose-binding lectins with molecular masses 16–34 kDa have been obtained from the eggs of *Aplysia kurodai* [11,20–22] and other *Aplysia* species (*A. depilans, A. dactylomela,* and *A. californica*) [23–25]. Primary structures of all these previously purified egg lectins possess a novel triple tandem repeating sequence consisting of 210–230 amino acids and show striking similarities to domain DUF3011 of some uncharacterized bacterial proteins [26].

These Aplysia egg lectins showed cytotoxic activities against certain tumor cells and were involved with organogenesis in early developmental stages. They might have played protective roles in marine organisms as well. Hasan et al. reported an α-galactose-binding 40 kDa lectin from eggs of AKL in 2014 that inhibited streptolysin O-induced hemolysis and growth of *Streptococcus pyogenes* [22].

The 40 kDa lectin from *Aplysia kurodai* eggs will be designated here as AKL-40, to differentiate it from other AKLs. We characterized this egg lectin and determined its N-terminal sequence. Previously, in vitro antiproliferative activity of two AKLs was reported [11,21]. In this study, antiproliferative activities of this particular lectin were determined for the first time in vivo, using Swiss albino mice. In addition, other biological properties, such as in vitro anticancer activity against different cancer cell lines and antibacterial potential of the lectin, was further evaluated along with its antifungal activity.

2. Results

2.1. Purification, Confirmation of the Molecular Mass and Hemagglutination Activity of Aplysia kurodai Egg Lectin (AKL-40)

Affinity chromatography showed different molecular masses of AKLs found in the eggs of *Aplysia kurodai* (Figure 1). After applying the crude protein sample, the melibiosyl-agarose affinity chromatography column was washed with TBS and bound lectins were eluted with 10 mM D-galacturonic acid or D-galactose-containing TBS. Each polypeptide was separated with gel filtration chromatography by using Sephacryl S-200HR, (Cytiva, Marlborough, MA, USA) (data not shown). In the reducing condition, polypeptides with molecular weights of 40 kDa, 32/30 kDa, and 16 kDa were denoted as AKL-40, AKL, and AKL-2, respectively, in this study (Figure 1: white triangle: AKL-40; black triangles: AKL and AKL-2). The minimum concentration of AKL-40 to agglutinate human and mice erythrocytes was found to be 24 and 32 μg/mL, respectively.

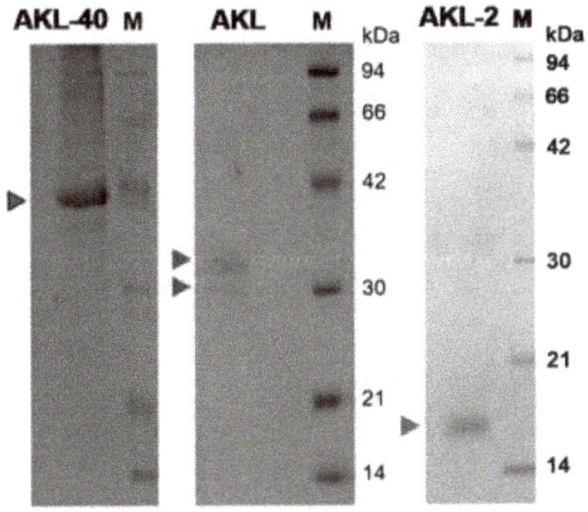

Figure 1. Different molecular masses of AKLs found in the eggs of *Aplysia kurodai*. AKL mixtures obtained from the melibiose column were combined and separated by using Sephacryl S-200 gel filtration column (black triangles show the polypeptides indicating AKLs). AKL-40: 40 kDa species (used in this study), AKL: 32/30 kDa species, and AKL-2: 16 kDa species. M: standard markers of Phosphorylase b (94 kDa), serum albumin (66 kDa), ovalbumin (42 kDa), carbonic anhydrase (30 kDa), trypsin inhibitor (21 kDa), and lysozyme (14 kDa). All samples are separated in the reducing condition and 15% polyacrylamide gels are used except AKL-2, which was separated by a 12% gel.

2.2. N-Terminal Amino Acid Sequence of AKL-40

The N-terminal region of AKL-40, including the first 35 amino acids of the polypeptide, was determined by Edman degradation with a repetitive yield of 84.19% (Supplementary Figure S1). This partial N-terminal sequence of AKL-40 fitted with those of AKL-a to -d and ADEL, showing similarities (Figure 2). Unlike others, AKL-40 possessed a forward extension comprising four additional amino acids.

```
              1       5         10        15        20        25        30        35
AKL-40        V S Y L P V N P I V x D V I R V E T W A Y K T L D D T F G x D N Q x L

AKL-a    1            N T D Q A M C D V I R V E S W S Y K Y A E K I V D G A S Y V L   31
AKL-b    1            S S D Q A M C D V I R V E S W N Y K Y A E K I V N G A S Y V L   31
AKL-c    1            N P A P E ▒ C D V I R V E S W S Y K Y A E K V V E D A S Y V L   31
AKL-d    1            C C D P D K C D T I R V E S W S Y K Y A E K V V E D A S W V L   31
ADEL                      D P D M C K T I R V E S W S Y K Y A E K V V E D A S Y V L   29
```

Figure 2. Comparison of N-terminal amino acid sequences of AKL-40 and other Aplysia egg lectins. Thirty-five amino acids of the N-terminal sequence of AKL-40 were indicated by the single-letter amino acid code. Identical amino acids in the N-terminal sequences of 32–30 kDa Aplysia egg lectins isolated as AKL-a to -d [25] and ADEL [23] were shown as bold letters. X: unidentified, Gray box: skipped.

2.3. Toxicity of AKL-40 against Brine Shrimp Artemia nauplii

A dose-dependent graph showed that 70% of shrimp nauplii died at a concentration of 160 µg/mL of AKL-40 (Figure 3). The LC_{50} value was determined to be 63.63 µg/mL indicating AKL-40 as a moderately toxic protein.

2.4. In Vivo Anticancer Activity of AKL-40 against Ehrlich Ascites Carcinoma Cells, Their Morphological Changes, and Expression of Apoptosis-Related Genes

After administering 1 mg/kg/day and 2 mg/kg/day of AKL-40 for five days, growth of EAC-cells in the lectin-treated mice (Groups B and C) was inhibited to 28.7% and 58.3%, compared to the untreated (or control) mice from group A (Figure 4A). Weight of the treated mice reduced significantly compared to the untreated mice (data not shown). EAC cells from untreated mice were spherical and regular-sized (Figure 4B, column 0).

Irregular-shaped kidney-bean-like EAC cells from the treated mice groups (Figure 4B, column 1 and 2). Upregulation of p53 gene expression was observed in EAC cells from group C, though it was a weak one. Expression of Bax gene was there in EAC cells from lectin-treated mice whereas no expression was found in EAC cells from untreated mice. Cells from AKL-40-treated mice showed no expression of the Bcl-X_L gene whereas, in EAC cells from control mice, it was visible. Expression of GAPDH gene in EAC cells from both untreated and lectin-treated mice confirmed the quality of mRNA isolated (Figure 4C).

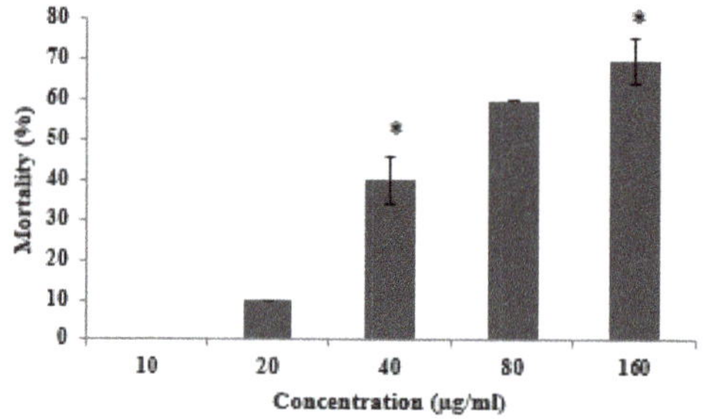

Figure 3. Toxicity of AKL-40 against shrimp nauplii at different concentrations. Data are expressed in mean ± SD (n = 10). The results were statistically significant (* $p < 0.05$, when mortality percentage of lectin-treated shrimps were compared to untreated shrimps).

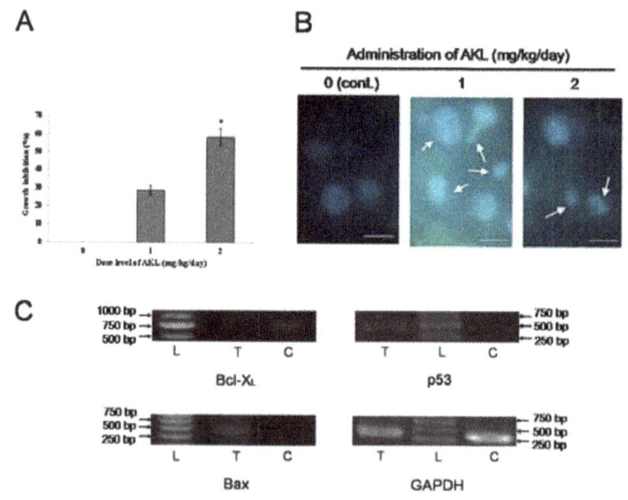

Figure 4. In vivo anticancer activity of AKL-40. (**A**). Growth inhibition in AKL-40 treated mice compared to the untreated mice. The lectin was administrated to mice at doses 0 (untreated), 1 and 2 mg/kg/day (treated). Data are expressed in mean ± SD (n = 6). The results were statistically significant (* $p < 0.05$, when cells from lectin-treated mice were compared to cells from untreated mice). (**B**). Change of morphology in Hoechst-stained EAC cells harvested from AKL-40 treated mice. EAC cells from untreated and treated (with 1 mg/kg/day and 2 mg/kg/day of AKL-40, for five days) mice were observed by a fluorescence microscope. White arrows show changes in the morphology of cells. Scale bar: 25 µm. (**C**). Expression of apoptosis-related genes (Bcl-X_L, p53 and Bax) and GAPDH. L, 1000 bp DNA ladder; T, RNA of EAC cells from AKL-40 treated mice; C, RNA of EAC cells from untreated (control) mice.

2.5. In Vitro Anticancer Activity of AKL-40 against Different Cancer Cells and Activation of Signal Transduction Molecules

AKL-40 significantly inhibited the growth of K562 cells at a concentration of 100 µg/mL or more (Figure 5A, orange bar). On the other hand, the lectin could not influence the growth of human B-lymphoma cells (Raji) and rat basophilic leukemia cells (RBL-1) (Figure 5A, blue and gray bars). During a period of 24 h, extracellular signal-regulated kinase $(ERK)_{1/2}$ and p38kinase molecules became phosphorylated in AKL-40 treated K562 cells, as shown by western blotting (Figure 5B).

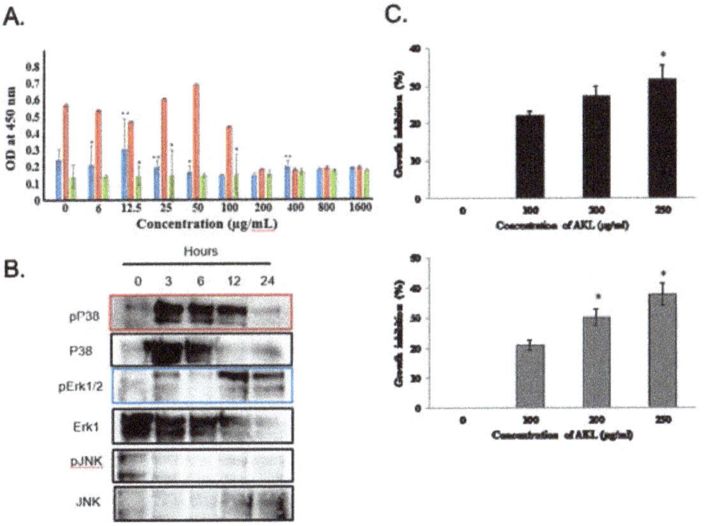

Figure 5. In vitro antiproliferative activity of AKL-40 against different cancer cell lines. (**A**). Antiproliferative activity of AKL-40 against erythroleukemia (K562), human B-lymphoma (Raji), and rat basophilic leukemia (RBL-1) cells. Orange, blue, and gray bars denote the growth of K562, Raji, and RBL-1 cells, respectively, at different concentrations of AKL-40. Data are expressed in mean ± SD. (**B**). Activation of MAPK pathway in AKL-40 treated K562 cells was observed during a period of 24 h. Phosphorylation of P38 was observed from a 3 to 12 h treatment and it diminished at 24 h (surrounded in red). Phosphorylation of Erk1/2 was also shown after the treatment for 12 to 24 h (surrounded in blue). No phosphorylation of JNK was observed. (**C**). Anticancer activity of AKL-40 against EAC and U937 cells. After treating with 100–250 µg/mL of AKL-40, percentages of growth inhibition were determined by MTT assay ($n = 3$, mean ± SD). Gray and black bars indicate EAC and U937 cells, respectively. The results were statistically significant (* $p < 0.05$, when lectin-treated cells were compared to untreated cells). The results were statistically significant (** $p < 0.01$ and * $p < 0.05$, when lectin-treated cells were compared to untreated cells).

Minimum agglutination concentrations of AKL-40 for EAC and U937 cells were 20 and 24 µg/mL, respectively. The dose-dependent effect of AKL-40 against U937 and EAC cells was observed by MTT assay. At lower concentrations (100 µg/mL), growth inhibition of both cell types was similar, but at higher concentrations, AKL-40 showed slightly higher activity against EAC cells. At the concentration of 250 µg/mL, the percentage of growth inhibition for EAC cells was 37.9%, compared to 31.8% for U937 cells (Figure 5C).

2.6. Antimicrobial Activity of AKL-40

AKL-40 showed antibacterial activity against *Staphylococcus aureus*, *Bacillus cereus*, and *Shigella sonnei*. No zone of inhibition was there for *Escherichia coli*. Slightly larger zones were observed around the disks soaked with higher doses (100 µg/disc) of AKL-40 in case of every susceptible bacterial species (Figure 6A). Maximum activity of AKL-40 was found against *Staphylococcus aureus* in terms of the inhibition zones formed. Growth of *Talaromyces verruculosus* was also very efficiently inhibited by AKL-40. The fungi rapidly grew and after two weeks, nearly covered the whole petri dish. However, when discs soaked with AKL-40 were placed in the fungal media, almost no growth of the mold was observed (Figure 6B).

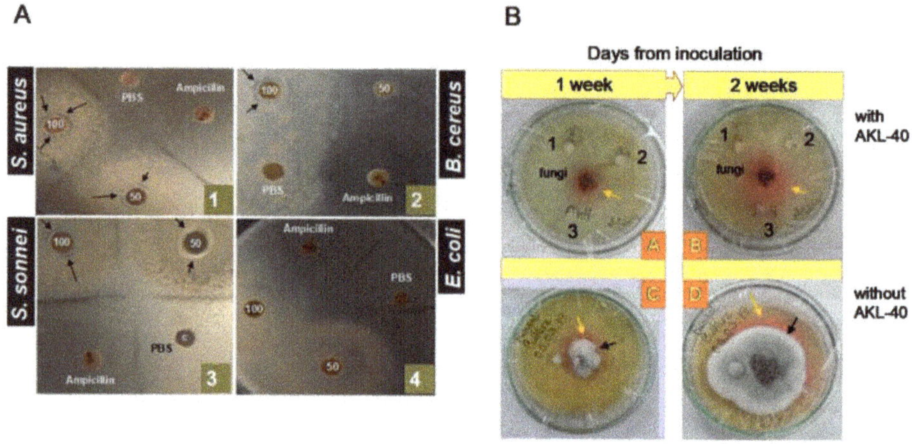

Figure 6. Antimicrobial activity of AKL-40. (**A**). Disc diffusion assay of the antibacterial activity of AKL-40; 1, 2, 3, and 4 shows the activity for *Staphylococcus aureus, Bacillus cereus, Shigella sonnei*, and *Escherichia coli*, respectively. Moreover, 50 and 100 on discs soaked with 50 and 100 μg/disc of AKL, respectively. PBS is the negative control (soaked with PBS) and Ampicillin indicated 15 μg of ampicillin (positive control). Black arrows mark the zone of inhibition. (**B**). AKL-40 showed time-dependent antifungal activity against *Talaromyces verruculosus*. A and B show the petri dishes containing three discs soaked with 400 μg/mL of AKL-40 after one week and two weeks of fungal inoculation, respectively). C and D are the petri dishes with no AKL-40 (control) after one week and two weeks of inoculation, respectively. Orange and black arrows indicate the pigment produced by the fungus and its subsequent growth, respectively.

3. Discussion

Following the purification of two lectins AKL and AKL-2 from *Aplysia kurodai* eggs, a third lectin with different molecular mass (40 kDa) had been purified [11,21,22]. This study focused on the biological activities of AKL-40 and the comparison of those activities to other marine lectins, specially isolated from mollusks. A number of lectin families have been reported in the phylum Mollusca whereas the Aplysia egg lectin family is structurally unique consisting of various isotypes. Galacturonic acid has been found in polysaccharides present in mollusks, such as cuttlefish [27]. Eggs of *Aplysia kurodai* also contained galactose/galacturonic acid-binding lectins as a mixture of 40, 32/30, and 16 kDa polypeptides (Figure 1). Such diversity of galactose/galacturonic acid-binding lectins in Aplysia eggs perhaps is a result of their response to changing marine environments. It might be interesting to find out the function of each molecular species of Aplysia egg lectins during each developmental stage of the embryo. The primary structure of the sea hare egg lectin family is different from other lectins because of the presence of homolog sequences that are also found in organisms, such as bacteria and brachiopods [26]. A partial but novel N-terminal sequence of AKL-40, consisting of 35 amino acids, indicated that this polypeptide was different from other variants (such as AKL-a to -d) by having a four-amino acid forward extension (Figure 2). This result also showed low sequence similarity of AKL-40 to ADEL.

Determining the level of toxicity of lectins could be important to study and predict their structures, physiological functions, and biological applications. With an LC_{50} value of 63.63, AKL-40 was moderately toxic to brine shrimp nauplii (Figure 3) whereas AKL-2, the lactose-binding counterpart, was more than three times toxic [28]. However, toxicity of lectins is not always related to cell regulatory effects as their interactions with glycans play vital roles in cell signaling. Despite having much lower LC_{50} value (384.53 μg/mL), MytiLec-1, a lectin from another marine mussel, could kill Burkitt lymphoma and U937 cells in vitro, as well as Ehrlich ascites carcinoma cells in vivo [3,29]. A variable level of toxicity with different LC_{50} values (850.1, 142.1, 9.5, and 6.4 μg/mL) was also observed in lectins purified from marine sponges and a sea cucumber [30,31].

Ehrlich ascites carcinoma cells originated from mammary tissues, are spontaneous, differentiated, transplantable, and aggressive in nature. These cells grow in certain mice strains and are approved worldwide as a standard mice model [32]. AKL-40 agglutinated EAC cells via galactosyl ligands present in their cell membrane [33,34]. In a previous report, another Molluscan D-Gal-binding lectin, MytiLec-1 strongly agglutinated EAC cells at a minimum concentration of 16 μg/mL and inhibited 28 and 49% of their growth at doses of 1 and 2 mg/kg/day, respectively [29]. In the present study, minimum agglutination concentration of AKL-40 for EAC cells was 20 μg/mL and, compared to MytiLec-1, similar growth inhibition activities (28.70% and 58.32% at doses of 1 and 2 mg/kg/day) have been observed (Figure 4A). Lectin-treated cancer cells presented Irregular shapes and nuclear condensation compared to the control (or untreated) cells (Figure 4B).

Appearance of a faint band of p53 indicated to the transcription of a pro-apoptotic gene like Bax in AKL-40 treated Ehrlich ascites carcinoma cells. Expression of Bcl-X_L, an anti-apoptotic gene had also become downregulated in the treated cells (Figure 4C). Similar expressions of p53, Bax, and Bcl-X_L genes have been found when MytiLec-1 was applied on EAC cells [29]. A previous study reported a lactose-binding lectin from the marine sponge *Cinachyrella apion* to induce the apoptotic death of HeLa cells through activation of Bax protein. The anti-apoptotic Bcl-2 protein showed no significant change in expression, compared to the control [35]. Evaluation of gene expression of MCF-7 cells revealed that marine red alga *Solieria filiformis* triggered caspase-dependent apoptosis. The anti-apoptotic gene Bcl-2 became down expressed, whereas the pro-apoptotic Bax gene underwent over-expression [36]. Downregulation of the anti-apoptotic factor Bcl-2 was determined also in the case of two other lectins from Sea bass (*Dicentrarchus labrax*) and sea urchin (*Strongylocentrotus purpuratus*) [37]. Therefore, it can be suggested that AKL-40 possibly calling the Bcl-2 apoptotic protein family into play marks its ability to promote apoptosis.

Marine invertebrate lectins have already been reported to exhibit antiproliferative properties, promote apoptosis, and block angiogenesis. In contrast, tumor-derived galactose-binding lectins, especially galectins, could compromise the anti-tumor immune response of CD8+ T cells to escape from the host immune surveillance [38,39]. AKL-40 significantly reduced the growth of erythroleukemia K562 cells compared to Raji and RBL-1 cells (Figure 5A). Phosphorylation of two major mitogen-activated protein kinases, p38 and ERK 1/2 became significantly upregulated by the administration of AKL-40, whereas no phosphorylation was observed for and c-Jun N-terminal kinase (JNK) (Figure 5B). It can be suggested that like other lectins previously reported, the lectin might have participated in a signaling cascade controlling cellular responses, probably leading to apoptosis [3,5,14]. In another in vitro study, AKL-40 inhibited the growth of two other cell lines, EAC and U937 (Figure 5B). U937 is a monocytic human myeloid leukemia cell line frequently used in biomedical research [40]. Compared to AKL-40, MytiLec-1 and Ricin inhibited growth of these cells in vitro at much lower concentrations, which also indicates the difference of glycan recognition for these three lectins [30,41,42].

In marine organisms, lectins recognize and bind to the surface polysaccharides of a variety of bacteria. They have ability to kill bacteria or inhibit their growth and thereby contribute to defense against infection as a part of the innate immune system [43,44]. Similar to AKL, AKL-40 also exhibited growth inhibitory activity against both Gram-positive and Gram-negative organisms (Figure 6A) [45]. *Staphylococcus aureus* were the most susceptible bacteria to both lectins. Unlike AKL, another egg lectin from *Aplysia dactylomela*

(ADEL) agglutinated *Staphylococcus aureus* cells, but could not inhibit their growth [24]. A mannose/galactose-binding C-type lectin isolated from bay scallop *Argopectenirradians* also displayed this property against Gram-positive *Staphylococcus aureus* and Gram-negative *Escherichia coli* and *Vibrio anguillarum* [46]. However, due to unknown reasons, AKL, AKL-40, and ADEL could not affect the growth of *Escherichia coli* [24]. On the other hand, CvL from marine sponge *Cliona varians* showed cytotoxic effect on Gram-positive bacteria, such as *Bacillus subtilis* and *Staphylococcus aureus*, but could not inhibit the growth of Gram-negative bacteria like *Escherichia coli* and *Pseudomonas aeruginosa* [15]. These results suggest that the bactericidal activity of lectins depend not only on glycan-binding specificities, but also on multiple systems, such as multivalency or binding constancy.

Antifungal properties are not very common in lectins isolated from marine invertebrates [47]. *Aplysia depilans* gonad lectin was previously used to study the distribution of galacturonic acids in the cell walls of some pathogenic fungi [48]. In this study, high concentration (400 μg/mL) of AKL-40 totally inhibited the growth of *Talaromyces verruculosus* (Figure 6B). A study by Ruperez et al. reported the presence of galactose sugars in the cell wall of *Talaromyces verruculosus*, which justified our findings [49]. Another antifungal protein (Aplysianin E) from *Aplysia kurodai* eggs completely suppressed the growth of *Saccharomyces cerevisiae* and *Candida albicans* at a concentration of 16 μg/mL [50]. AKL also repressed the mycelial growth of *Curvularia lunata* at 100 μg/mL whereas growth inhibitory effects of two other lectins from Japanese black sponge (*Halichondria okadai*) and mussel (*Crenomytilus grayanus*) were found against *Aspergillus niger* and *Pichia pastoris*, respectively [5,45,51]. It can be postulated that certain polysaccharides present in fungal cell walls interacted with these lectins and inhibited their growth by disturbing spore germination, growth of mycelium, and synthesis of chitin, to alter the fungal cell wall [52]. Like eggs from other organisms, sea hare eggs are found lying exposed in the seashore, vulnerable to the attack of predators and microbes [53]. Therefore, egg lectins might have a role to serve as protective molecules.

4. Materials and Methods

4.1. Preparation of the Crude Extract

The eggs of sea hare were gathered from the Zushi coast, Kanagawa, Japan. Eggs were crushed in a mortar, solubilized with Tris-buffered saline or TBS (10 mM Tris(hydroxymethyl) aminomethane-HCl, pH 7.4, with 150 mM NaCl). Then, crushed eggs were blended with 10 volumes of TBS containing a protease inhibitor (10 mM, Protease Inhibitor Cocktail 100X, Wako Pure Chemical Corp, Osaka, Japan).The homogenized sample was filled up in 500-mL centrifuge bottles and centrifuged at 14,720× *g* for 1 h at 4 °C. A Suprema 21 centrifuge equipped with an NA-18HS rotor (TOMY Co. Ltd., Tokyo, Japan) was used for this step. After repeated centrifugation of the supernatant at the same speed for the same duration, a clear solution was prepared and stored as the crude extract.

4.2. Purification of the Lectin

The lectin was purified according to a previously described procedure [22]. The crude extract was twice centrifuged at 27,500× *g* for 1 h at 4°C and was administered to a 5 mL melibiose-agarose affinity column (J-Oil Mills Inc., Tokyo, Japan). This column was connected to a 5 mL Sephadex G-75 pre-column. After loading on the crude protein sample, the column was washed well with TBS. Lectins bound to the column were eluted with 10 mM D-galacturonic acid or D-galactose-containing TBS. The mixture of AKLs was separated by using a gel filtration chromatography of Sephacryl S-200 (Cytiva, Marlborough, MA, USA) connected to a fraction collector (FRC-10A, Shimadzu Corporation, Tokyo, Japan). Molecular weights of the lectins were confirmed by SDS-PAGE using standard marker proteins [54]. The 40 kDa polypeptide species was denoted as 'AKL-40′ in this study.

4.3. Determination of N-Terminal Partial Amino Acid Sequence of AKL-40

The N-terminal sequence of the 35 amino acids of AKL-40 was determined with automated Edman degradation by using a protein/peptide sequencer PPSQ (Shimadzu Corporation, Kyoto, Japan) [55].

4.4. Determination of the Toxicity of AKL-40 by Brine Shrimp Nauplii Lethality Assay

The assay was performed according to a method reported earlier [29]; 4 mL of artificial seawater was taken in test tubes. Ten brine shrimp nauplii were taken in each vial and AKL-40 was added to these vials to adjust its concentrations from 10 to 160 μg/mL. There were control vials containing only seawater and nauplii, but no lectin. The experiment was repeated thrice at 30 °C for 24 h with 6 h of light exposure. Percentages of mortality of the nauplii were determined for each concentration and the LC_{50} value of AKL-40 was also determined according to the method of Finney [56].

4.5. In Vivo Anticancer Activity of AKL-40 against Ehrlich Ascites Carcinoma Cells Grown in Swiss Albino Mice

The in vivo experiment was approved by the Institutional Animal, Medical Ethics, Biosafety and Bio-security Committee (IAMEBBC) for experimentations on animals, human, microbes, and living natural sources, Institute of Biological Sciences (IBSc), University of Rajshahi, Bangladesh (memo no. 102(6)/320-IAMEBBC/IBSc). Four to six weeks old Swiss albino mice (weight range 25–30 g) of both genders were collected from ICDDR'B (International Center for Diarrheal Diseases Research, Bangladesh) and EAC cells were propagated into these mice by a bi-weekly intraperitoneal transformation. Cells in ascitic fluid were drawn from a donor mouse bearing 6–7 days old tumor cells and with the help of a hemocytometer, adjusted to 2×10^6 cells/mL. Normal saline was used for the dilution. Viability of tumor cells was checked by 0.4% trypan blue assay.

The mice were randomly distributed into three groups, consisting of six mice in each group. These groups were denoted as 'A' or control, 'B' or lectin-treated with lower dose (1 mg/kg/day)' and 'C' or lectin-treated with higher dose (2 mg/kg/day). Moreover, 0.1 mL of cellular suspension containing viable EAC cells was injected intraperitoneally to each Swiss albino mouse. After 24 h, both lectin-treated groups (B and C) were treated for five days with an intraperitoneal injection of AKL-40 at doses of 1.0 and 2.0 mg/kg/day. All mice in groups A, B, and C were weighed to check the rate of tumor growth. Mice were sacrificed on the sixth day and EAC cells were collected from the ascitic fluid. The total number of viable EAC cells in every mouse of the treated groups (B and C) was compared to those of the control group (A) using the following formula:

$$\text{Percentage of inhibition} = 100 - \{(\text{cells from AKL-40 treated mice}/\text{cells from control mice}) \times 100\}$$

4.6. Morphological Observation of AKL-40 Treated EAC Cells by Fluorescence Microscope

Morphological changes of control (or untreated) and AKL-40 treated EAC cells were observed using fluorescence microscopy (Olympus iX71, Seoul, Korea). EAC cells from mice treated with and without AKL-40 for five consecutive days were collected. After washing with phosphate buffer saline (PBS), the cells were stained with 0.1 μg/mL of Hoechst-33342 at 37 °C for 20 min in dark, washed again with PBS, and observed in the microscope.

4.7. RNA Isolation and Checking the Expression of Apoptosis-Related Genes from Ehrlich Ascites Carcinoma Cells

EAC cells from mice treated with and without AKL-40 were collected and the total RNA was isolated using a reagent kit (Tiangen Biotech Co., Beijing, China). Concentration and purity of the isolated RNA were checked at 260 and 280 nm using a spectrophotometer. cDNA samples were prepared following the protocol of Applied Biosystems, Waltham, MA, USA, and bands for all PCR reactions were visualized in 1.4% agarose gel with a gel documentation system (Cleaver Scientific Ltd., Rugby, UK). Ten μg/mL of ethidium bromide solution was used to stain the gel. GAPDH was used as a housekeeping gene to compare with the standard. GeneRuler 1000 bp DNA ladder (Fermentas, Waltham, MA, USA) was used as marker. Specific oligonucleotides (Integrated DNA Technologies or IDT, Singapore) like p53, Bax, Bcl-X_L and GAPDH generated 458 bp, 477 bp, 780 bp, and 475 bp amplification products, respectively. Primer sequences of these genes under study are provided in Table 1.

Table 1. Primer constructions for apoptosis related and housekeeping genes.

Primer	Forward	Reverse
p53	5′-GCGTCTTAGAGACAGTTGCCT-3′	5′-GGATAGGTCGGCGGTTCATGC-3′
Bax	5′-GGCCCACCAGCTCTGAGCAGA-3′	5′-GCCACGTGGGCGTCCCAAAGT-3′
Bcl-X_L	5′-TTGGACAATGGACTGGTTGA-3′	5′-GTAGAGTGGATGGTCAGTG-3′
GAPDH	5′-GTGGAAGGACTCATGACCACAG-3′	5′-CTGGTGCTCAGTGTAGCCCAG-3′

For gene amplification, a program was set in a thermal cycler (Gene, Atlas 482, Tokyo, Japan) at 95 °C for 3 min, followed by 35 cycles of 95 °C for 30 s, 55 °C for 30 s, 72 °C for 50 s, finally at 72 °C for 10 min, and then was eventually held at 20 °C. In case of Bax and Bcl-XL, the annealing temperature was 54 °C instead of 55 °C.

4.8. *Determination of Cytotoxic Activity of AKL-40 against Cancer Cell Lines and Detection of Activated Signal Transduction Molecules*

Cytotoxic activity of AKL-40 against cancer cell lines was evaluated according to a previous report [5]. Three leukemia cell lines K562, Raji, and RBL-1 (2×10^5 cells) were seeded into a 96-well titer plate and treated with different concentrations of AKL-40 for 24 h at 37 °C. Ten micro liter of WST-8 solution was added to each well and incubated for 4 h at the same temperature. The absorbance was measured at 450 nm to assay the reduction in proportion of living cells by a GloMax Multi Detection System (Promega, Madison, WI, USA).

K562, Raji, and RBL-1 (2×10^5 cells) were cultured with AKL-40 (0–1600 μg/mL) for 24 h, and lysed with 200 μL cell lysis buffer M. The cell lysate was separated by SDS-PAGE and electroblotted onto PVDF membrane. Primary antibodies used were directed to phospho-ERK1/2 (extracellular signal-regulated kinase), phospho-P38 (P38 MAP kinase), and phospho-JNK (c-Jun N-terminal kinase) by using each mouse mAb (Becton Dickinson, Franklin Lakes, NJ, USA) at the dilution of 1/3000. Membrane was masked with TBS containing 1% BSA, soaked with 2% Triton X-100 at RT, incubated with HRP-conjugated goat anti-mouse IgG for mouse mAb (Tokyo Chemical Industry Co., Tokyo, Japan) for 1 h, and colored with EzWestBlue (ATTO Corp., Tokyo, Japan).

4.9. *Determination of the Minimum Agglutination Concentration of EAC and U937 Cells by AKL-40*

U937 cells (ATCC CRL-3253) were collected from Yokohama City University, Japan. Fifty μL of 20 mM Tris-HCl buffer saline containing 10 mM CaCl2 (pH 7.8) was taken in each well of two U-bottomed 96-well microtiter plates (one for EAC and the other for U937 cells). AKL-40 (50 μL) was added to the titer plates through serial dilution. The number of EAC and U937 cells in RPMI-1640 media was counted using a hemocytometer (Hirschmann EM Techcolor, Eberstadt, Germany) and around 5×10^5 cells were seeded in each well of the two plates. The plates were agitated for 5 min in a microshaker, kept at room temperature for 30 min, and agglutination titers for both cell types were recorded.

4.10. *Anticancer Activity of AKL-40 in Vitro against U937 and Ehrlich Ascites Carcinoma Cells*

One hundred μL of RPMI-1640 media was taken in two 96-well flat bottom titer plates (one for EAC and the other for U937 cells). AKL-40 was added to the wells at final concentrations of 0, 100, 200, and 250 μg/mL. 100 μL of EAC cells (collected from mice) and U937 cells were added to each well (5×10^5 cells/well). There were three control wells containing only cancer cells. After an incubation period of 24 h in 5% CO_2 incubator at 37 °C, the clear supernatant was carefully removed from each well. Then 180 µL of PBS and 20 µL of 3-(4,5-dimethylthiazol-2-yl)-2,5-diphenyl tetrazolium bromide or MTT (5 mg/mL) were added and both plates were kept in the incubator for 8 h at 37 °C. The supernatant was removed again from each well, 200 µL of acidic isopropanol was added, and absorbance of each well was recorded by a titer plate reader at 570 nm. Percentages of cell proliferation inhibition were calculated by the following equation:

$$\text{Proliferation inhibition ratio (\%)} = (A - B) \times 100/A$$

where A is the OD_{570} nm of the cellular homogenate from control wells and B is the OD_{570} nm of the cellular homogenate from wells treated with AKL-40.

4.11. Bactericidal Activity of AKL against Pathogenic Bacteria

Antibacterial activity of AKL-40 was determined by agar disc diffusion method against four bacteria—two Gram-positive (*Staphylococcus aureus* and *Bacillus cereus*) and two Gram-negative (*Shigella sonnei* and *Escherichia coli*). Petri dishes containing lysogeny broth (LB) media for each bacterial species were prepared and four paper discs (control, high dose, low dose, and antibiotic) were placed on each petri dish. Moreover, two hundred and one hundred µg of AKL-40 were used as high and low doses, respectively, whereas the antibiotic disc contained 15 µg of ampicillin. Control disc was soaked only with LB media. After keeping at 37 °C for 24 h in an incubator, zones of inhibition around the discs were observed in each petri dish.

4.12. Antifungal Activity of AKL-40 against Talaromyces verruculosus

Talaromyces verruculosus, a fungus from the division ascomycota, was cultured in petri dishes with potato dextrose agar (HiMedia Laboratories Pvt. Ltd., Mumbai, India) as a medium following the standard procedure. In the 'Test' petri dish, three discs soaked with 400 µg/mL of AKL-40 were placed around whereas the 'Control' petri dish had only one disc, soaked only with media. After keeping at 25 °C for 1 week in an incubator, growth of the fungi in these petri dishes was observed. After another week, there was a second observation to find any growth inhibition.

4.13. Statistical Analysis

All experiments of this research work were performed in three replicates. Experimental data were expressed as mean ± SD. To test differences between experimental conditions, one way ANOVA and Dunnett's post-test correction was used. The results were statistically significant when ** $p < 0.01$ and * $p < 0.05$.

5. Conclusions

This study might be the start of isolating and sequencing amino acids in the specific peptides/proteins accountable for the observed antiproliferative effects of AKL-40. Abolition of biological activities by AKL-40 was not checked in the presence of inhibiting sugars, such as D-galacturonic acid or D-galactose, which is a limitation of this study. Determination of MIC and MBC values for the bacteria and fungi could also be performed. Additional investigations are required to investigate the association of lectin in infection and pathogenesis of different bacteria, as well as to recognize the molecular mechanism of signaling pathways activated during infection, which could help in the development of new therapeutic approaches. It is quite interesting to predict that a protein domain from bacteria was evolved in the course of time to exert antimicrobial effects in marine organisms. Such predictions could be evaluated in the future via a combined study of functional genomics, transcriptomics, and glycobiology of Aplysia lectins.

Supplementary Materials: The following are available online at https://www.mdpi.com/article/10.3390/md19070394/s1. Figure S1: N-terminal region of AKL-40, including the first 35 amino acids of the polypeptide determined by Edman degradation.

Author Contributions: Conceptualization, Y.O. (Yasuhiro Ozeki) and I.H.; methodology, R.R.S., Y.O. (Yasuhiro Ozeki), and I.H.; software, A.K.M.A.; validation, S.M.A.K. and S.R.; formal analysis, A.K.M.A. and Y.F.; investigation, R.R.S., N.A., S.R., Y.F., K.H., N.K., M.Y. and I.H.; resources, S.R.K., K.H., N.K., and Y.O. (Yukiko Ogawa); data curation, I.H.; writing—original draft preparation, I.H.; writing—review and editing, I.H. and Y.O. (Yasuhiro Ozeki); visualization, Y.O. (Yasuhiro Ozeki) and I.H.; supervision, I.H.; project administration, I.H and Y.O. (Yasuhiro Ozeki); funding acquisition, Y.F., Y.O. (Yasuhiro Ozeki) and I.H. All authors have read and agreed to the published version of the manuscript.

Funding: This research was funded in part by Rajshahi University research projects (A-1179/5/52/R.U/Science-22/19-20), University of Rajshahi, Bangladesh. This work was also supported by JSPS Grant-in-Aid for Scientific Research-KAKENHI under grant no. JP19K06239.

Institutional Review Board Statement: The study was conducted according to the guidelines of the Declaration of Helsinki, and approved by the Institutional Animal, Medical Ethics, Bio-safety and Bio-security Committee (IAMEBBC) for Experimentations on Animal, Human, Microbes and Living Natural Sources (memo no. 102(6)/320/IAMEBBC/IBSc, date of approval: 7 July 2020), Institute of Biological Sciences, University of Rajshahi, Bangladesh. As of 2020, collections of sea hare eggs on the coasts are not restricted from the view of biological preservation in Japan (Ministry of Environment, Government of Japan).

Acknowledgments: The authors would like to thank Abu Reza, Department of Genetic Engineering and Biotechnology, University of Rajshahi, Bangladesh, for providing the fungal strain and Ruhul Amin, Senior Scientific Officer, Bangladesh Council of Scientific and Industrial Research (BCSIR), Rajshahi, Bangladesh for providing laboratory support.

Conflicts of Interest: The authors declare no conflict of interest. The funders had no role in the design of the study; in the collection, analyses, or interpretation of data; in the writing of the manuscript, or in the decision to publish the results.

References

1. Fujii, Y.; Dohmae, N.; Takio, K.; Kawsar, S.M.A.; Matsumoto, R.; Hasan, I.; Koide, Y.; Kanaly, R.A.; Yasumitsu, H.; Ogawa, Y.; et al. A lectin from the mussel *Mytilus galloprovincialis* has a highly novel primary structure and induces glycan-mediated cytotoxicity of globotriaosylceramide-expressing lymphoma cells. *J. Biol. Chem.* **2012**, *287*, 44772–44783. [CrossRef] [PubMed]
2. Hirabayashi, J.; Dutta, S.K.; Kasai, K. Novel galactose-binding proteins in Annelida. Characterization of 29-kDa tandem repeat-type lectins from the earthworm *Lumbricus terrestris*. *J. Biol. Chem.* **1998**, *273*, 14450–14460. [CrossRef] [PubMed]
3. Wang, J.H.; Kong, J.; Li, W.; Molchanova, V.; Chikalovets, I.; Belogortseva, N.; Lukyanov, P.; Zheng, Y.T. A beta-galactose-specific lectin isolated from the marine worm *Chaetopterus variopedatus* possesses anti-HIV-1 activity. *Comp. Biochem. Physiol. C Toxicol. Pharmacol.* **2006**, *142*, 111–117. [CrossRef] [PubMed]
4. Gowda, N.M.; Goswami, U.; Khan, M.I. Purification and characterization of a T-antigen specific lectin from the coelomic fluid of a marine invertebrate, sea cucumber (*Holothuria scabra*). *Fish Shellfish Immunol.* **2008**, *24*, 450–458. [CrossRef]
5. Hasan, I.; Ozeki, Y. Histochemical localization of N-acetylhexosamine-binding lectin HOL-18 in *Halichondria okadai* (Japanese black sponge), and its antimicrobial and cytotoxic anticancer effects. *Int. J. Biol. Macromol.* **2019**, *124*, 819–827. [CrossRef]
6. Ohtsubo, K.; Marth, J.D. Glycosylation in cellular mechanisms of health and disease. *Cell* **2006**, *126*, 855–867. [CrossRef]
7. Pang, X.; Li, H.; Guan, F.; Li, X. Multiple Roles of Glycans in Hematological Malignancies. *Front. Oncol.* **2018**, *8*, 364. [CrossRef]
8. Stowell, S.R.; Ju, T.; Cummings, R.D. Protein glycosylation in cancer. *Annu. Rev. Pathol.* **2015**, *10*, 473–510. [CrossRef] [PubMed]
9. Powlesland, A.S.; Hitchen, P.G.; Parry, S.; Graham, S.A.; Barrio, M.M.; Elola, M.T.; Mordoh, J.; Dell, A.; Drickamer, K.; Taylor, M.E. Targeted glycoproteomic identification of cancer cell glycosylation. *Glycobiology* **2009**, *19*, 899–909. [CrossRef] [PubMed]
10. Queiroz, A.F.S.; Silva, R.A.; Moura, R.M.; Dreyfuss, J.L.; Paredes-Gamero, E.J.; Souza, A.C.S.; Tersariol, I.L.S.; Santos, E.A.; Nader, H.B.; Justo, G.Z.; et al. Growth inhibitory activity of a novel lectin from *Cliona varians* against K562 human erythroleukemia cells. *Cancer Chemother. Pharmacol.* **2009**, *63*, 1023–1033. [CrossRef]
11. Kawsar, S.M.A.; Matsumoto, R.; Fujii, Y.; Matsuoka, H.; Masuda, N.; Iwahara, C.; Yasumitsu, H.; Kanaly, R.A.; Sugawara, S.; Hosono, M.; et al. Cytotoxicity and glycan-binding profile of a D-galactose-binding lectin from the eggs of a Japanese sea hare (*Aplysia kurodai*). *Protein J.* **2011**, *30*, 509–519. [CrossRef]
12. Chernikov, O.; Kuzmich, A.; Chikalovets, I.; Molchanova, V.; Hua, K.F. Lectin CGL from the sea mussel *Crenomytilus grayanus* induces Burkitt's lymphoma cells death via interaction with surface glycan. *Int. J. Biol. Macromol.* **2017**, *104*, 508–514. [CrossRef]

13. Zurga, S.; Nanut, M.P.; Kos, J.; Sabotic, J. Fungal lectin MpL enables entry of protein drugs into cancer cells and their subcellular targeting. *Oncotarget* **2017**, *8*, 26896–26910. [CrossRef] [PubMed]
14. Fujii, Y.; Sugawara, S.; Araki, D.; Kawano, T.; Tatsuta, T.; Takahashi, K.; Kawsar, S.M.; Matsumoto, R.; Kanaly, R.A.; Yasumitsu, H.; et al. MRP1 expressed on Burkitt's lymphoma cells was depleted by catfish egg lectin through Gb3-glycosphingolipid and enhanced cytotoxic effect of drugs. *Protein J.* **2012**, *31*, 15–26. [CrossRef] [PubMed]
15. Moura, R.M.; Queiroz, A.F.; Fook, J.M.; Dias, A.S.; Monteiro, N.K.; Ribeiro, J.K.; Moura, G.E.; Macedo, L.L.; Santos, E.A.; Sales, M.P. CvL, a lectin from the marine sponge *Cliona varians*: Isolation, characterization and its effects on pathogenic bacteria and *Leishmania* promastigotes. *Comp. Biochem. Physiol. A Mol. Integr. Physiol.* **2006**, *145*, 517–523. [CrossRef]
16. Schroder, H.C.; Ushijima, H.; Krasko, A.; Gamulin, V.; Thakur, N.L.; Diehl-Seifert, B.; Muller, I.M.; Muller, W.E. Emergence and disappearance of an immune molecule, an antimicrobial lectin, in basal metazoa. A tachylectin-related protein in the sponge *Suberitesdomuncula*. *J. Biol. Chem.* **2003**, *278*, 32810–32817. [CrossRef] [PubMed]
17. Frirdich, E.; Whitfield, C. Lipopolysaccharide inner core oligosaccharide structure and outer membrane stability in human pathogens belonging to the Enterobacteriaceae. *J. Endotoxin Res.* **2005**, *11*, 133–144. [CrossRef] [PubMed]
18. Cox, A.D.; Li, J.; Richards, J.C. Identification and localization of glycine in the inner core lipopolysaccharide of *Neisseria meningitidis*. *Eur. J. Biochem.* **2002**, *269*, 4169–4175. [CrossRef]
19. Lodowska, J.; Wolny, D.; Jaworska-Kik, M.; Kurkiewicz, S.; Dzierzewicz, Z.; Weglarz, L. The chemical composition of endotoxin isolated from intestinal strain of *Desulfovibrio desulfuricans*. *Sci. World J.* **2012**, *2012*, 647352. [CrossRef]
20. Kamiya, H.; Shimizu, Y. A natural agglutinin inhibitable by d-galacturonic acid in the sea hare *Aplysia* eggs: Characterization and purification. *Nippon. Suisan Gakk.* **1981**, *47*, 255–259. [CrossRef]
21. Kawsar, S.M.A.; Matsumoto, R.; Fujii, Y.; Yasumitsu, H.; Dogasaki, C.; Hosono, M.; Nitta, K.; Hamako, J.; Matsui, T.; Kojima, N.; et al. Purification and biochemical characterization of D-galactose binding lectin from Japanese sea hare (*Aplysia kurodai*) eggs. *Biochemistry* **2009**, *74*, 709–716. [CrossRef]
22. Hasan, I.; Watanabe, M.; Ishizaki, N.; Sugita-Konishi, Y.; Kawakami, Y.; Suzuki, J.; Dogasaki, C.; Rajia, S.; Kawsar, S.M.A.; Koide, Y.; et al. A galactose-binding lectin isolated from *Aplysia kurodai* (sea hare) eggs inhibits streptolysin-induced hemolysis. *Molecules* **2014**, *19*, 13990–14003. [CrossRef] [PubMed]
23. Gilboa-Garber, N.; Susswein, A.J.; Mizrahi, L.; Avichezer, D. Purification and characterization of the gonad lectin of *Aplysia depilans*. *FEBS Lett.* **1985**, *181*, 267–270. [CrossRef]
24. Carneiro, R.F.; Torres, R.C.; Chaves, R.P.; de Vasconcelos, M.A.; de Sousa, B.L.; Goveia, A.C.; Arruda, F.V.; Matos, M.N.; Matthews-Cascon, H.; Freire, V.N.; et al. Purification, biochemical characterization, and amino acid sequence of a novel type of lectin from *Aplysia dactylomela* eggs with antibacterial/antibiofilm potential. *Mar. Biotechnol.* **2017**, *19*, 49–64. [CrossRef]
25. Wilson, M.P.; Carrow, G.M.; Levitan, I.B. Modulation of growth of Aplysia neurons by an endogenous lectin. *J. Neurobiol.* **1992**, *23*, 739–750. [CrossRef]
26. Motohashi, S.; Jimbo, M.; Naito, T.; Suzuki, T.; Sakai, R.; Kamiya, H. Isolation, amino acid sequences, and plausible functions of the galacturonic acid-binding egg lectin of the Sea hare *Aplysia kurodai*. *Mar. Drugs.* **2017**, *15*, 161. [CrossRef] [PubMed]
27. Jridi, M.; Nasri, R.; Marzougui, Z.; Abdelhedi, O.; Hamdi, M.; Nasri, M. Characterization and assessment of antioxidant and antibacterial activities of sulfated polysaccharides extracted from cuttlefish skin and muscle. *Int. J. Biol. Macromol.* **2019**, *123*, 1221–1228. [CrossRef]
28. Kawsar, S.M.A.; Aftabuddin, S.; Yasumitsu, H.; Ozeki, Y. The cytotoxic activity of two D-galactose-binding lectins purified from marine invertebrates. *Arch. Biol. Sci.* **2010**, *62*, 1027–1034. [CrossRef]
29. Hasan, I.; Asaduzzaman, A.K.M.; Swarna, R.R.; Fujii, Y.; Ozeki, Y.; Uddin, M.B.; Kabir, S.R. MytiLec-1 shows glycan-dependent toxicity against brine shrimp *Artemia* and induces apoptotic death of Ehrlich ascites carcinoma cells in vivo. *Mar Drugs.* **2019**, *17*, 502. [CrossRef]
30. Moura, R.M.; Melo, A.A.; Carneiro, R.F.; Rodrigues, C.R.; Delatorre, P.; Nascimento, K.S.; Saker-Sampaio, S.; Nagano, C.S.; Cavada, B.S.; Sampaio, A.H. Hemagglutinating/Hemolytic activities in extracts of marine invertebrates from the Brazilian coast and isolation of two lectins from the marine sponge *Cliona varians* and the sea cucumber *Holothuria grisea*. *An. Acad. Bras. Cienc.* **2015**, *87*, 973–984. [CrossRef]
31. Carneiro, R.F.; de Melo, A.A.; Nascimento, F.E.; Simplicio, C.A.; Nascimento, K.S.; Rocha, B.A.; Saker-Sampaio, S.; Moura, R.M.; Mota, S.S.; Cavada, B.S.; et al. Halilectin 1 (H-1) and Halilectin 2 (H-2): Two new lectins isolated from the marine sponge *Haliclona caerulea*. *J. Mol. Recognit.* **2013**, *26*, 51–58. [CrossRef]
32. Rahman, M.S.; Alam, M.B.; Choi, Y.H.; Yoo, J.C. Anticancer activity and antioxidant potential of *Aponogeton undulatus* against Ehrlich ascites carcinoma cells in Swiss albino mice. *Oncol. Lett.* **2017**, *14*, 3169–3176. [CrossRef] [PubMed]
33. Eckhardt, A.E.; Goldstein, I.J. Occurrence of alpha-D-galactosyl-containing glycoproteins on Ehrlich tumor cell membranes. *Biochemistry* **1983**, *22*, 5280–5289. [CrossRef]
34. Sakakibara, F.; Kawauchi, H.; Takayanagi, G.; Ise, H. Egg lectin of *Rana japonica* and its receptor glycoprotein of Ehrlich tumor cells. *Cancer Res.* **1979**, *39*, 1347–1352.
35. Rabelo, L.; Monteiro, N.; Serquiz, R.; Santos, P.; Oliveira, R.; Oliveira, A.; Rocha, H.; Morais, A.H.; Uchoa, A.; Santos, E. A lactose-binding lectin from the marine sponge *Cinachyrellaapion* (Cal) induces cell death in human cervical adenocarcinoma cells. *Mar. Drugs.* **2012**, *10*, 727–743. [CrossRef] [PubMed]

36. Chaves, R.P.; Silva, S.R.D.; Neto, L.G.N.; Carneiro, R.F.; Silva, A.L.C.D.; Sampaio, A.H.; Sousa, B.L.D.; Cabral, M.G.; Videira, P.A.; Teixeira, E.H.; et al. Structural characterization of two isolectins from the marine red alga *Solieria filiformis* (Kützing) P.W. Gabrielson and their anticancer effect on MCF-7 breast cancer cells. *Int. J. Biol. Macromol.* **2018**, *107*, 1320–1329. [CrossRef]
37. Wu, L.; Yang, X.; Duan, X.; Cui, L.; Li, G. Exogenous expression of marine lectins DlFBL and SpRBL induces cancer cell apoptosis possibly through PRMT5-E2F-1 pathway. *Sci. Rep.* **2014**, *4*, 4505. [CrossRef] [PubMed]
38. Timoshenko, A.V. Towards molecular mechanisms regulating the expression of galectins in cancer cells under microenvironmental stress conditions. *Cell Mol. Life Sci.* **2015**, *72*, 4327–4340. [CrossRef]
39. Cedeno-Laurent, F.; Dimitroff, C.J. Galectins and their ligands: Negative regulators of anti-tumor immunity. *Glycoconj. J.* **2012**, *29*, 619–625. [CrossRef]
40. Chanput, W.; Peters, V.; Wichers, H. THP-1 and U937 Cells. In *The Impact of Food Bioactives on Health*, 1st ed.; Verhoeckx, K., Cotter, P., Lopez-Exposito, I., Kleiveland, C., Lea, T., Mackie, A., Requena, T., Swiatecka, D., Wichers, H., Eds.; Springer International Publishing: New York, NY, USA, 2015; pp. 147–149. [CrossRef]
41. Oda, T.; Iwaoka, J.; Komatsu, N.; Muramatsu, T. Involvement of N-acetylcysteine-sensitive pathways in ricin-induced apoptotic cell death in U937 cells. *Biosci. Biotechnol. Biochem.* **1999**, *63*, 341–348. [CrossRef]
42. Hasegawa, N.; Kimura, Y.; Oda, T.; Komatsu, N.; Muramatsu, T. Isolated ricin B-chain-mediated apoptosis in U937 cells. *Biosci. Biotechnol. Biochem.* **2000**, *64*, 1422–1429. [CrossRef] [PubMed]
43. Sahly, H.; Keisari, Y.; Crouch, E.; Sharon, N.; Ofek, I. Recognition of bacterial surface polysaccharides by lectins of the innate immune system and its contribution to defense against Infection: The case of pulmonary pathogens. *Infect. Immun.* **2008**, *76*, 1322–1332. [CrossRef] [PubMed]
44. Hasan, I.; Gerdol, M.; Fujii, Y.; Rajia, S.; Koide, Y.; Yamamoto, D.; Kawsar, S.M.; Ozeki, Y. cDNA and gene structure of MytiLec-1, a bacteriostatic R-type lectin from the mediterranean mussel (*Mytilus galloprovincialis*). *Mar Drugs.* **2016**, *14*, 92. [CrossRef]
45. Kawsar, S.M.A.; Mamun, S.M.A.; Rahman, M.S.; Yasumitsu, M.S.; Ozeki, Y. Growth inhibitory effects on microorganisms by a D-galactose-binding lectin purified from the sea hare (*Aplysia kurodai*) eggs: An in vitro study. *Nat. Sci.* **2010**, *8*, 82–89. [CrossRef]
46. Huang, M.; Song, X.; Zhao, J.; Mu, C.; Wang, L.; Zhang, H.; Zhou, Z.; Liu, X.; Song, L. A C-type lectin (AiCTL-3) from bay scallop *Argopecten irradians* with mannose/galactose binding ability to bind various bacteria. *Gene* **2013**, *531*, 31–38. [CrossRef]
47. Cheung, R.C.; Wong, J.H.; Pan, W.; Chan, Y.S.; Yin, C.; Dan, X.; Ng, T.B. Marine lectins and their medicinal applications. *Appl. Microbiol. Biotechnol.* **2015**, *99*, 3755–3773. [CrossRef]
48. Benhamou, N. Ultrastructural study of galacturonic acid distribution in some pathogenic fungi using gold-complexed *Aplysia depilans* gonad lectin. *Can. J. Microbiol.* **1989**, *35*, 349–358. [CrossRef]
49. Ruperez, P.; Moya, A.; Leal, J.A. Cell wall polysaccharides from *Talaromyces* species. *Arch. Microbiol.* **1986**, *146*, 250–255. [CrossRef]
50. Iijima, R.; Kisugi, J.; Yamazaki, M. Antifungal activity of Aplysianin E, a cytotoxic protein of sea hare (*Aplysia kurodai*) eggs. *Dev. Comp. Immunol.* **1995**, *19*, 13–19. [CrossRef] [PubMed]
51. Chikalovets, I.V.; Chernikov, O.V.; Pivkin, M.V.; Molchanova, V.I.; Litovchenko, A.P.; Li, W.; Lukyanov, P.A. A lectin with antifungal activity from the mussel *Crenomytilus grayanus*. *Fish Shellfish Immunol.* **2015**, *42*, 503–507. [CrossRef] [PubMed]
52. Gomes, B.S.; Siqueira, A.B.S.; Maia, R.C.C.; Giampaoli, V.; Teixeira, E.H.; Arruda, F.V.S.; Nascimento, K.S.; Lima, A.N.; Souza-Motta, C.M.; Cavada, B.S.; et al. Antifungal activity of lectins against yeast of vaginal secretion. *Braz. J. Microbiol.* **2012**, *43*, 770–778. [CrossRef] [PubMed]
53. Ituarte, S.; Brola, T.R.; Fernandez, P.E.; Mu, H.; Qiu, J.-W.; Heras, H.; Dreon, M.S. A lectin of a non-invasive apple snail as an egg defense against predation alters the rat gut morphophysiology. *PLoS ONE* **2018**, *13*, e0198361. [CrossRef]
54. Laemmli, U.K. Cleavage of structural proteins during the assembly of the head of bacteriophage T4. *Nature* **1970**, *227*, 680–685. [CrossRef] [PubMed]
55. Ozeki, Y.; Matsui, T.; Suzuki, M.; Titani, K. Amino acid sequence and molecular characterization of a D-galactoside-specific lectin purified from sea urchin (*Anthocidaris crassispina*) eggs. *Biochemistry* **1991**, *30*, 2391–2394. [CrossRef]
56. Finney, D.J. *Probit Analysis*, 3rd ed.; Cambridge University Press: London, UK, 1971; p. 333.

marine drugs

MDPI

Review

Pharmacological Activities of Sulfated Fucose-Rich Polysaccharides after Oral Administration: Perspectives for the Development of New Carbohydrate-Based Drugs

Roberto J. C. Fonseca [1,2,*] and Paulo A. S. Mourão [1,3]

1 Laboratório de Tecido Conjuntivo, Hospital Universitário Clementino Fraga Filho, Rio de Janeiro 21941-913, Brazil; pmourao@hucff.ufrj.br
2 Centro de Ciências da Saúde, Instituto de Ciências Biomédicas, Universidade Federal do Rio de Janeiro, Rio de Janeiro 21941-913, Brazil
3 Centro de Ciências da Saúde, Instituto de Bioquímica Médica Leopoldo de Meis, Universidade Federal do Rio de Janeiro, Rio de Janeiro 21941-913, Brazil
* Correspondence: robertofonseca@hucff.ufrj.br

Citation: Fonseca, R.J.C.; Mourão, P.A.S. Pharmacological Activities of Sulfated Fucose-Rich Polysaccharides after Oral Administration: Perspectives for the Development of New Carbohydrate-Based Drugs. *Mar. Drugs* **2021**, *19*, 425. https://doi.org/10.3390/md19080425

Academic Editor: Orazio Taglialatela-Scafati

Received: 16 June 2021
Accepted: 23 July 2021
Published: 27 July 2021

Abstract: Marine organisms are a source of active biomolecules with immense therapeutic and nutraceutical potential. Sulfated fucose-rich polysaccharides are present in large quantities in these organisms with important pharmacological effects in several biological systems. These polysaccharides include sulfated fucan (as fucoidan) and fucosylated chondroitin sulfate. The development of these polysaccharides as new drugs involves several important steps, among them, demonstration of the effectiveness of these compounds after oral administration. The oral route is the more practical, comfortable and preferred by patients for long-term treatments. In the past 20 years, reports of various pharmacological effects of these polysaccharides orally administered in several animal experimental models and some trials in humans have sparked the possibility for the development of drugs based on sulfated polysaccharides and/or the use of these marine organisms as functional food. This review focuses on the main pharmacological effects of sulfated fucose-rich polysaccharides, with an emphasis on the antidislipidemic, immunomodulatory, antitumor, hypoglycemic and hemostatic effects.

Keywords: sulfated fucose-rich polysaccharides; sulfated fucan; fucosylated chondroitin sulfate; fucoidan; oral administration; anticoagulant activity

1. Introduction

Sulfated fucose-rich polysaccharides have been described in seaweed for approximately a century and are denominated as fucoidan [1]. The structural complexity of these polysaccharides resulted in contradictory reports about their molecular structure since the analytical methods did not allow their detailed characterization. More recently, with the advance of new analytical methodologies, especially high resolution nuclear magnetic resonance, the structure of these polysaccharides has been elucidated [2,3]. Although these studies are restricted to a limited number of species, a high variability is observed among them [4]. Figure 1 shows examples of fucoidans already characterized. One of them contains alternating α (1→3)- and α (1→4)-linked fucose units, while the other is composed exclusively by α (1→3) units (Figure 1a). In both cases, the polysaccharide possesses a heterogeneous sulfation pattern and branches of sulfated and non-sulfated fucose. In addition to fucose, many other sugars are present in these fucoidans, such as galactose, xylose, mannose and uronic acid. It is not possible to clarify whether these sugars are part of the fucoidan molecule or a result of incomplete purification. In terms of the chemical structure, fucoidan could be designated as sulfated fucan (SF). However, "fucoidan" is a traditional denomination and also expresses the heterogeneous composition of these molecules.

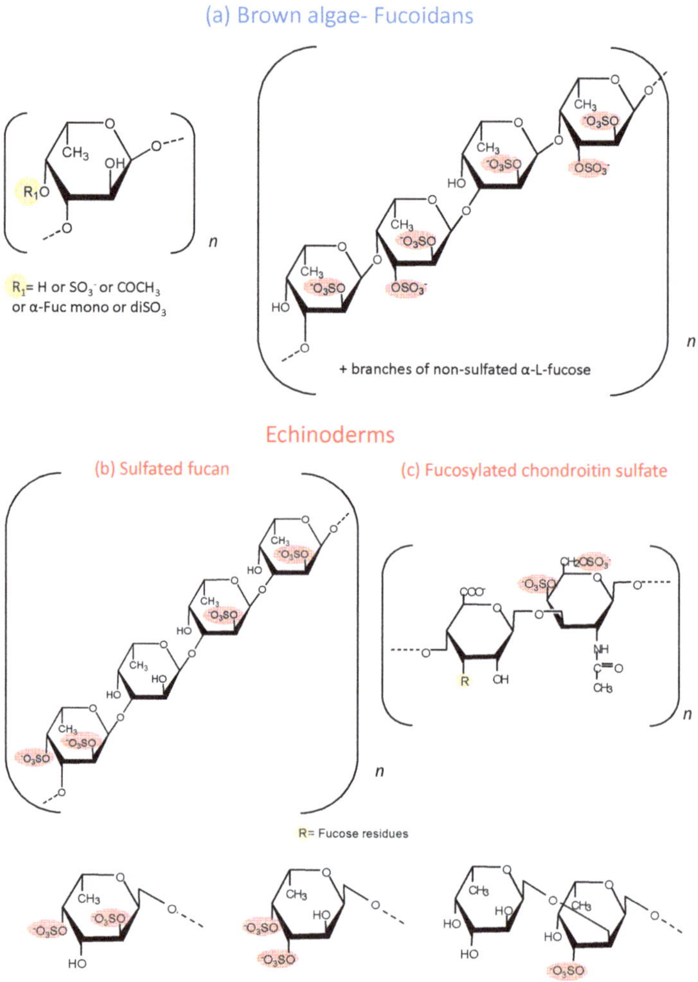

Figure 1. Structure of the sulfated fucose-rich polysaccharides from brown algae and echinoderms. (**a**) Fucoidans from brown algae are composed of α (1→3)-linked fucose units or alternating α (1→3)- and α (1→4)-linked fucose units. Mannose, galactose, xylose, uronic acid and branches of other monosaccharides make this polysaccharide highly variable and with complex structures. (**b**) SFs from echinoderms are made up of a repetitive tetrasaccharide units, formed by α (1→3) units and with a regular sulfation pattern at positions 2 and 4. (**c**) Structure of a fucCS from sea cucumbers. This polysaccharide has a chondroitin sulfate-like backbone, with branches of α-fucose linked to position 3 of the β-glucuronic acid of the central core. These fucose branches varies among species. In the specie *L. grisea*, for example, three types of branches are observed: α-Fuc-2,4diSO_4, α-Fuc-3,4diSO_4 and disaccharides composed of α-Fuc1→2-α-Fuc-3SO_4→. The sulfated fucose-rich polysaccharides from echinoderms have a more regular and repetitive structures compared with brown algae polysaccharides.

In the last 30 years, the study of polysaccharides rich in sulfated fucose was extended to echinoderms (sea urchins and cucumbers) [5–7]. In clear contrast with the fucoidans from marine algae, SFs from echinoderms have regular and repetitive structures [8]. Initially, these studies were concentrated on SFs from sea urchins, which are involved in the fertilization process of the invertebrate [9,10]. These SFs are obtained in very small quantities and cannot be tested in in vivo experimental models that require high doses. However, it soon became clear that sea cucumbers also have these SFs, which are present in more expressive quantities. Sea cucumbers contain SFs made up of repetitive tetrasaccharide units, formed by α (1→3) units and with a regular sulfation pattern at positions 2 and 4, which varies among species [11]. Figure 1b shows a representative structure of one of these SFs. In the case of echinoderms, the term SF is appropriate since it denominates chemically homogeneous molecules, distinct from fucoidan.

Moreover, sea cucumbers possess another polysaccharide rich in sulfated fucose, denominated as fucosylated chondroitin sulfate (fucCS) [12]. This compound has a central chain similar to chondroitin sulfate from vertebrates, but it has branches of fucose linked to position 3 of glucuronic acid of the central core. The structure of these fucose branches varies between species. Figure 1c shows the structure of this sulfated polysaccharide. Tables 1 and 2 show the structural characteristics of fucoidans and fucCS cited in this manuscript. It is important to emphasize that fucoidans can show structural variation according to the source, extraction method and time of the year [13].

These sulfated polysaccharides from echinoderms allowed a significant advance in the attempts to correlate structure and biological activities of these molecules, which is difficult to establish with fucoidans [14]. Many studies report the biological effects of the fucose-rich sulfated polysaccharides administered intravascularly, subcutaneously or intraperitoneally [15–19]. More recently, studies have emerged reporting several pharmacological effects of these polysaccharides after oral administration [20,21]. The observation that these compounds have a therapeutic effect after their oral administration is very significant since it opens the perspective of the use of these molecules for the development of new drugs. The oral route is more practical, comfortable and preferred by patients for long term treatments.

The purpose of this review was to conduct a systematic analysis of the effects observed after oral administration of the sulfated fucose-rich polysaccharides. We distinguished the effects observed with the complex fucoidans from algae from those obtained with the echinoderm polysaccharides. The use of fucCS and SFs is an important pharmacological tool to define structure versus therapeutic effects of polysaccharides rich in fucose as a basis for the development of new drugs.

Table 1. Structural characteristics of fucoidans with pharmacological activities after oral administration.

Species	Structure	Sugar and Sulfate Content	Mw (kDa)	Ref
A. nodosum	1→3)-α-L-Fucp and a few (1→4)-α-L-Fucp with →3-α-L-(2 and/or 4 Fucp)	The carbohydrate and sulfate content of the fraction A3 were 74.7% and 12.0%, respectively.	97.52	[22]
S. henslowianum	→3)-α-L-Fuc*p*(2 SO_3^-)-(1→3)-α-L-Fuc*p* (4 SO_3^-)-(1→	Fucose and glucose as main sugars. Sulfate content: 25.20%.	ND	[23]
F. evanescens	→3)-α-L-Fuc*p*(2 SO_3^-)-(1→4)-α-L-Fuc*p*(2 SO_3^-)-(1→	Fucose, sulfate and acetyl groups at a molar ratio of 1:1.23:0.36 and trace amounts of galactose and xylose.	10–100	[24]
F. vesiculosus	→3)-α-L-Fuc*p*(2 SO_3^-)-(1→4)-α-L-Fuc*p*(2,3-SO_3^-)-(1→3)	55.9% of carbohydrates, 27.0% of sulfate residues and 5.7% of uronic acid. Carbohydrates were represented mainly by fucose (38%), galactose (3.5%), xylose (2.7%).	20.7	[25]
C. okamuranus	→3)-α-L-Fuc*p* (SO_3^-)-(1→3)-α-L-Fuc*p*(4 SO_3^-)-(1→	The glucuronic acid residues are linked to the C-2 positions of the fucose residues, which are not substituted by a sulfate group. Sulfate content ~15%.	92.1	[26]
U. pinnatifida	→3)-α-L-Fuc*p*(2 SO_3^-)-(1→4)-α-L-Fuc*p*(2,3-di SO_3^-)-(1→3)	This sulphated galactofucan is composed of: galactose 44.6% and fucose 50.9%. Xylose (4.2%), mannose (0.3%). Sulfate content 15%. A significant number of *O*-acetyl groups.	378	[27]
S. japonica	→3)-α-L-Fuc*p*(2 SO_3^-)-(1→4)-α-L-Fuc*p*(2,3-di SO_3^-)-(1→3)	79.49% of fucose and 16.76% of galactose. Sulfate content ~30.72%.	30	[28]

Table 1. *Cont.*

Species	Structure	Sugar and Sulfate Content	Mw (kDa)	Ref
Mozuku *(High molecular weight fraction)*	ND	Sulfate content: 13%.	240	[29]
L. japonica	→3)-α-L-Fuc*p*(4 SO_3^-)-(1→	46.5% fucoxanthin, 8.01% lipids and 45.4% carbohydrates of mostly cellulose. Sulfate content: 13%.	300	[30]

ND: not determined.

Table 2. Structural characteristics of fucCS with pharmacological activities after oral administration.

Species	Proportions of the Branching Sulfated Fucose Units	Mw (kDa)	Ref
P. graeffei	81.6% α-Fuc-4SO_4, 18.4% α-Fuc-2,4diSO_4	49 kDa	[31]
I. badionotus	4.1% α-Fuc-4SO_4, 95.9% α-Fuc-2,4diSO_4	70.4 kDa	[32]
L. grisea	~27% α-Fuc-2,4diSO_4; ~20% α-Fuc3,4diSO_4 and ~53% disaccharides composed of α-Fuc1→2-α-Fuc-3SO_4→	40 kDa	[20]
C. frondosa	The chemical composition contained mainly glucuronic acid, galactosamine and fucose in the molar ratio of 1:1.50:1.16, with 30.07% sulfate content.	14.76 kDa	[33]

2. Antidislipidemic Effect

Dyslipidemia refers to a spectrum of metabolic disorders characterized by either an excess or a deficiency of lipoprotein particles, resulting in elevated plasma concentrations of total cholesterol (TC), low-density lipoprotein cholesterol (LDL-C) or triglyceride (TG) and/or depressed high-density lipoprotein cholesterol (HDL-C). Blood levels of different lipoproteins are strongly associated with the risk of cardiovascular diseases [34]. Thus, the management of blood lipid levels has an enormous significance for the control of these diseases. However, many patients fail to reach target levels of lipids with currently available drugs and still experience adverse clinical evolution [35]. Thus, additional pharmaceutical strategies are required to fill these gaps in efficacy and tolerability. There is some research reporting the lipid-lowering effect of sulfated polysaccharide administered by oral route (Table 3).

Initial studies of the antidislipidemic effect of sulfated fucose-rich polysaccharides were performed with sulfated polysaccharide from sea cucumber. A hypolipidemic effect was observed on rats fed with a cholesterol-rich diet and simultaneously received orally SF and fucCS for 6 weeks [36]. The echinoderm polysaccharides significantly decreased TC, LDL-C and the atherogenic index. The authors proposed that these effects may be due to inhibition of Hydroxymethylglutaryl-CoA reductase and/or increased lipoprotein lipase activity, although they did not provide any data regarding this mechanism.

Oral administration of fucoidan from *Ascophyllum nodosum* for 4 weeks improves reverse cholesterol transport in mice [22]. Plasma levels of TC and triglycerides were reduced, as well as fat pad index. The proposed mechanism is related to improvement of the hepatic lipids uptake by activating scavenger receptor B1 and LDL Receptor (LDLR), thus decreasing plasma LDL levels. Another study from the same group showed that oral administration of fucoidan ameliorated atherosclerotic lesion and lipid profiles in a dose-dependent manner in the apolipoprotein (apo) E-deficient mice fed with a high-fat diet [37]. Oil red staining revealed a decrease in the lesion/lumen ratio and in the liver lipid deposition with oral fucoidan-treated apo$E^{-/-}$ mice. Moreover, animals treated with the high dose of fucoidan showed reduction of the morphological changes of the kidney induced by high-fat diet. It also reduced triacylglycerol levels and plasma alanine transaminase, suggesting reduction in the high-fat-induced toxicity. Moreover, oral fucoidan increases plasma lipoprotein lipase (LPL) activity, apoA1 and peroxisome proliferator-activated receptor (PPAR) α/β levels. The combination of these effects can improve fatty acid oxidation and lower triglycerides. Another study showed that oral administration of fucoidan from *Sargassum henslowianum* decreased TC, triglyceride and LDL-C levels on obese mice [23].

Subsequent studies attempted to correlate the effect of sulfated fucose-rich polysaccharides on lipid levels with their molecular dynamics using compounds from echinoderms with well-defined chemical structure [38]. FucCS from *Isostichopus badionotus* and SF from *Pearsonothuria graeffei* showed potent effects on triglyceride lowering after oral administration. In contrast, SF from *I. badionotus* showed only weak effects. The distinct effects of these polysaccharides were correlated with their dynamics in solution: fucCS and SF from *P. graeffei* and fucCS from *I. badionotus* form random linear chains in solution with a few spherical aggregations, while SF from *I. badionotus* assumes a spherical conformation in solution and exhibited high viscosity. This study shows a new perspective to explore the structure versus pharmacological effect of sulfated fucose-rich polysaccharides at a molecular level.

Table 3. Antidislipidemic effects of sulfated fucose-rich polysaccharides after oral administration.

Polysaccharide	Dosage Regimen and Species	Major Observations and Mechanism Proposed	Ref.
Fucoidan from *A. nodosum*	100 mg/kg/day, 4 weeks, Mice	Improvement of reverse cholesterol transport and bile acid synthesis related genes expression. Reduction of plasma TC (~23.2%) and triglyceride (~48.7%) levels.	[22]
Fucoidan from *A. nodosum*	50 and 100 mg/kg/day, 8 weeks, ApoE$^{-/-}$ mice	Reduction of hepatotoxicity induced by high-fat diet; increased plasma LPL activity, apoA1 level and protein expression of PPARα/β ($\sim$2-fold), improved fatty acid oxidation and TG lowering ($\sim$24.5%).	[37]
Fucoidan from *S. henslowianum*	100 mg/kg/day, 4 weeks, Obese mice	Decreased cholesterol and LDL levels by ~23% and 18%, respectively.	[23]
Glycosaminoglycans from *M. scabra*	5, 10, 20 and 50 mg/kg, 6 weeks, Rats	Inhibition of HMG-CoA reductase and/or increased lipoprotein lipase activity and metabolism of cholesterol.	[36]
FucCS and sulfated fucan from *P. graeffei* and from *I. badionotus*	40 mg/kg, 8 days, Rats on high-fat diet	Hypolipidemic activity of sulfated polysaccharides is determined by the molecular dynamics of the sulfated polysaccharide.	[38]

LPL: lipoprotein lipase; PPAR: peroxisome proliferator-activated receptor; TG: triglycerides; HMG-CoA: 3-hidroxi-3-methyl-glutaril-*CoA* reductase. Results obtained with fucoidans from marine brown algae are in blue while those with polysaccharides from echinoderms are in red.

3. Anticancer Effect

Cancer is a leading cause of death, along with cardiovascular diseases. The hallmark of cancer treatment has been conventional chemotherapy. Chemotherapeutic drugs target rapidly dividing cells, such as cancer cells; however, these drugs also target normal cells, such as intestinal epithelium, bone marrow and hair follicles. In an attempt to target only cancer cells, a new generation of anticancer drugs arises using specific monoclonal antibodies, small molecule inhibitors and immunotoxins [39]. However, side effects and emerging resistance are still an issue, which increase the demand for new compounds that could act as adjuvant therapy and/or increase efficacy [40]. Regarding this issue, some data in the literature explore the anticancer effect of sulfated polysaccharides after oral administration. Table 4 summarizes the major observations.

An example of the beneficial effect of these polysaccharides as anticancer drugs is the observation that oral doses of fucoidan from *Fucus vesiculosus* delayed tumor growth in a xenograft model and increased cytolytic activity of natural killer cells [41]. Athymic mice were pre-treated with fucoidan daily for 2 weeks and then a human acute promyelocytic leukemia cell line was injected subcutaneously. Significant antitumor activity was observed without any sign of toxicity. Tumor development was clearly slower in the oral fucoidan-treated mice than in the control group. An enhancement of the cytotoxic activity of splenic natural killer cells in mice that were orally treated with fucoidan was also observed, which could be in part responsible for its pharmacological effect. Interestingly, when fucoidan from the same specie were orally administered for 21 days, starting on the seventh day post-tumor implantation, significant reduction in tumor volume and tumor weight was observed when compared with the control group [42]. In vitro assays showed that fucoidan could induce G_0/G_1 cell cycle arrest and caspase-dependent apoptosis in diffuse large B-cell lymphoma culture. This indicates that oral fucoidan administration can inhibit tumor growth and development.

Another observation of the anticancer effect of orally administered fucoidan was obtained with Lewis lung carcinoma cells (LLC) [43]. These cells were inoculated into the hypodermic dorsum of mice, and the tumor growth rate was assessed over 21 days. A marked dose-dependent reduction in tumor volume and weight was observed in the fucoidan-treated group, with the maximum effect observed with 144 mg/kg daily oral dose. Expression of growth factors receptors showed a decrease in fucoidan-treated mice compared with the control group. No signs of liver toxicity due to fucoidan administration were observed. Continuous oral administration of fucoidan has a greater efficacy in suppressing tumorigenesis than discontinuous doses, as expected.

In another xenograft model using human prostate carcinoma cells, oral administration of fucoidan for 28 days significantly hindered the tumor growth and tumor vascular density, as indicated by hemoglobin quantification assay [44]. The mRNA expression level of CD31 and CD105, biomarkers of endothelium, also declined. Analysis of the protein expression and gene promoters related to angiogenesis showed that their levels were reduced after oral fucoidan treatment, suggesting that fucoidan hindered tumor growth by inhibiting the formation of new blood vessels.

The anticancer effect of fucoidan was also observed using the polysaccharide from brown alga *Fucus evanescens* using a xenograft model. Colon cancer cells were inoculated into athymic nude mice [24]. Oral treatment with fucoidan for 21 days inhibited tumor growth compared with the vehicle-treated group. The antitumor effect of fucoidan was associated with its inhibition of lymphokine-activated killer T-cell-originated protein kinase (TOPK), highly expressed in many cancers. Tissues from each group were analyzed for phosphorylation of TOPK downstream targets, and the expression of these markers was decreased after 20 days of oral fucoidan treatment. Additional in vitro assays showed that this fucoidan modulates EGF-induced neoplastic transformation of mouse epidermal cells in a concentration-dependent manner. This pathway is related to the machinery that controls fundamental cellular processes, such as growth, proliferation, differentiation, migration and apoptosis. The polysaccharide also binds and decreases TOPK kinase activity in vitro, although a high concentration is required for this effect. The antitumoral activity of the echinoderm polysaccharides has not been tested so far after oral administration. These well-defined structures may help to clarify the effect of sulfated polysaccharides on cancer cells.

Table 4. Anticancer effects of sulfated fucose-rich polysaccharides after oral administration.

Polysaccharide	Dosage Regimen and Species	Major Observations and Mechanism Proposed	Ref.
Fucoidan from *F. evanescens*	1–50 mg/kg, 3 times/week/ up to 21 days, Rats	Inhibition of lymphokine-activated killer T-cell-originated protein kinase (TOPK) (64% at 400 µg/mL) and EGF-downstream signaling. ↓Tumor growth 72% at 50 mg/kg.	[24]
Fucoidan from *F. vesiculosus*	150 mg/kg/body weight, 2 weeks, Athymic mice	Enhancement of the cytotoxic activity of splenic NK cells (~2.3 fold).	[41]
Fucoidan from *F. vesiculosus*	100 mg/kg, 21 days starting on the seventh day pos tumor implantation, Mice	Induces G_0/G_1 cell cycle arrest (2–10%) and caspase-dependent apoptosis.	[42]
Fucoidan from *F. vesiculosus*	144 mg/kg, 26 days, Mice	Reduction of Transforming Growth Factor Receptor (TGFR) levels (↓~50%) and its downstream signaling pathways. Enhancement of TGFR degradation.	[43]
Fucoidan from *F. vesiculosus*	20 mg/kg, 28 days, Athymic mice	Inhibition of angiogenesis by decreasing mRNA expression level of angiogenesis related markers (↓~70%) and gene promoters.	[44]

TOPK: Lymphokine-activated killer T-cell-originated protein kinase; EGF: epidermal growth factor; NK: natural killer; TGFR: transforming growth factor receptor. Results obtained with fucoidans from marine brown algae are in blue.

4. Immunomodulatory Effect

Immunomodulatory drugs can act at different levels of the immune system. Therefore, different kinds of drugs have been developed that selectively either inhibit or intensify the specific populations of immune responsive cells, i.e., lymphocytes, macrophages, neutrophils, natural killer cells, and cytotoxic T lymphocytes. Immunomodulators affect the cells systems by producing soluble mediators such as cytokines. Therefore, the rational use of drugs with anti-inflammatory effects is necessary to avoid excessive inflammation triggered by external agents or autoimmune diseases and drugs with immunostimulatory effects to increase the immune response such as the production of specific antibodies. In this context, oral administration of sulfated fucose-rich polysaccharides has also shown some interesting effects. A summary of these effects is shown in Table 5.

Oral administration of fucoidan from *Cladosiphon okamuranus* had an antifibrogenesis effect in an N-nitrosodiethylamine-induced liver fibrosis model in rats [26]. Two fractions of fucoidan with distinct molecular weight were tested on this model after oral administration for 12 weeks. A high-molecular-weight fraction of fucoidan prevents liver fibrosis, as indicated by histological examination and hydroxyproline measurement. It also prevents the increase in plasma levels of bilirubin, which occurs as a consequence of liver damage. A low-molecular-weight fraction of fucoidan had only a modest effect on hydroxyproline and bilirubin levels. This observation indicates that the biological effects of fucoidan may differ depending on the molecular weight of the molecule. TGF-β1 appears to play a major role in liver fibrosis and that the mRNA expression of this cytokine is upregulated in this experimental model. The expression of this cytokine decreases significantly in oral fucoidan-treated animals. Furthermore, a chemokine ligand, denominated CXCL12, is markedly stained in the liver epithelium after the induction of experimental fibrosis. Oral treatment with fucoidan prevents the increase of this chemokine expression.

Fucoidan from *F. vesiculosus* was tested on a model of alcohol-induced hepatic dysfunction [45]. Seven days of oral administration of this polysaccharide to mice prevents the increase of transaminase levels. It also prevents the expression of TGF-β1 and COX-2, both in the liver from the animal experimental model and in the hepatic cells in culture. Oral fucoidan also decreases mRNA expressions of hepatic inflammatory matrix metalloproteinase-2. Histopathological evaluation showed that macrovesicular steatosis of hepatocytes and focal hepatic necrosis associated with inflammatory cells infiltration induced by high-fat diet was clearly reduced in rats treated with oral fucoidan.

The therapeutic effect of oral fucoidan on non-alcoholic fatty liver disease was tested in rats using an experimental model induced by a high-fat diet [46]. Oral administration of fucoidan for 4 weeks resulted in a decrease of body and liver index and aminotransferase levels when compared with the non-treated group. Total cholesterol and triglycerides also decreased in serum and liver, as well as serum fasting glucose, insulin levels and liver inflammation of the animals fed with fucoidan.

Prevention of arthritis encompasses a variety of the immunomodulatory effects of sulfated polysaccharides. In this particular event, oral administration of fucoidan from *Undaria pinnatifida* for 25 days showed an anti-arthritic effect in a carrageenan-induced paw edema model in rats [27]. Animals treated with fucoidan exhibited significant reduction in paw edema, compared with a standard anti-inflammatory drug, although the doses required to achieve similar protection differ significantly (150 vs 10 mg/kg body weight). Histological analysis revealed that oral fucoidan and standard drug-treated groups exhibited protective effects on joint architecture, such as less edema, cell infiltrations and cartilage destruction. Furthermore, the increase in several biochemical parameters were ameliorated by oral fucoidan administration, and this polysaccharide showed no signs of toxicity at doses up to 1000 mg/kg. Fucoidan demonstrated concentration-dependent antioxidant and anti-inflammatory activities in various in vitro assays, suggesting that its anti-arthritic properties might be related to suppression of prostaglandin production and other inflammatory mediators.

Another study showed that oral administration of fucoidan from *F. vesiculosus* for 13 weeks to Goto-Kakizaki rats, which spontaneously develop mild hyperglycemia and hyperinsulinemia, protected the animals from diabetes nephropathy [47]. The increased fasting blood glucose, urea, serum creatinine and urine protein levels observed in positive control animals were significantly decreased in GK rats that received fucoidan orally at both doses. Fucoidan diminished levels of collagen IV in the renal cortex and decreased expression of TGF-β1 and fibronectin in the renal cortex and in the glomerular mesangial cells. Histopathological analysis revealed vacuolation of renal tubular epithelial cells and inflammatory cell infiltration in the renal interstitium of the kidneys from the diabetic rats compared with those from the control and fucoidan-treated rats. The increased expression of NF-κB in the nuclei of glomerular mesangial cells was also attenuated significantly by the oral administration of fucoidan, suggesting that this pathway is involved in the nephropathy and in the anti-inflammatory activity of the polysaccharide.

Suppression of allergic symptoms is another event related to immunomodulatory activity of fucoidan [28]. The effect of the polysaccharide in this particular event was investigated using fucoidan from *S. japonica* orally administered for 4 days to mice submitted to a passive cutaneous anaphylaxis reaction. The ear edema was evaluated 2 h after antigen challenge. Fucoidan showed an inhibitory effect only after oral but not intraperitoneal administration. The mechanism proposed is related to an increase in galectin-9 expression in intestinal epithelial cells and in the blood of mice fed with fucoidan. In fact, administration of anti-galectin 9 antibody suppressed fucoidan effects on ear edema. Moreover, this polysaccharide prevented the interaction of IgE and mast cells, an important event that mediates allergic responses. These data suggest that dietary intake of fucoidan from *Saccharina japonica* may prevent allergic symptoms.

The impact of the molecular weight on the anti-inflammatory activity of sulfated fucan from the sea cucumber *Acaudina molpadioides* with varying degrees of polymerization was also reported [48]. The SF was tested on an animal model of intestinal mucositis after oral administration for 26 days. Histological analysis revealed that morphology of the intestinal mucosa of fucoidan-treated animals was similar to the healthy group and that this effect was more pronounced with the high-molecular-weight fractions. Interestingly, these fractions regulated Th1/Th2 immune balance processes by altering IFN-γ/IL-4 ratio, while the oral administration of intact fucoidan had no effect. Intact SF and the high-molecular-weight fractions enhanced IgA protein expression levels in intestinal mucosa and strengthened intestinal adaptive immunity. Another interesting aspect of this work was the analysis of plasma concentration achieved by the oral administration of the SF. The low-molecular-weight fractions achieved high plasma levels when compared with unfractioned polysaccharide; therefore, the absorption and bioavailability of SF are likely to depend on the molecular size of the polysaccharide.

Another approach investigated the anti-inflammatory effect of oral administration of fucCS from the sea cucumber *I. badionotus* for 7 days. In contrast with the highly heterogeneous fucoidan from brown algae, this polysaccharide has a regular repetitive structure containing mostly 2,4 disulfated fucose units, as shown in Figure 1c. When tested on an experimental model of colitis induced by dextran sulfate, oral fucCS attenuated the body weight loss, expression of colonic TNF-α gene and colon shortening caused by experimental colitis. The authors proposed that this protective effect might be due to downregulation of NF-kB and downstream genes such as COX-2 and TNF-α and a benefic profile on gut microbiota [49].

Table 5. Immunomodulatory effects of sulfated fucose-rich polysaccharides after oral administration.

Immunomodulatory Effect	Polysaccharide	Dosage Regimen and Species	Major Observations and Mechanism Proposed	Ref.
Antifibrotic effect	Fucoidan from *C. okamuranus*	Free access to drinking water containing 2% low (28.8 kDa) or high (41.4 kDa) MW fractions, 12 weeks, Rats	↓TGF-β1 mRNA expression and the levels of chemokine ligand CXCL12 in the liver (~3 fold).	[26]
Hepatoprotection	Fucoidan from *F. vesiculosus*	30 or 60 mg/kg, 7 days, Mice	↓expression of liver TGF-β1(~40%) and COX-2, ↑antioxidant pathways.	[45]
Hepatoprotection	Fucoidan from *F. vesiculosus*	100 mg/kg, 4 weeks, Rats on high-fat diet	↓TNF-α, IL-1β and MMP-2 mRNA expressions (~50–70%). Prevention of the increase in serum lipids and glucose levels induced by HFD.	[46]
Nephroprotection	Fucoidan from *F. vesiculosus*	50 and 75 mg/kg, 13 weeks, Rats	Decreased levels of collagen IV, NF-κB, TGF-β1 and fibronectin in the renal cortex and in the glomerular mesangial cells.	[47]
Anti-arthritic and antioxidant effects	Fucoidan from *U. pinnatifida*	50 or 150 mg/kg, 25 days, Rats	Downregulation of COX-2 and other inflammatory mediators (68% inhibition of in vivo inflammation).	[27]
Immunostimulatory effects	Fucoidan from *U. pinnatifida*	300 mg daily,20 weeks, Human	Higher immunogenicity of influenza trivalent vaccine than control group and increase of natural killer cell activity.	[50]
Suppression of allergic symptoms	Fucoidan from *S. japonica*	100–400 μg/day,4 days, Rats	Prevention of the interaction of IgE and mast cells via an increase in galectin-9 mRNA expression (↑~50%) in intestinal epithelial cells.	[28]
Anti-inflammatory effect	FucCS from *I. badionotus*	80 m/kg, 7 days, Rats	Downregulation of NF-kB and downstream genes such as COX-2 and TNF-α and a benefic effect on gut microbiota.	[49]
Anti-inflammatory effect	Sulfated fucan from *A. molpadioides* with varying degrees of polymerization (10–500 kDa)	50 mg/kg, 26 days, Mice	Regulation of IFN-γ/IL-4 ratio (0.53 to 0.70) and Th1/Th2 response, IL-6 and IL-10 levels, enhanced IgA protein expression levels (~35%) in intestinal mucosa.	[48]

CXCL12: C-X-C motif chemokine ligand 12; TNF-α: tumor necrosis factor; TGF-β: transforming growth factor beta; NF-kB: nuclear factor kappa B; COX-2: ciclooxigenase 2; IFN-γ: interferon gamma; IgA: immunoglobulin A; HFD: high-fat diet. Results obtained with fucoidans from marine brown algae are in blue while those with polysaccharides from echinoderms are in red.

5. Effects on Diabetes

Diabetes is a highly prevalent disease characterized by high levels of blood sugar, due to deficiency of insulin concentration and/or activity. Pharmacological therapy may be required in order to maintain normal level of blood glucose and to delay or prevent the development of diabetes-related health problems. The first choice in type 2 diabetes is oral hypoglycemic drugs, but side effects, toxicity and unwanted drug–drug interactions can compromise the effectiveness of the treatment [51]. Nevertheless, the idea of a diet rich in sulfated fucose-rich polysaccharides with hypoglycemic effect as adjuvant therapy may be an interesting alternative. A summary of these effects is shown in Table 6.

Oral administration of fucoidan from *F. vesiculosus* for 13 weeks to Goto-Kakizaki rats reduced high blood glucose and recovers serum insulin levels [52]. Moreover, histopathological analysis of the pancreas also demonstrated that fucoidan markedly reduced islet atrophy, fibrosis and inflammation. Additional in vitro assays showed that treatment with the phosphodiesterase inhibitor significantly increased fucoidan-induced insulin secretion, whereas treatment with the adenylyl cyclase inhibitor significantly decreased fucoidan-induced insulin secretion. These results suggested that the cAMP signaling pathway may be important in the antidiabetic effect of fucoidan. A further study showed that the polysaccharide inhibits dipeptidyl peptidase-IV, which prolongs the action of incretins, reduces glucose and increases insulin production. This is another possible mechanism involved in the antihyperglycemic effect of fucoidan [25].

A detailed study about the effect of oral sulfated fucose-rich polysaccharides on diabetes employed an experimental model in mice inducing type 2 diabetes by high fat/sucrose diet [53]. The authors tested a fucCS from the sea cucumber *Cucumaria frondose*. Oral administration for 19 weeks stimulated insulin-dependent glucose uptake in skeletal muscle cells and improved insulin sensitivity. Oral fucCS treatment promoted insulin-stimulated phosphorylation of phosphoinositide 3-kinase and protein kinase B, the major regulators of glucose uptake response to insulin in skeletal muscle and increased GLUT4 translocation. It also increased mRNA expression levels of these regulators in the skeletal muscle of oral polysaccharide-treated mice. Furthermore, fucCS increased hepatic glycogen content and restored the activities of key enzymes for glucose metabolism in the liver to near-control levels [54]. Therefore, oral fucCS can promote hepatic glycogen synthesis by regulating gene expression.

A further study attempted to investigate the mechanisms involved in the favorable effect of fucCS on experimental diabetes [33]. Animals were submitted to a high-fat/sucrose diet, which disrupts insulin signaling and thus results in endoplasmic reticulum stress and inflammation. After oral administration of oral fucCS for 19 weeks, several cytokines and inflammatory markers were reduced in the serum and in the liver of treated animals. Analysis of mRNA expression showed that the polysaccharide attenuates the increase of several markers of liver endoplasmic reticulum stress, inhibits important inflammatory signaling pathways and improves insulin sensitivity in the liver.

The antidiabetic effect of sulfated fucose-rich polysaccharides extracted from 10 low-edible-value sea cucumber species was tested after oral administration for 8 weeks using a classic experimental model of diabetes induced by streptozotocin in rats [32]. A variety of effects were observed, such as reduced polyphagia and loss of body weight, decreased fasting blood glucose level and improved glucose tolerance by increasing insulin secretion and enhancement of its sensitivity. A significant improvement of antioxidant enzymes was also observed indicating a decrease in inflammatory status and oxidative stress. The sulfated polysaccharides decrease the levels of transaminases, suggesting a repair of liver damage associated with the experimental model. They also restored normal levels of TNF-α content in the serum and enhanced synthesis of liver glycogen to decrease blood glucose level. Furthermore, they reduced levels of serum triacylglycerol, TC and LDL-C and increased HDL-C/LDL-C values, which indicates that oral administration of sulfated fucose-rich polysaccharides can alleviate dyslipidemia resulting from diabetes. In this study, the authors did not show a clear correlation between the structure of the polysaccharide and its biological effect. Nevertheless, the sulfated polysaccharides from *C. frondosa* and *Thelenota ananás* seem to show more potent effects.

Table 6. Hypoglycemic effects of sulfated fucose-rich polysaccharides after oral administration.

Polysaccharide	Dosage Regimen and Species	Major Observations and Mechanism Proposed	Ref.
Fucoidan from *F. vesiculosus*	75 m/kg, 13 weeks, Rats	Reduced islet atrophy, fibrosis and inflammation mediated by cAMP signaling pathway. Inhibition of dipeptidyl peptidase-IV.	[25,52]
High molecular weight fucoidan from Mozuku (*C. okamuranus*)	1620 mg, 12 weeks, Human	Alterations in GLP-1 (from 6.42 ± 3.52 to 4.93 ± 1.88 pmol/L) and hemoglobin A1c levels (from 6.73 ± 1.00 to 6.59 ± 1.00).	[29]
Fucoidan extract from *Laminaria* ssp.	500 mg, 3 months, Human	Decrease in diastolic blood pressure and LDL-C (↓13%) with increase in insulin levels (↑ 30%), HOMA β-cell, and HOMA IR.	[29,55]
FucCS from *C. frondosa*	20 or 80 mg/kg, 19 weeks, Mice	↑insulin-stimulated phosphorylation of PI3K and PKB; ↑GLUT4 translocation ↑glycogen synthesis-related gene expression; ↓liver ER stress markers, ROS, TNF-α and other inflammatory markers levels in serum and liver; ↓inflammatory signaling pathways in the liver.	[33,53,54]

Table 6. *Cont.*

Polysaccharide	Dosage Regimen and Species	Major Observations and Mechanism Proposed	Ref.
Sulfated polysaccharides from 10 sea cucumber species	200 or 400 mg/kg, 8 weeks, Rats	↓TNF-α, ↑antioxidant enzymes; ↑glucose metabolism related gene signaling pathway.	[32]

GLP-1: glucagon-like peptide 1; PI3K: phosphatidylinositol 3-kinase; PKB: protein kinase B; GLUT4: glucose transporter 4; ER: endoplasmic reticulum; ROS: reactive oxygen species. Results obtained with fucoidans from marine brown algae are in blue while those with polysaccharides from echinoderms are in red.

6. Thrombosis and Hemostasis

Thromboembolic events are expanding due to the aging of the population and a more precise diagnosis. Heparin is the classic anticoagulant used in the treatment and prevention of thrombosis, but its use is limited to the intravenous or subcutaneous route, and it has significant adverse effects [56–58]. New oral anticoagulants are available, but bleeding is still a concern [59]. Therefore, there is a demand for new antithrombotic drugs.

The antithrombotic effects were the first significant pharmacological effects reported for the sulfated fucose-rich polysaccharides [60–62]. Several authors addressed the parenteral use of fucoidan and echinoderm polysaccharides in experimental models of venous and arterial thrombosis [63–65]. The initial studies associate the mechanism of action of these molecules with heparin, the most traditional anticoagulant sulfated polysaccharide. However, recent studies using sea cucumber fucCS showed that the anticoagulant mechanism of this compound is serpin-independent, inhibiting the assembly of the tenase and prothrombinase complexes and the generation of thrombin and factor Xa [66]. In addition to the distinct mechanism of action, the preserved antithrombotic effect after oral administration has made this sulfated polysaccharide an interesting candidate for the development of new drugs [20,67]. Table 7 summarizes the effects of sulfated fucose-rich polysaccharide in hemostasis.

An initial study about the antithrombotic effect of oral fucoidan employed a low-molecular-weight fraction obtained by chemical degradation of the native polysaccharide from *L. japonica* [68]. After oral administration for 30 days to rats, the polysaccharide prolonged aPTT and TT values, increased TFPI and suppressed thromboxane levels in rat plasma. It also inhibited thrombin-induced platelet aggregation and enhanced fibrinolysis. The antithrombotic effect was tested in an arterial thrombosis model induced by electrical stimulus. The low-molecular-weight fucoidan prolonged the time for formation of the thrombus. Unlike aspirin, the low-molecular-weight fucoidan did not decrease platelet number and fibrinogen level after oral administration for 30 days, which suggest a safe antithrombotic profile.

The first report of the antithrombotic effect of an echinoderm polysaccharide after oral administration employed a fucCS from the sea cucumber *Ludwigothurea grisea* [20]. The polysaccharide increased aPTT and TT values and decreased thrombin residual activity. A dose-dependent antithrombotic effect is observed using a vena cava and an arterial shunt thrombosis models in rats. After removal of the fucose branches, the antithrombotic activity of the polysaccharide was abolished. The dose necessary to achieve complete inhibition of the thrombus formation was 50 mg/kg administered in aqueous solution. A great achievement was the encapsulation of the polysaccharide on gastro-resistant tablets, which prevents the degradation in the acid juice fluid [67]. This approach allowed the dose of fucCS to be decreased to 25 mg/kg and to still observe the same anticoagulant and antithrombotic effects. FucCS does not alter bleeding tendency or arterial pressure after oral administration, which is the major concern with this polysaccharide due to activation of the contact system and the release of bradykinin [69].

Recently, oligosaccharides containing 6→18 units were obtained by controlled depolymerization of fucCS from the sea cucumber *P. graeffei* [70]. These oligosaccharides were the active ingredient of gastro-resistant microcapsules using a chitosan-coated alginate system and were orally administered to rats in a single dose of 10 or 50 mg/kg. Microcapsules containing the oligosaccharides prolonged aPTT values with a stronger intensity compared with microcapsules containing native fucCS. In a venous thrombosis model, oral administration of 50 mg/kg fucCS oligomers delivered by aqueous solution exhibited a weaker antithrombotic effect than observed with gastro-resistant microcapsules, probably due to the partial removal of sulfated fucose branches in the acid gastric fluid. No bleeding tendency was observed for fucCS oligomers tested on gastro-resistant microcapsules. Using an intestinal Caco-2 cell system, the authors confirmed that fucCS oligomers showed higher absorption than native polysaccharide.

A very curious observation comes from a study involving fucoidan from *F. vesiculosus* and *Laminaria japonica* orally administered twice daily in a multiweek dose-escalation study of dogs with hemophilia A [71]. A dose-dependent decrease in bleeding time score and improved clotting dynamics was observed, indicating a procoagulant effect of these polysaccharides after oral administration. In vitro assays showed that this fucoidan inhibited exogenous TFPI activity and accelerated the clotting time of human hemophilia A and B plasma. Current methods of hemophilia treatment are expensive, challenging and involve regular administration of clotting factors. While gene therapy is expensive and still under investigation, additional therapeutic options have already explored heparin-like sulfated polysaccharides, including pentosan polysulfate and fucoidan, with unique procoagulant activity for bleeding disorders [72]. These results explore another aspect of the effects of sulfated polysaccharides on the coagulation system. Interestingly, sulfated polysaccharides from red algae have already shown a dual effect on coagulation either as a pro- or anticoagulant drug [73].

Table 7. Effects on hemostasis of sulfated fucose-rich polysaccharides after oral administration.

Polysaccharide	Dosage Regimen and Species	Major Observations and Mechanism Proposed	Ref.
Low molecular weight fucoidan (Mw7.6 kDa) from *L. japonica*	400 and 800 mg/kg 30 days, Rats	↑TFPI (4.5 to 110.2 U/mL) and 6-keto-PGF1α levels (32.8 to 50.4 U/mL). ↑Fibrinolysis (tPA and PAI-1 levels) ↓Thromboxane A2 levels.	[68]
Fucoidan from *L. japonica*	400 mg for 5 weeks to humans	↑6-keto-PGF1a (44 to 113 ng/L) ↑fibrinolysis.	[30]
Fucoidans from *F. vesiculosus* and *L. japonica*	5–20 mg/kg, Twice daily in a multiweek escalation dose, Dogs	Procoagulant effect, Inhibition of TFPI activity.	[71]
Native and gastro-resistant tablets of fucCS from *L. grisea*	5–50 mg/kg, Single dose or 5 days, Rats	Serpin-independent anticoagulant effect by inhibiting the formation of factor Xa and/or IIa through the procoagulants tenase and prothrombinase complexes. Antithrombotic effects at 50 mg/kg: ~85% vs. 55% inhibition of the venous and arterial thrombus weight, respectively.	[20,66,67]
Gastro-resistant tablets containing FucCS oligomers (6 to 18 saccharide units, Mw 3,4 kDa) from *P. graeffei*	10 or 50 mg/kg Single dose, Rats	Anticoagulant and antithrombotic effects (82% of venous thrombosis inhibition at 50 mg/kg).	[70]

TFPI: tissue factor pathway inhibitor; *6-keto* PGF1α: *6-keto* prostaglandin F1α; tPA: tissue plasminogen activator; PAI-1: plasminogen activator inhibitor. Results obtained with fucoidans from marine brown algae are in blue while those with polysaccharides from echinoderms are in red.

7. Clinical Trials

Preclinical studies using animal models are important to assess the effectiveness of fucose-rich polysaccharides in different pathologies and to elucidate the mechanisms involved in their mechanism of action. Clinical trials are the next step for the development of these polysaccharides as new drugs and/or using marine organisms as a food supplement, and some studies in the literature address this issue. These aspects are also under investigation, and we describe the major observations in this review.

Very few studies report the anticancer effect of fucoidan in humans. Some clinical trials report an improvement in the quality of life of patients who have used fucoidan orally as an adjuvant therapy. Cancer patients receiving oral fucoidan for 4 weeks showed reduced levels of proinflammatory cytokines, including IL-1β, IL-6 and TNF-α [74]. Interestingly, the responsiveness of IL-1β was significantly correlated with overall survival, suggesting that this might be a useful prognostic biomarker for advanced cancer patients receiving fucoidan. Another study examined the effects of fucoidan extracted from *C. okamuranus* on natural killer cell activity in cancer survivors [75]. Male patients treated with 3 g of fucoidan for 6 months showed an enhanced activation of natural killer cells. Fucoidan also has the potential for adjuvant therapy and may also reduce chemotherapy toxicity for cancer patients [76,77].

Another study in humans involves the immunogenicity response to influenza trivalent vaccine in the elderly, whose antibody production is generally attenuated [50]. Oral intake of fucoidan from seaweed *U. pinnatifida* for 20 weeks (300 mg daily) increased antibody titers, which is most evident against influenza B strain. This effect may be related to NK cell activity. This suggest that popular seaweeds containing fucoidan that are eaten daily in Japan could have immunostimulatory effects in enhancing vaccination efficacy. Another study showed no decrease of osteoarthritis symptoms after a 300 mg daily oral dose of *F. vesiculosus* extract (85% fucoidan) over a 12-week period [78].

Studies of the effect of sulfated polysaccharides on diabetes after oral administration were also reported in humans [29]. Thirty patients with type 2 diabetes were selected for oral intake of a high-molecular-weight fucoidan from Mozuku seaweed for a 12-week period. Oral fucoidan altered hemoglobin A1c and levels of glucagon-like peptide-1 and increased the number of bowel movements and stool frequency. These effects were associated with a beneficial control of diabetes. Another randomized, double-blind, placebo-controlled clinical trial was carried out with 25 overweight volunteers to evaluate the effect of fucoidan administration on insulin secretion and sensitivity [55]. A total of 13 patients received an oral dose of 500 mg of fucoidan once daily before breakfast and 12 patients received placebo for 3 months. A significant decrease in diastolic blood pressure and LDL levels with an increase in insulin levels were observed after oral fucoidan administration. There were no significant adverse events associated with the long-term intake of fucoidan in both studies.

Human studies reporting the effect of sulfated polysaccharides on hemostasis are scarce, as in the case of other biological effect. In one study, oral administration of capsules containing 400 mg fucoidan from *L. japonica* to healthy participants for 5 weeks resulted in increased fibrinolysis and antiplatelet effects [30]. Fucoidan was not detected in the plasma, probably due to low polysaccharide concentrations and/or the sensitivity of the method used. This is one of the challenges associated with assessing the pharmacokinetics of orally administered sulfated polysaccharides.

8. Future Perspectives: Pharmacokinetics Studies and Prebiotic Effects

This review summarizes the therapeutic effects achieved after oral administration of sulfated polysaccharides in a variety of pathological processes. These observations were obtained mainly using animal experimental models, although some preliminary data have already been reported in humans. These results are not limited to the therapeutic effect but also highlight the proposed molecular mechanisms involved in the pharmacological action of these polysaccharides. Further studies are necessary to further understand their pharmacokinetic and the modulating effect on the intestinal microbiota.

In the case of heparin, a paradigm of an anticoagulant drug with carbohydrate structure, the transition from intravenous to subcutaneous administration was associated with the development of low-molecular-weight heparin. This led to the development of new analytical methods to study its pharmacodynamics, resulting in the now widespread methods to determine the plasma concentration of heparins based on anti-FXa and anti-FIIa assays [56]. Likewise, there is a need to develop sensitive methods for the study of the pharmacokinetics/pharmacodynamics of fucose-rich polysaccharides after their oral administration.

A similar approach was employed for oral administration of fucCS. The plasma concentration of the polysaccharide was determined using ex vivo coagulation assays and purified proteases [67]. The results allowed the correlation of the anticoagulant with its antithrombotic effects. However, it was not possible to evaluate tissue distribution, minor structural modifications and urinary elimination of the compound. Such analyses are limited by the very low concentration achieved by the polysaccharide after oral administration. We need to validate new methods for fucCS labeling and quantification as a critical step for the pharmacokinetic studies.

A partially depolymerized fucCS, administered as a single oral dose of 50 mg/kg, was detected in plasma between 0.5 and 7.5 h using a chromatographic method. Only 0.1% of the dose was detected in the urine accumulated during 24 h [79]. It is very challenging to measure the plasma levels of these polysaccharides analytically because of the heterogeneous molecular weight, branched structure and similarity in monosaccharide composition to mammalian polysaccharides.

Some studies report other sensitive methods to assess the pharmacodynamics of fucose-rich polysaccharides after oral administration. In one study, the authors employed an antibody against fucoidan extracted from *C. okamuranus* and developed a sensitive ELISA method for the measurement of its serum and urinary concentration after a single oral dose of fucoidan (1 g) in ten healthy volunteers [80]. The anti-fucoidan antibody specifically recognized fucoidan from *C. okamuranus* and *F. vesiculosus* with different specificities, with low cross-reactivity with heparin and heparin-like substances. Fucoidan concentration in serum and urine was detectable 3 h and mostly 6 h after its administration. The time and peak concentrations varied among individuals, suggesting a high variability of fucoidan absorption in the intestine. The concentration of fucoidan was higher in the urine than in the serum. The molecular weight of the ingested fucoidan remained unchanged in the serum, whereas the fucoidan excreted in the urine showed an expressive decrease in size. Possibly, fucoidan degradation occurs mostly in the excretory system but not during its absorption through the gastrointestinal tract by local microbiota. Using the same method, a further study confirmed fucoidan in the urine of Japanese volunteers after 100 g of oral intake of seaweed *C. okamuranus* [81].

Another work using fucoidan antibody revealed that the polysaccharide accumulated in jejunal epithelial cells, mononuclear cells in the jejunal lamina propria and sinusoidal non-parenchymal cells in the liver of rats fed standard chow containing 2% fucoidan for one or two weeks [82]. Fucoidan was detected in the sinusoids of hepatic lobules, which suggested its internalization by macrophages. The intestinal absorption was also observed using an intestinal Caco-2 cells system in vitro.

One of the major challenges associated with assessing the bioavailability of orally administered fucoidan has been the lack of a sensitive and accurate analytical method that can quantify fucoidans in the blood since this polysaccharide exhibits low anti-FXa and anti-FIIa activities compared with heparin. However, one study evaluated the pharmacokinetics and tissue distribution of fucoidan in rats after a single-dose oral administration of 100 mg/kg of fucoidan from *F. vesiculosus* based on its anti-Xa activity. The C_{max} in plasma was observed at 4 h after oral administration. Fucoidan accumulated mainly in the kidney and was also present in liver and spleen and showed a relatively long absorption time and extended circulation in the blood [83]. Different analytical methods are reported for the evaluation of pharmacokinetic parameters of marine-derived drugs [84].

The classic mechanism of action proposed for the sulfated fucose-rich polysaccharides after oral administration is summarized in Figure 2A. The polysaccharides are absorbed through the gastrointestinal tract, probably by endocytosis due to their high molecular weights [85], reach appropriate plasma concentration and exert their therapeutic action. Subsequently, the polysaccharides are distributed to different tissues, metabolized and excreted. Structural modifications might occur during these processes.

Moreover, there is evidence for another mechanism involved in the therapeutic effect of fucose-rich sulfated polysaccharides administered orally, which involves modification of the gut microbiota induced by the polysaccharides [86]. Probiotics are important microorganisms in the intestinal microflora. When colonized in adequate amounts, they confer a health benefit to the host by modulating several physiological activities [87]. Gut microbiota degrade polysaccharides and produce short-chain fatty acids, which might play an important role in maintaining the epithelial barrier function, regulating the immune responses and metabolic processes as well as inhibiting tumor development [88–91]. Models of gut microbe cultivation in vitro provide a convenient way to study the structural modifications of polysaccharides during digestion and absorption in the gastrointestinal tract and have already demonstrated a large number of applications in the field of intestinal fermentation of polysaccharides and oligosaccharides [92,93].

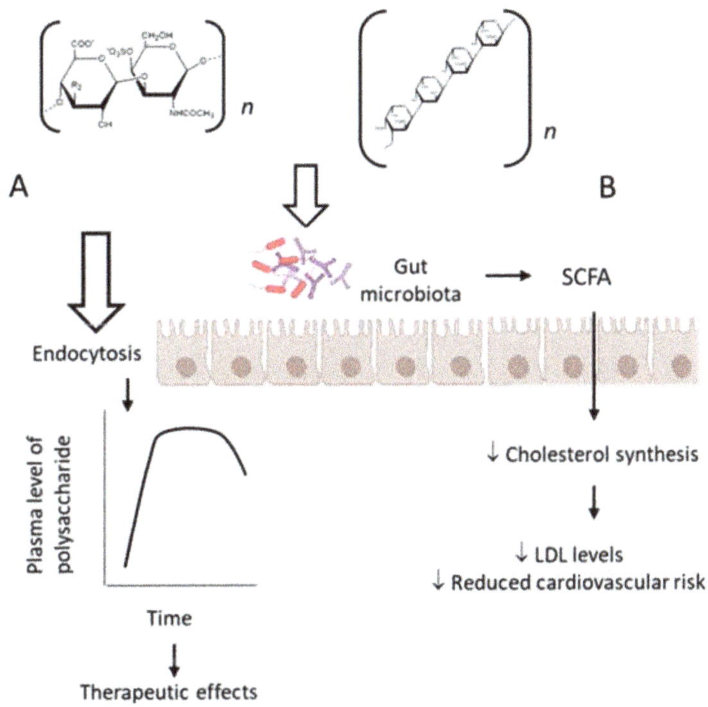

Figure 2. Pharmacological effects of sulfated polysaccharides after oral administration. (**A**) As the polysaccharides pass through the gastrointestinal tract, they are absorbed, probably by endocytosis due to their high-molecular-weight, and reach the bloodstream. Subsequently, they are distributed among various tissues and excreted unchanged and/or metabolized, as classically described for orally active drugs. (**B**) Alternatively, polysaccharides can exert a prebiotic effect by modulating the intestinal microbiota, which will produce short-chain fatty acids that can pass the intestinal mucosa by passive diffusion and reach the bloodstream, inhibiting cholesterol synthesis and reducing cardiovascular risk.

This particular aspect was investigated using a fucCS from the sea cucumber *P. graefei* [31]. The incubation of the polysaccharide with human intestinal flora in a simulated intestinal digestion model in vitro induced changes of intestinal microflora and degradation of the polysaccharide. Three samples of these bacteria utilize fucCS as a carbon source for their growth and produced short-chain fatty acids that decreased the pH of the media. A high content of acetate, propionate and butyrate was observed, indicating that they were the major products of microbial metabolism. Propionic acid can inhibit cholesterol synthesis in the liver, promote redistribution of cholesterol in plasma and liver, inhibit lipogenesis enzymes and reduce plasma lipid levels [94]. This activity could be responsible, in part, for the hypolipidemic effect of fucCS after oral administration. *Bifi-*

dobacterium, Bacteriodes prevotella and three species of *Clostridium* seemed to be involved in the metabolism of fucCS. These observations are summarized in Figure 2B.

Another study reported that fucCS from *S. japonicas* was not broken down under salivary and gastrointestinal digestion [95]. Due to the inhibition of pancreatic lipase in vitro by increasing concentrations of fucCS, the authors hypothesized that fucCS may work at the level of the gastrointestinal tract itself and/or after its absorption for the hypolipidemic effect.

The intestinal flora of part of the human population can ferment fucoidan to afford low-molecular-weight oligosaccharides [96], while another research had a different conclusion [97]. This suggests that the consumption of non-sterile marine foods with associated bacteria may have been the route by which these novel enzymes were acquired in human gut. Interestingly, the consumption of *C. okamuranus* algae by Japanese volunteers was associated with increased oral absorption of fucoidan contained in algae [98,99]; therefore, these contradictory results may be due to individual differences between species and strain level, which results in different metabolic capabilities of the microbiota to hydrolyze the molecules. Obviously, it also may depend on the polysaccharide structure. Further research is necessary to assess whether these changes in both bacterial composition and sulfated polysaccharide degradation also occur in vivo, and critically, whether such changes are responsible for some of the biological effects of these sulfated fucose-rich polysaccharides after oral administration.

9. Conclusions

Sulfated fucose-rich polysaccharides from marine organisms are unique molecules with various pharmacological effects. They might have promising therapeutic applications in different diseases. There has been an increasing interest in the therapeutic use of natural products for treatment of chronic cardiovascular and/or inflammatory diseases. The fact that these sulfated polysaccharides preserve their pharmacological effect after oral administration opens the perspective for the development of new drugs and/or the use of marine organisms as a source of functional food. High doses of the orally administered sulfated polysaccharides from marine organisms are employed in most studies. This limits their therapeutic use. Further larger trials are required to establish the role of fucoidan in several diseases. More efficient techniques of labeling sulfated polysaccharides will help to understand the pharmacokinetic parameters of these molecules. In recent years, much attention has been given to the prebiotic function of bioactive polysaccharides. Elucidating the linkage between the sulfated polysaccharides and gut microbiota will help to understand the biological effects of these molecules after oral administration.

Funding: This research was funded by Conselho Nacional de Desenvolvimento Científico e Tecnológico (CNPq) and Fundação de Amparo à Pesquisa do Estado do Rio de Janeiro (FAPERJ).

Conflicts of Interest: The authors declare no conflict of interest.

References

1. Springer, G.F.; Wurzel, H.A.; Mcneal, G.M.; Ansell, N.J.; Doughty, M.F. Isolation of Anticoagulant Fractions from Crude Fucoidin. *Proc. Soc. Exp. Biol. Med.* **1957**, *94*, 404–409. [CrossRef]
2. Pomin, V.H. NMR Structural Determination of Unique Invertebrate Glycosaminoglycans Endowed with Medical Properties. *Carbohydr. Res.* **2015**, *413*, 41–50. [CrossRef]
3. Soares, P.A.G.; Queiroz, I.N.L.; Pomin, V.H. NMR Structural Biology of Sulfated Glycans. *J. Biomol. Struct. Dyn.* **2017**, *35*, 1069–1084. [CrossRef]
4. Fitton, J.H.; Stringer, D.N.; Karpiniec, S.S. Therapies from Fucoidan: An Update. *Mar. Drugs* **2015**, *13*, 5920–5946. [CrossRef] [PubMed]
5. Mourão, P.A.S.; Bastos, I.G. Highly Acidic Glycans from Sea Cucumbers: Isolation and Fractionation of Fucose-rich Sulfated Polysaccharides from the Body Wall of Ludwigothurea Grisea. *Eur. J. Biochem.* **1987**, *166*, 639–645. [CrossRef]
6. Mourão, P.A.S.; Perlin, A.S. Structural Features of Sulfated Glycans from the Tunic of Styela Plicata (Chordata-Tunicata): A Unique Occurrence of L-galactose in Sulfated Polysaccharides. *Eur. J. Biochem.* **1987**, *166*, 431–436. [CrossRef] [PubMed]

7. Santos, J.A.; Mulloy, B.; Mourão, P.A.S. Structural Diversity among Sulfated A-L-galactans from Ascidians (Tunicates): Studies on the Species Ciona Intestinalis and Herdmania Monus. *Eur. J. Biochem.* **1992**, *204*, 669–677. [CrossRef]
8. Pereira, M.S.; Mulloy, B.; Mourão, P.A.S. Structure and Anticoagulant Activity of Sulfated Fucans. Comparison between the Regular, Repetitive, and Linear Fucans from Echinoderms with the More Heterogeneous and Branched Polymers from Brown Algae. *J. Biol. Chem.* **1999**, *274*, 7656–7667. [CrossRef] [PubMed]
9. Vilela-Silva, A.C.E.S.; Alves, A.P.; Valente, A.P.; Vacquier, V.D.; Mourão, P.A.S. Structure of the Sulfated α-L-Fucan from the Egg Jelly Coat of the Sea Urchin Strongylocentrotus Franciscanus: Patterns of Preferential 2-O- and 4-O-Sulfation Determine Sperm Cell Recognition. *Glycobiology* **1999**, *9*, 927–933. [CrossRef]
10. Alves, A.P.; Mulloy, B.; Diniz, J.A.; Mourão, P.A.S. Sulfated Polysaccharides from the Egg Jelly Layer Are Species-Specific Inducers of Acrosomal Reaction in Sperms of Sea Urchins. *J. Biol. Chem.* **1997**, *272*, 6965–6971. [CrossRef]
11. Ribeiro, A.C.; Vieira, R.P.; Mourão, P.A.S.; Mulloy, B. A Sulfated α-l-Fucan from Sea Cucumber. *Carbohydr. Res.* **1994**, *255*, 225–240. [CrossRef]
12. Pacheco, R.G.; Vicente, C.P.; Zancan, P.; Mourão, P.A.S. Different Antithrombotic Mechanisms among Glycosaminoglycans Revealed with a New Fucosylated Chondroitin Sulfate from an Echinoderm. *Blood Coagul. Fibrinolysis* **2000**, *11*, 563–573. [CrossRef] [PubMed]
13. Jiao, G.; Yu, G.; Zhang, J.; Ewart, H.S. Chemical Structures and Bioactivities of Sulfated Polysaccharides from Marine Algae. *Mar. Drugs* **2011**, *9*, 196–233. [CrossRef] [PubMed]
14. Pereira, M.S.; Melo, F.R.; Mourão, P.A.S. Is There a Correlation between Structure and Anticoagulant Action of Sulfated Galactans and Sulfated Fucans? *Glycobiology* **2002**, *12*, 573–580. [CrossRef]
15. Mourão, P.A.S. Perspective on the Use of Sulfated Polysaccharides from Marine Organisms as a Source of New Antithrombotic Drugs. *Mar. Drugs* **2015**, *13*, 2770–2784. [CrossRef]
16. Pomin, V.H. Holothurian Fucosylated Chondroitin Sulfate. *Mar. Drugs* **2014**, *12*, 232–254. [CrossRef]
17. Borsig, L.; Wang, L.; Cavalcante, M.C.M.; Cardilo-Reis, L.; Ferreira, P.L.; Mourão, P.A.S.; Esko, J.D.; Pavão, M.S.G. Selectin Blocking Activity of a Fucosylated Chondroitin Sulfate Glycosaminoglycan from Sea Cucumber: Effect on Tumor Metastasis and Neutrophil Recruitment. *J. Biol. Chem.* **2007**, *282*, 14984–14991. [CrossRef]
18. Lee, Y.E.; Kim, H.; Seo, C.; Park, T.; Lee, K.B.; Yoo, S.Y.; Hong, S.C.; Kim, J.T.; Lee, J. Marine Polysaccharides: Therapeutic Efficacy and Biomedical Applications. *Arch. Pharmacal Res.* **2017**, *40*, 1006–1020. [CrossRef]
19. Wang, Y.; Xing, M.; Cao, Q.; Ji, A.; Liang, H.; Song, S. Biological Activities of Fucoidan and the Factors Mediating Its Therapeutic Effects: A Review of Recent Studies. *Mar. Drugs* **2019**, *17*, 183. [CrossRef] [PubMed]
20. Fonseca, R.J.C.; Mourão, P.A.S. Fucosylated Chondroitin Sulfate as a New Oral Antithrombotic Agent. *Thromb. Haemost.* **2006**, *96*, 822–829. [CrossRef]
21. Richards, C.; Williams, N.A.; Fitton, J.H.; Stringer, D.N.; Karpiniec, S.S.; Park, A.Y. Oral Fucoidan Attenuates Lung Pathology and Clinical Signs in a Severe Influenza a Mouse Model. *Mar. Drugs* **2020**, *18*, 246. [CrossRef] [PubMed]
22. Yang, Z.; Yin, J.; Wang, Y.; Wang, J.; Xia, B.; Li, T.; Yang, X.; Hu, S.; Ji, C.; Guo, S. The Fucoidan A3 from the Seaweed Ascophyllum Nodosum Enhances RCT-Related Genes Expression in Hyperlipidemic C57BL/6J Mice. *Int. J. Biol. Macromol.* **2019**, *134*, 759–769. [CrossRef] [PubMed]
23. Cuong, H.D.; Thuy, T.T.T.; Huong, T.T.; Ly, B.M.; Van, T.T.T. Structure and Hypolipidaemic Activity of Fucoidan Extracted from Brown Seaweed Sargassum Henslowianum. *Nat. Prod. Res.* **2015**, *29*, 411–415. [CrossRef]
24. Vishchuk, O.S.; Sun, H.; Wang, Z.; Ermakova, S.P.; Xiao, J.J.; Lu, T.; Xue, P.P.; Zvyagintseva, T.N.; Xiong, H.; Shao, C.; et al. PDZ-Binding Kinase/T-LAK Cell-Originated Protein Kinase Is a Target of the Fucoidan from Brown Alga Fucus Evanescens in the Prevention of EGF-Induced Neoplastic Cell Transformation and Colon Cancer Growth. *Oncotarget* **2016**, *7*, 18763–18773. [CrossRef]
25. Pozharitskaya, O.N.; Obluchinskaya, E.D.; Shikov, A.N. Mechanisms of Bioactivities of Fucoidan from the Brown Seaweed Fucus Vesiculosus L. Of the Barents Sea. *Mar. Drugs* **2020**, *18*, 275. [CrossRef]
26. Nakazato, K.; Takada, H.; Iha, M.; Nagamine, T. Attenuation of N-Nitrosodiethylamine-Induced Liver Fibrosis by High-Molecular-Weight Fucoidan Derived from Cladosiphon Okamuranus. *J. Gastroenterol. Hepatol.* **2010**, *25*, 1692–1701. [CrossRef]
27. Phull, A.R.; Majid, M.; Haq, I.U.; Khan, M.R.; Kim, S.J. In Vitro and in Vivo Evaluation of Anti-Arthritic, Antioxidant Efficacy of Fucoidan from Undaria Pinnatifida (Harvey) Suringar. *Int. J. Biol. Macromol.* **2017**, *97*, 468–480. [CrossRef]
28. Tanino, Y.; Hashimoto, T.; Ojima, T.; Mizuno, M. F-Fucoidan from Saccharina Japonica Is a Novel Inducer of Galectin-9 and Exhibits Anti-Allergic Activity. *J. Clin. Biochem. Nutr.* **2016**, *59*, 25–30. [CrossRef] [PubMed]
29. Sakai, C.; Abe, S.; Kouzuki, M.; Shimohiro, H.; Ota, Y.; Sakinada, H.; Takeuchi, T.; Okura, T.; Kasagi, T.; Hanaki, K. A Randomized Placebo-Controlled Trial of an Oral Preparation of High Molecular Weight Fucoidan in Patients with Type 2 Diabetes with Evaluation of Taste Sensitivity. *Yonago Acta Med.* **2019**, *62*, 14–23. [CrossRef]
30. Ren, R.; Azuma, Y.; Ojima, T.; Hashimoto, T.; Mizuno, M.; Nishitani, Y.; Yoshida, M.; Azuma, T.; Kanazawa, K. Modulation of Platelet Aggregation-Related Eicosanoid Production by Dietary F-Fucoidan from Brown Alga Laminaria Japonica in Human Subjects. *Br. J. Nutr.* **2013**, *110*, 880–890. [CrossRef] [PubMed]
31. Wei, C.Y.; Liao, N.B.; Zhang, Y.; Ye, X.Q.; Li, S.; Hu, Y.Q.; Liu, D.H.; Linhardt, R.J.; Wang, X.; Chen, S.G. In Vitro Fermentation Behaviors of Fucosylated Chondroitin Sulfate from Pearsonothuria Graeffei by Human Gut Microflora. *Int. J. Biol. Macromol.* **2017**, *102*, 1195–1201. [CrossRef]

32. Zhu, Q.; Lin, L.; Zhao, M. Sulfated Fucan/Fucosylated Chondroitin Sulfate-Dominated Polysaccharide Fraction from Low-Edible-Value Sea Cucumber Ameliorates Type 2 Diabetes in Rats: New Prospects for Sea Cucumber Polysaccharide Based-Hypoglycemic Functional Food. *Int. J. Biol. Macromol.* **2020**, *159*, 34–45. [CrossRef] [PubMed]
33. Li, S.; Jiang, W.; Hu, S.; Song, W.; Ji, L.; Wang, Y.; Cai, L. Fucosylated Chondroitin Sulphate from Cusumaria Frondosa Mitigates Hepatic Endoplasmic Reticulum Stress and Inflammation in Insulin Resistant Mice. *Food Funct.* **2015**, *6*, 1547–1556. [CrossRef] [PubMed]
34. Soppert, J.; Lehrke, M.; Marx, N.; Jankowski, J.; Noels, H. Lipoproteins and Lipids in Cardiovascular Disease: From Mechanistic Insights to Therapeutic Targeting. *Adv. Drug Deliv. Rev.* **2020**, *159*, 4–33. [CrossRef]
35. Penson, P.E.; Banach, M. The Role of Nutraceuticals in the Optimization of Lipid-Lowering Therapy in High-Risk Patients with Dyslipidaemia. *Curr. Atheroscler. Rep.* **2020**, *22*, 1–9. [CrossRef]
36. Liu, H.H.; Ko, W.C.; Hu, M.L. Hypolipidemic Effect of Glycosaminoglycans from the Sea Cucumber Metriatyla Scabra in Rats Fed a Cholesterol-Supplemented Diet. *J. Agric. Food Chem.* **2002**, *50*, 3602–3606. [CrossRef]
37. Yin, J.; Wang, J.; Li, F.; Yang, Z.; Yang, X.; Sun, W.; Xia, B.; Li, T.; Song, W.; Guo, S. The Fucoidan from the Brown Seaweed: Ascophyllum Nodosum Ameliorates Atherosclerosis in Apolipoprotein E-Deficient Mice. *Food Funct.* **2019**, *10*, 5124–5139. [CrossRef] [PubMed]
38. Li, S.; Li, J.; Zhi, Z.; Wei, C.; Wang, W.; Ding, T.; Ye, X.; Hu, Y.; Linhardt, R.J.; Chen, S. Macromolecular Properties and Hypolipidemic Effects of Four Sulfated Polysaccharides from Sea Cucumbers. *Carbohydr. Polym.* **2017**, *173*, 330–337. [CrossRef]
39. Kumar, M.; Nagpal, R.; Hemalatha, R.; Verma, V.; Kumar, A.; Singh, S.; Marotta, F.; Jain, S.; Yadav, H. Targeted Cancer Therapies: The Future of Cancer Treatment. *Acta Biomed.* **2012**, *83*, 220–233. [PubMed]
40. Rejhová, A.; Opattová, A.; Čumová, A.; Slíva, D.; Vodička, P. Natural Compounds and Combination Therapy in Colorectal Cancer Treatment. *Eur. J. Med. Chem.* **2018**, *144*, 582–594. [CrossRef]
41. Atashrazm, F.; Lowenthal, R.M.; Woods, G.M.; Holloway, A.F.; Karpiniec, S.S.; Dickinson, J.L. Fucoidan Suppresses the Growth of Human Acute Promyelocytic Leukemia Cells In Vitro and In Vivo. *J. Cell. Physiol.* **2016**, *231*, 688–697. [CrossRef]
42. Yang, G.; Zhang, Q.; Kong, Y.; Xie, B.; Gao, M.; Tao, Y.; Xu, H.; Zhan, F.; Dai, B.; Shi, J.; et al. Antitumor Activity of Fucoidan against Diffuse Large B Cell Lymphoma in Vitro and in Vivo. *Acta Biochim. Sinica* **2015**, *47*, 925–931. [CrossRef]
43. Hsu, H.Y.; Lin, T.Y.; Wu, Y.C.; Tsao, S.M.; Hwang, P.A.; Shih, Y.W.; Hsu, J. Fucoidan Inhibition of Lung Cancer in Vivo and in Vitro: Role of the Smurf2-Dependent Ubiquitin Proteasome Pathway in TGFβ Receptor Degradation. *Oncotarget* **2014**, *5*, 7870–7885. [CrossRef] [PubMed]
44. Rui, X.; Pan, H.F.; Shao, S.L.; Xu, X.M. Anti-Tumor and Anti-Angiogenic Effects of Fucoidan on Prostate Cancer: Possible JAK-STAT3 Pathway. *BMC Complement. Altern. Med.* **2017**, *17*, 1–8. [CrossRef] [PubMed]
45. Lim, J.D.; Lee, S.R.; Kim, T.; Jang, S.A.; Kang, S.C.; Koo, H.J.; Sohn, E.; Bak, J.P.; Namkoong, S.; Kim, H.K.; et al. Fucoidan from Fucus Vesiculosus Protects against Alcohol-Induced Liver Damage by Modulating Inflammatory Mediators in Mice and Hepg2 Cells. *Mar. Drugs* **2015**, *13*, 1051–1067. [CrossRef] [PubMed]
46. Heeba, G.H.; Morsy, M.A. Fucoidan Ameliorates Steatohepatitis and Insulin Resistance by Suppressing Oxidative Stress and Inflammatory Cytokines in Experimental Non-Alcoholic Fatty Liver Disease. *Environ. Toxicol. Pharmacol.* **2015**, *40*, 907–914. [CrossRef]
47. Wang, Y.; Nie, M.; Lu, Y.; Wang, R.; Li, J.; Yang, B.; Xia, M.; Zhang, H.; Li, X. Fucoidan Exerts Protective Effects against Diabetic Nephropathy Related to Spontaneous Diabetes through the NF-KB Signaling Pathway in Vivo and in Vitro. *Int. J. Mol. Med.* **2015**, *35*, 1067–1073. [CrossRef]
48. Zuo, T.; Li, X.; Chang, Y.; Duan, G.; Yu, L.; Zheng, R.; Xue, C.; Tang, Q. Dietary Fucoidan of Acaudina Molpadioides and Its Enzymatically Degraded Fragments Could Prevent Intestinal Mucositis Induced by Chemotherapy in Mice. *Food Funct.* **2015**, *6*, 415–422. [CrossRef] [PubMed]
49. Olivera-Castillo, L.; Grant, G.; Kantún-Moreno, N.; Barrera-Pérez, H.A.; Montero, J.; Olvera-Novoa, M.A.; Carrillo-Cocom, L.M.; Acevedo, J.J.; Puerto-Castillo, C.; Solís, V.M.; et al. A Glycosaminoglycan-Rich Fraction from Sea Cucumber Isostichopus Badionotus Has Potent Anti-Inflammatory Properties in Vitro and in Vivo. *Nutrients* **2020**, *12*, 1698. [CrossRef] [PubMed]
50. Negishi, H.; Mori, M.; Mori, H.; Yamori, Y. Supplementation of Elderly Japanese Men and Women with Fucoidan from Seaweed Increases Immune Responses to Seasonal Influenza Vaccination. *J. Nutr.* **2013**, *143*, 1794–1798. [CrossRef] [PubMed]
51. Ganesan, K.; Sultan, S. *Oral Hypoglycemic Medications*; StatPearls Publishing: Treasure Island, FL, USA, 2019.
52. Jiang, X.; Yu, J.; Ma, Z.; Zhang, H.; Xie, F. Effects of Fucoidan on Insulin Stimulation and Pancreatic Protection via the CAMP Signaling Pathway in Vivo and in Vitro. *Mol. Med. Rep.* **2015**, *12*, 4501–4507. [CrossRef] [PubMed]
53. Hu, S.; Chang, Y.; He, M.; Wang, J.; Wang, Y.; Xue, C. Fucosylated Chondroitin Sulfate from Sea Cucumber Improves Insulin Sensitivity via Activation of PI3K/PKB Pathway. *J. Food Sci.* **2014**, *79*, H1424–H1429. [CrossRef]
54. Hu, S.; Chang, Y.; Wang, J.; Xue, C.; Li, Z.; Wang, Y. Fucosylated Chondroitin Sulfate from Sea Cucumber in Combination with Rosiglitazone Improved Glucose Metabolism in the Liver of the Insulin-Resistant Mice. *Biosci. Biotechnol. Biochem.* **2013**, *77*, 2263–2268. [CrossRef]
55. Hernández-Corona, D.M.; Martínez-Abundis, E.; González-Ortiz, M. Effect of Fucoidan Administration on Insulin Secretion and Insulin Resistance in Overweight or Obese Adults. *J. Med. Food* **2014**, *17*, 830–832. [CrossRef] [PubMed]
56. Mulloy, B.; Hogwood, J.; Gray, E.; Lever, R.; Page, C.P. Pharmacology of Heparin and Related Drugs. *Pharmacol. Rev.* **2015**, *68*, 76–141. [CrossRef]

57. Bleeding Associated with Antithrombotic—PubMed. Available online: https://pubmed.ncbi.nlm.nih.gov/6160618/ (accessed on 13 June 2021).
58. Heparin-Induced Thrombocytopenia: Molecular Pathogenesis—PubMed. Available online: https://pubmed.ncbi.nlm.nih.gov/10605736/ (accessed on 13 June 2021).
59. Nutescu, E.A.; Dager, W.E.; Kalus, J.S.; Lewin, J.J.; Cipolle, M.D. Management of Bleeding and Reversal Strategies for Oral Anticoagulants: Clinical Practice Considerations. *Am. J. Health-Syst. Pharm.* **2013**, *70*, 1914–1929. [CrossRef] [PubMed]
60. Mourão, P.A.S.; Pereira, M.S. Searching for Alternatives to Heparin: Sulfated Fucans from Marine Invertebrates. *Trends Cardiovasc. Med.* **1999**, *9*, 225–232. [CrossRef]
61. Mourao, P. Use of Sulfated Fucans as Anticoagulant and Antithrombotic Agents: Future Perspectives. *Curr. Pharm. Des.* **2005**, *10*, 967–981. [CrossRef]
62. Mourao, P.A.S.; Pereira, M.S.; Pavo, M.S.G.; Mulloy, B.; Tollefsen, D.M.; Mowinckel, M.C.; Abildgaard, U. Structure and Anticoagulant Activity of a Fucosylated Chondroitin Sulfate from Echinoderm. Sulfated Fucose Branches on the Polysaccharide Account for Its High Anticoagulant Action. *J. Biol. Chem.* **1996**, *271*, 23973–23984. [CrossRef]
63. Thorlacius, H.; Vollmar, B.; Seyfert, U.T.; Vestweber, D.; Menger, M. The Polysaccharide Fucoidan Inhibits Microvascular Thrombus Formation Independently from P- and L-Selectin Function in Vivo. *Eur. J. Clin. Investig.* **2000**, *30*, 804–810. [CrossRef]
64. Zancan, P.; Mourão, P.A. Venous and Arterial Thrombosis in Rat Models. *Blood Coagul. Fibrinolysis* **2004**, *15*, 45–54. [CrossRef]
65. Fonseca, R.J.C.; Santos, G.R.C.; Mourão, P.A.S. Effects of Polysaccharides Enriched in 2,4-Disulfated Fucose Units on Coagulation, Thrombosis and Bleeding: Practical and Conceptual Implications. *Thromb. Haemost.* **2009**, *102*, 829–836. [CrossRef]
66. Glauser, B.F.; Pereira, M.S.; Monteiro, R.Q.; Mourão, P.A.S. Serpin-Independent Anticoagulant Activity of a Fucosylated Chondroitin Sulfate. *Thromb. Haemost.* **2008**, *100*, 420–428. [CrossRef] [PubMed]
67. Fonseca, R.J.C.; Sucupira, I.D.; Oliveira, S.N.M.C.G.; Santos, G.R.C.; Mourão, P.A.S. Improved Anticoagulant Effect of Fucosylated Chondroitin Sulfate Orally Administered as Gastro-Resistant Tablets. *Thromb. Haemost.* **2017**, *117*, 662–670. [CrossRef] [PubMed]
68. Zhao, X.; Guo, F.; Hu, J.; Zhang, L.; Xue, C.; Zhang, Z.; Li, B. Antithrombotic Activity of Oral Administered Low Molecular Weight Fucoidan from Laminaria Japonica. *Thromb. Res.* **2016**, *144*, 46–52. [CrossRef] [PubMed]
69. Fonseca, R.J.C.; Oliveira, S.N.M.C.G.; Pomin, V.H.; Mecawi, A.S.; Araujo, I.G.; Mourão, P.A.S. Effects of Oversulfated and Fucosylated Chondroitin Sulfates on Coagulation: Challenges for the Study of Anticoagulant Polysaccharides. *Thromb. Haemost.* **2010**, *103*, 994–1004. [CrossRef]
70. Yan, L.; Zhu, M.; Wang, D.; Tao, W.; Liu, D.; Zhang, F.; Linhardt, R.J.; Ye, X.; Chen, S. Oral Administration of Fucosylated Chondroitin Sulfate Oligomers in Gastro-Resistant Microcapsules Exhibits a Safe Antithrombotic Activity. *Thromb. Haemost.* **2021**, *121*, 15–26. [CrossRef] [PubMed]
71. Prasad, S.; Lillicrap, D.; Labelle, A.; Knappe, S.; Keller, T.; Burnett, E.; Powell, S.; Johnson, K.W. Efficacy and Safety of a New-Class Hemostatic Drug Candidate, AV513, in Dogs with Hemophilia A. *Blood* **2008**, *111*, 672–679. [CrossRef]
72. Improved Coagulation in Bleeding Disorders by Non-Anticoagulant Sulfated Polysaccharides (NASP)—PubMed. Available online: https://pubmed.ncbi.nlm.nih.gov/16543964/ (accessed on 13 June 2021).
73. Fonseca, R.J.C.; Oliveira, S.N.M.C.G.; Melo, F.R.; Pereira, M.G.; Benevides, N.M.B.; Mourão, P.A.S. Slight Differences in Sulfation of Algal Galactans Account for Differences in Their Anticoagulant and Venous Antithrombotic Activities. *Thromb. Haemost.* **2008**, *99*, 539–545. [CrossRef]
74. Takahashi, H.; Kawaguchi, M.; Kitamura, K.; Narumiya, S.; Kawamura, M.; Tengan, I.; Nishimoto, S.; Hanamure, Y.; Majima, Y.; Tsubura, S.; et al. An Exploratory Study on the Anti-Inflammatory Effects of Fucoidan in Relation to Quality of Life in Advanced Cancer Patients. *Integr. Cancer Ther.* **2018**, *17*, 282–291. [CrossRef] [PubMed]
75. Nagamine, T.; Kadena, K.; Tomori, M.; Nakajima, K.; Iha, M. Activation of NK Cells in Male Cancer Survivors by Fucoidan Extracted from Cladosiphon Okamuranus. *Mol. Clin. Oncol.* **2020**, *12*, 81–88. [CrossRef]
76. Hsu, H.; Hwang, P. Clinical Applications of Fucoidan in Translational Medicine for Adjuvant Cancer Therapy. *Clin. Transl. Med.* **2019**, *8*, 15. [CrossRef]
77. Ikeguchi, M.; Yamamoto, M.; Arai, Y.; Maeta, Y.; Ashida, K.; Katano, K.; Miki, Y.; Kimura, T. Fucoidan Reduces the Toxicities of Chemotherapy for Patients with Unresectable Advanced or Recurrent Colorectal Cancer. *Oncol. Lett.* **2011**, *2*, 319–322. [CrossRef] [PubMed]
78. Myers, S.P.; Mulder, A.M.; Baker, D.G.; Robinson, S.R.; Rolfe, M.I.; Brooks, L.; Fitton, J.H. Effects of Fucoidan from Fucus Vesiculosus in Reducing Symptoms of Osteoarthritis: A Randomized Placebo-Controlled Trial. *Biol. Targets Ther.* **2016**, *10*, 81–88. [CrossRef]
79. Imanari, T.; Washio, Y.; Huang, Y.; Toyoda, H.; Suzuki, A.; Toida, T. Oral Absorption and Clearance of Partially Depolymerized Fucosyl Chondroitin Sulfate from Sea Cucumber. *Thromb. Res.* **1999**, *93*, 129–135. [CrossRef]
80. Irhimeh, M.R.; Fitton, J.H.; Lowenthal, R.M.; Kongtawelert, P. A Quantitative Method to Detect Fucoidan in Human Plasma Using a Novel Antibody. *Methods Find. Exp. Clin. Pharmacol.* **2005**, *27*, 705–710. [CrossRef]
81. Tokita, Y.; Hirayama, M.; Nakajima, K.; Tamaki, K.; Iha, M.; Nagamine, T. Detection of Fucoidan in Urine after Oral Intake of Traditional Japanese Seaweed, Okinawa Mozuku (Cladosiphon Okamuranus Tokida). *J. Nutr. Sci. Vitaminol.* **2017**, *63*, 419–421. [CrossRef] [PubMed]
82. Nagamine, T.; Nakazato, K.; Tomioka, S.; Iha, M.; Nakajima, K. Intestinal Absorption of Fucoidan Extracted from the Brown Seaweed, Cladosiphon Okamuranus. *Mar. Drugs* **2015**, *13*, 48–64. [CrossRef]

83. Pozharitskaya, O.N.; Shikov, A.N.; Faustova, N.M.; Obluchinskaya, E.D.; Kosman, V.M.; Vuorela, H.; Makarov, V.G. Pharmacokinetic and Tissue Distribution of Fucoidan from Fucus Vesiculosus after Oral Administration to Rats. *Mar. Drugs* **2018**, *16*, 132. [CrossRef]
84. Shikov, A.N.; Flisyuk, E.V.; Obluchinskaya, E.D.; Pozharitskaya, O.N. Pharmacokinetics of Marine-Derived Drugs. *Mar. Drugs* **2020**, *18*, 557. [CrossRef]
85. Conte, A.; Volpi, N.; Palmieri, L.; Bahous, I.; Ronca, G. Biochemical and Pharmacokinetic Aspects of Oral Treatment with Chondroitin Sulfate. *Arzneim. Forsch. Drug Res.* **1995**, *45*, 918–925.
86. Zhang, T.; Wu, S.; Ai, C.; Wen, C.; Liu, Z.; Wang, L.; Jiang, L.; Shen, P.; Zhang, G.; Song, S. Galactofucan from Laminaria Japonica Is Not Degraded by the Human Digestive System but Inhibits Pancreatic Lipase and Modifies the Intestinal Microbiota. *Int. J. Biol. Macromol.* **2021**, *166*, 611–620. [CrossRef]
87. Mueller, M.; Ganesh, R.; Bonnes, S. Gut Health = Mental Health? The Impact of Diet and Dietary Supplements on Mood Disorders. *Curr. Nutr. Rep.* **2020**, *9*, 361–368. [CrossRef]
88. Dronkers, T.M.G.; Ouwehand, A.C.; Rijkers, G.T. Global Analysis of Clinical Trials with Probiotics. *Heliyon* **2020**, *6*, e04467. [CrossRef] [PubMed]
89. Liao, W.; Chen, C.; Wen, T.; Zhao, Q. Probiotics in the Prevention of Antibiotic-Associated Diarrhea in Adults: A Meta-Analysis of Randomized Placebo-Controlled Trials. *J. Clin. Gastroenterol.* **2021**, *55*, 469–480. [CrossRef] [PubMed]
90. Fakhri, S.; Yarmohammadi, A.; Yarmohammadi, M.; Farzaei, M.H.; Echeverria, J. Marine Natural Products: Promising Candidates in the Modulation of Gut-Brain Axis towards Neuroprotection. *Mar. Drugs* **2021**, *19*, 165. [CrossRef] [PubMed]
91. Toumazi, D.; Constantinou, C. A Fragile Balance: The Important Role of the Intestinal Microbiota in the Prevention and Management of Colorectal Cancer. *Oncology* **2020**, *98*, 593–602. [CrossRef] [PubMed]
92. Clavel, T.; Lagkouvardos, I.; Stecher, B. From Complex Gut Communities to Minimal Microbiomes via Cultivation. *Curr. Opin. Microbiol.* **2017**, *38*, 148–155. [CrossRef]
93. Arnold, J.W.; Roach, J.; Azcarate-Peril, M.A. Emerging Technologies for Gut Microbiome Research. *Trends Microbiol.* **2016**, *24*, 887–901. [CrossRef]
94. Wiciński, M.; Gębalski, J.; Gołębiewski, J.; Malinowski, B. Probiotics for the Treatment of Overweight and Obesity in Humans—A Review of Clinical Trials. *Microorganisms* **2020**, *8*, 1148. [CrossRef] [PubMed]
95. Zhu, Z.; Dong, X.; Yan, C.; Ai, C.; Zhou, D.; Yang, J.; Zhang, H.; Liu, X.; Song, S.; Xiao, H.; et al. Structural Features and Digestive Behavior of Fucosylated Chondroitin Sulfate from Sea Cucumbers Stichopus Japonicus. *J. Agric. Food Chem.* **2019**, *67*, 10534–10542. [CrossRef]
96. Hehemann, J.H.; Correc, G.; Barbeyron, T.; Helbert, W.; Czjzek, M.; Michel, G. Transfer of Carbohydrate-Active Enzymes from Marine Bacteria to Japanese Gut Microbiota. *Nature* **2010**, *464*, 908–912. [CrossRef]
97. Strain, C.R.; Collins, K.C.; Naughton, V.; McSorley, E.M.; Stanton, C.; Smyth, T.J.; Soler-Vila, A.; Rea, M.C.; Ross, P.R.; Cherry, P.; et al. Effects of a Polysaccharide-Rich Extract Derived from Irish-Sourced Laminaria Digitata on the Composition and Metabolic Activity of the Human Gut Microbiota Using an in Vitro Colonic Model. *Eur. J. Nutr.* **2020**, *59*, 309–325. [CrossRef] [PubMed]
98. Imbs, T.I.; Zvyagintseva, T.N.; Ermakova, S.P. Is the Transformation of Fucoidans in Human Body Possible? *Int. J. Biol. Macromol.* **2020**, *142*, 778–781. [CrossRef] [PubMed]
99. Kadena, K.; Tomori, M.; Iha, M.; Nagamine, T. Absorption Study of Mozuku Fucoidan in Japanese Volunteers. *Mar. Drugs* **2018**, *16*, 254. [CrossRef] [PubMed]

marine drugs

Article

The Toxicology of Native Fucosylated Glycosaminoglycans and the Safety of Their Depolymerized Products as Anticoagulants

Lisha Lin [1,2], Sujuan Li [1,2], Na Gao [3], Weili Wang [1,2], Taocui Zhang [1,2], Lian Yang [1], Xingzhi Yang [1], Dan Luo [4], Xu Ji [5,*] and Jinhua Zhao [3,*]

[1] State Key Laboratory of Phytochemistry and Plant Resources in West China, Kunming Institute of Botany, Chinese Academy of Sciences, Kunming 650201, China; linlisha@mail.kib.ac.cn (L.L.); lisujuan@mail.kib.ac.cn (S.L.); wangweili@mail.kib.ac.cn (W.W.); zhangtaocui@mail.kib.ac.cn (T.Z.); yanglian@mail.kib.ac.cn (L.Y.); yangxingzhi@mail.kib.ac.cn (X.Y.)
[2] College of Life Sciences, University of Chinese Academy of Sciences, Beijing 100049, China
[3] School of Pharmaceutical Sciences, South-Central University for Nationalities, Wuhan 430074, China; gn2008.happy@163.com
[4] College of Traditional Chinese Medicine, Yunnan University of Chinese Medicine, Kunming 650201, China; luodan@mail.kib.ac.cn
[5] School of Chemical Science and Technology, Yunnan University, Kunming 650201, China
* Correspondence: jixu@ynu.edu.cn (X.J.); zhao.jinhua@yahoo.com (J.Z.)

Abstract: Fucosylated glycosaminoglycan (FG) from sea cucumber is a potent anticoagulant by inhibiting intrinsic coagulation tenase (iXase). However, high-molecular-weight FGs can activate platelets and plasma contact system, and induce hypotension in rats, which limits its application. Herein, we found that FG from *T. ananas* (TaFG) and FG from *H. fuscopunctata* (HfFG) at 4.0 mg/kg (i.v.) could cause significant cardiovascular and respiratory dysfunction in rats, even lethality, while their depolymerized products had no obvious side effects. After injection, native FG increased rat plasma kallikrein activity and levels of the vasoactive peptide bradykinin (BK), consistent with their contact activation activity, which was assumed to be the cause of hypotension in rats. However, the hemodynamic effects of native FG cannot be prevented by the BK receptor antagonist. Further study showed that native FG induced in vivo procoagulation, thrombocytopenia, and pulmonary embolism. Additionally, its lethal effect could be prevented by anticoagulant combined with antiplatelet drugs. In summary, the acute toxicity of native FG is mainly ascribed to pulmonary microvessel embolism due to platelet aggregation and contact activation-mediated coagulation, while depolymerized FG is a safe anticoagulant candidate by selectively targeting iXase.

Keywords: fucosylated glycosaminoglycan; anticoagulant; platelet aggregation; contact activation; hypotension; pulmonary embolism

Citation: Lin, L.; Li, S.; Gao, N.; Wang, W.; Zhang, T.; Yang, L.; Yang, X.; Luo, D.; Ji, X.; Zhao, J. The Toxicology of Native Fucosylated Glycosaminoglycans and the Safety of Their Depolymerized Products as Anticoagulants. *Mar. Drugs* **2021**, *19*, 487. https://doi.org/10.3390/md19090487

Academic Editors: Yuki Fujii, Marco Gerdol and Yasuhiro Ozeki

Received: 31 July 2021
Accepted: 25 August 2021
Published: 27 August 2021

1. Introduction

Fucosylated glycosaminoglycan (FG) is a sulfated polysaccharide isolated from sea cucumber (*Echinodermata*, *Holothuroidea*), a nutritious food widely consumed in Asia. FG has a unique chemical structure comprised of the backbone of mammal chondroitin sulfate and side chains of sulfated fucoses [1]. Previous study showed that it's a potent serpin-independent anticoagulant by targeting intrinsic coagulation factor Xase (iXase, FIXa-FVIIIa complex), the last and rate-limiting enzyme in the intrinsic coagulation pathway [2–4]. This novel anticoagulant target justifies FG to be a lead compound for safe and effective antithrombosis, the detailed information can be found in a recent review published in Blood Reviews [5]. After depolymerization, the obtained low-molecular-weight FGs (Mw 6~12 kDa) are a selective iXase inhibitor with potent anticoagulant and antithrombotic activities, without obvious bleeding risk [6,7]. Particularly, dHG-5 (Mw 5 kDa), depolymerized from the natural FG from *Holothuria Fuscopunctata* (HfFG) by β-elimination, is an anticoagulant candidate showing superior pharmacological activity compared to enoxaparin in preclinical studies [8–10]. Moreover, the clinical trial application of dHG-5 is recently approved by the Food and Drug Administration (IND 153953).

However, native FGs with high molecular weights (Mw > 50 kDa) have multiple pharmacological targets. Among others, they also have the contradictory activities of plasma contact activation (FXII activation) and platelet aggregation [11,12]. These contrary activities may offset their anticoagulant potency, and the lack of selectivity may cause undesired and severe off-target side effects. For instance, a natural FG from *Thelenota ananas* (TaFG) exhibited weaker in vivo antithrombotic activity than its depolymerized product in a stasis-induced venous thrombosis rat model [7]. Native FGs from *Ludwigothurea grisea*, *Isostichopus badionotus*, and *Pearsonothuria graeffei* could cause obvious hypotension and reduced heart rate in rats after intravenous injection [13,14]. However, the underlying mechanism is not clear, and how the undesired activities of FG are related to its in vivo toxicology remains to be studied.

In 2008, heparin contaminated with oversulfated chondroitin sulfate (OSCS) was found to cause an adverse reaction of hypotension in the clinic [15–17]. Investigation showed that OSCS induced severe anaphylactoid reactions by activating the plasma contact system, which led to the production of vasoactive bradykinin (BK) and anaphylatoxins (C3a, C5a) [17]. The plasma contact system contains the coagulation factor XII (FXII) and prekallikrein reciprocal activation loop. Contact activation can result in coagulation and inflammation, mediated by coagulation factor XI activation and the kallikrein-kinin pathway, respectively [18]. The contact system can be activated by negatively charged macromolecules such as sulfated polysaccharide [19,20]. Both OSCS and native FG are sulfated polysaccharide and both can potently enhance plasma contact activation; thus, it is assumed that the native FG has the same mechanism as OSCS in inducing hypotension in rats.

In this work, the in vivo toxicology of TaFG, HfFG, and their depolymerized products dTaFG13 (13 kDa) and dHG-5 were studied. Our results reveal that the systemic treatment of native FGs causes severe acute toxicity in rats and the main toxic mechanism was clarified. dTaFG13 had only a very weak effect on rat blood pressure and the anticoagulant candidate dHG-5 showed no side effect in all the assays. Although native FGs have various bioactivities such as antithrombosis, anti-inflammation, anticancer, antivirus, and pro-angiogenesis [1], according to these results, the application of any of their pharmacological activity should be approached cautiously. By contrast, the safety of low-molecular-weight FG as a promising novel anticoagulant is further demonstrated.

2. Results and Discussion

2.1. *In Vivo Effects of FGs in Rats*

It is reported that high-molecular-weight FG can potently activate the plasma contact system and cause hypotension in rats, similar to OSCS [13,14,16]. Herein, the in vivo effect of TaFG, HfFG, and their depolymerized products were studied. Both TaFG (65.8 kDa) and HfFG (61.1 kDa) have high molecular weights but have different sulfation pattern: 58% of the fucose branches of TaFG are di-sulfated (Fuc2S4S), while 85% of HfFG are di-sulfated (Fuc3S4S) (Figure 1). Unexpectedly, after intravenously injection of TaFG or HfFG (4.0 mg/kg) into the rats, respiration apnea followed by tachypnea could be observed macroscopically, and even death occurred within 5 min. Whereas, the depolymerized FGs caused no obvious macroscopic change. By continuously monitoring some physiological parameters, we found that native FGs caused severe and acute cardiovascular dysfunction and respiratory failure.

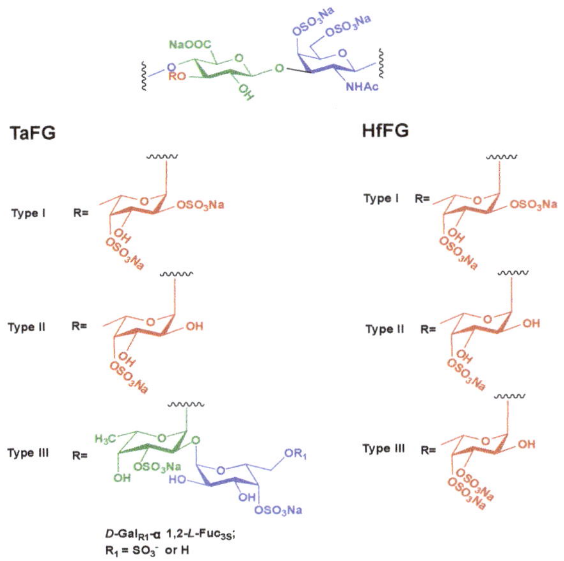

Figure 1. The chemical structures of TaFG and HfFG.

2.1.1. Effects of FGs on Rat Blood Pressure

TaFG and HfFG at 4.0 mg/kg injected into the rats caused an immediate drop of arterial pressure (AP) by 55% and 63%, respectively, and their activity was stronger than OSCS, which resulted in hypotension by 35% (Figures 2 and 3 and Table S1). The dTaFG13 injection caused mild and a delayed drop of rat blood pressure, while dHG-5 and heparin (UFH) had no obvious effects. The dose–effect relationship of HfFG was also studied (Figure 3 and Table S1). Rats treated with HfFG (0.125~4.0 mg/kg) showed severe hypotension with the similar peak effect at about 1 min, while the recovery time was increased with the dose. HfFG showed the lethal effect at 1 mg/kg and above, and had no obvious effect on blood pressure at 0.0625 mg/kg.

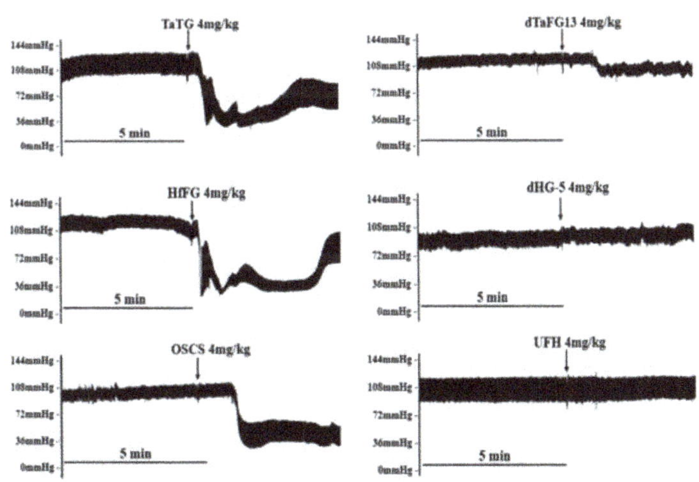

Figure 2. Rat typical arterial pressure response after treatment. Data were recorded by a multi-channel physiological signal-collecting instrument before and after treatment. Arrows indicate the administration of the test compound. Representative graphs are shown, n $\geq$ 8.

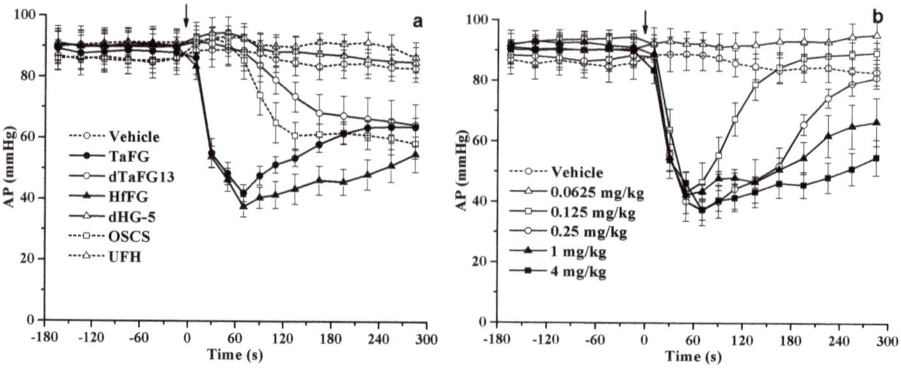

Figure 3. Effects of FG on rat arterial pressure. Effects of different polysaccharides on rat blood pressure (**a**). Effects of different doses of HfFG on rat blood pressure (**b**). After the 10-min adaptation period, rats were treated with a single intravenous bolus of the test compound. The arterial pressure and heart rate were detected for 10 min before and 30 min after treatment. Arrows indicate the administration of the test compound. Mean ± SEM, n ≥ 8. Abbreviation: AP, arterial pressure.

The results are unexpected as such severe side effects of high molecular FG have not been reported by others. Fonseca et al. reported that an FG from sea cucumber *L. grisea* caused a ~40% drop of AP and a slight decrease of heart rate, which was not statistically significant in Wistar rats that received an intravenous dose of 3.5 mg/kg and showed no hypotensive effect when below 3.0 mg/kg [13]. These differences are possibly ascribed to the different types of FG and animal species. Additionally, TaFG and HfFG induced much more acute and severe responses than OSCS, which was reported to induce the human hypotension and anaphylactoid reaction in clinic, mediated by plasma contact system activation [16,17].

2.1.2. Effects of FGs on the Rat Cardiac Function

Then, we investigated the effects of FGs on the rat cardiac function. After TaFG or HfFG (4.0 mg/kg) injection, the rat heart rate (HR), left ventricular pressure (LVP), and maximal rate of rise of ventricular pressure (+dp/dt$_{max}$, indicating the cardiac contractility) were dramatically decreased, reaching to the maximal effect at about 30 s (Table 1 and Figure S1). These changes occurred before the blood pressure drop (described above). A representative graph is shown in Figure 4, TaFG and HfFG reduced LVP by 61% and 68%, respectively, and reduced the heart rate by more than 70% (Table 1). Compared with native FGs, their depolymerized products had no obvious effects on rat cardiac function. OSCS and UFH also had no such effects, to confirm the result, the dose of OSCS was increased up to 8 mg/kg and no obvious change in rat heart rate was observed (data not shown).

Table 1. Effects of FGs on the rat cardiac function (mean ± SEM, n ≥ 8).

Treatment (mg/kg)	mLVP$_0$ (mmHg)	mLVP$_1$ (mmHg)	(+dP/dt$_{max}$)$_0$ (10^3 mmHg/s)	(+dP/dt$_{max}$)$_1$ (10^3 mmHg/s)	ΔHR (%)
Vehicle	47.2 ± 3.3	48.7 ± 2.3	4.8 ± 0.3	4.9 ± 0.3	−0.7 ± 1.4
TaFG (4.0)	47.4 ± 2.3	18.3 ± 3.2 ***	3.8 ± 0.4	1.4 ± 0.2 ***	−71.0 ± 5.8
dTaFG13 (4.0)	39.5 ± 2.5	44.9 ± 3.0	4.3 ± 0.6	4.5 ± 0.5	−3.9 ± 2.1
HfFG (4.0)	47.8 ± 2.3	15.4 ± 2.5 ***	3.8 ± 0.4	1.4 ± 0.2 ***	−72.9 ± 5.0
dHG-5 (4.0)	42.4 ± 2.2	42.5 ± 2.4	4.0 ± 0.4	4.2 ± 0.5	3.8 ± 2.3
OSCS (4.0)	46.6 ± 3.2	35.0 ± 2.3 *	3.3 ± 0.3	3.2 ± 0.4	14.1 ± 7.1
UFH (4.0)	43.5 ± 2.4	47.3 ± 2.3	3.7 ± 0.4	4.3 ± 0.5	5.3 ± 7.4
HfFG (1.0)	46.3 ± 3.2	16.5 ± 1.8 ***	3.7 ± 0.4	1.2 ± 0.1 ***	−77.7 ± 2.7
HfFG (0.25)	45.1 ± 2.4	17.1 ± 2.1 ***	3.1 ± 0.4	1.4 ± 0.2 **	−63.7 ± 6.6

* $p \leq 0.05$, ** $p \leq 0.01$, and *** $p \leq 0.001$ vs. pretreatment, two-tail *t*-test. The subscript 0 and 1 indicate the value before and after treatment, respectively. Abbreviations: mLVP, mean left ventricular pressure; +dP/dt$_{max}$, the maximal rates of rise of ventricular pressure; and ΔHR, the change of heart rate.

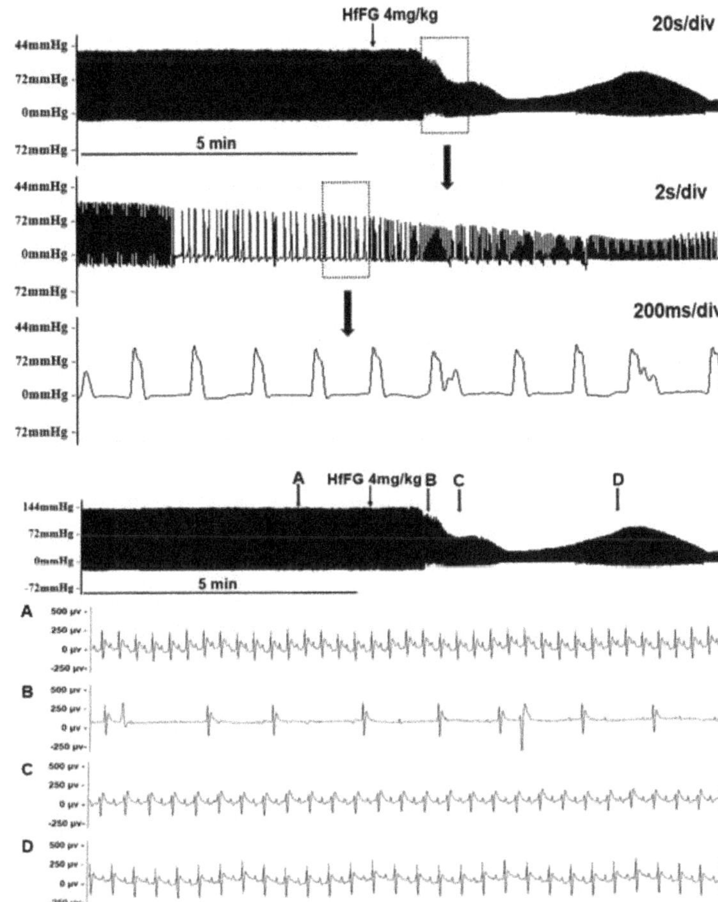

Figure 4. Effect of HfFG (i.v.) on rat left ventricular pressure (**upper**) and electrocardiogram (**below**). Data were recorded by a multi-channel physiological signal-collecting instrument before and after treatment. For the left ventricular pressure, different time scales were shown. For the electrocardiogram, different time points (**A–D**) in the left ventricular pressure recording were shown, time scale of (**A–D**) was 80 ms/div. Arrows indicate the administration of the test compound. Representative graphs are shown, n $\geq$ 8.

2.1.3. Effects of Native FGs on the Rat Respiration

The effects of TaFG and HfFG on the rat respiratory function were detected simultaneously with the ECG detection. The results showed that after injection of TaFG or HfFG, rats had accelerated respiration, followed immediately by respiratory failure or apnea before the heart rate reduction (Figure 5). This indicates that after native FGs injection, both rat hypotension and cardiac dysfunction occurred secondarily to respiratory dysfunction.

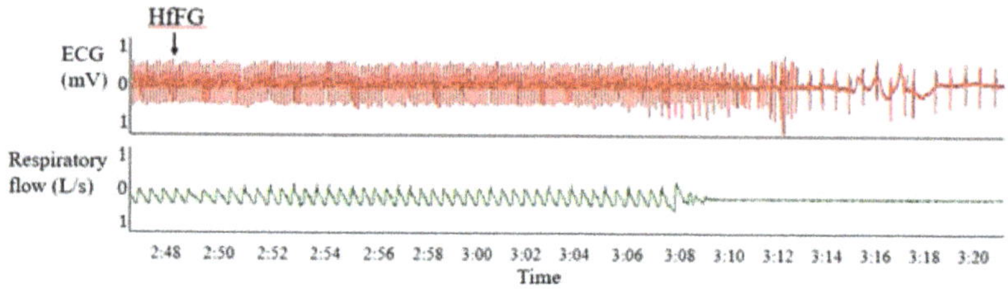

Figure 5. Effects of HfFG (i.v.) on rat ECG and the respiratory wave. Data were recorded by a multi-channel physiological signal-collecting instrument before and after treatment. HfFG was at 4.0 mg/kg and arrows indicate the administration of the test compound. Representative graph is shown and the experiment was repeated three times. Abbreviation: ECG, electrocardiogram.

2.2. Native FGs Activated the Plasma Contact System in Rats

Previously, in vitro assays showed that native FGs exhibit potent activity in plasma contact activation and platelet aggregation [11,12]. As expected, the injection of TaFG and HfFG significantly increased the kallikrein activity of rat plasma, and their activities were comparable to OSCS (Figure 6a). Compared with the control, dTaFG13 slightly enhanced the activation of the contact system in rat plasma, while UFH and dHG-5 caused no significant change. Once the contact system is activated, the reciprocal activation between prekallikrein and FXII produces more kallikrein [18]. Kallikrein can cleave high molecular weight kininogen (HMWK) to generate the vasoactive peptide BK, and mediate the generation of anaphylatoxins C3a and C5a [21–23]. These downstream products were detected, plasmas from rats treated with OSCS, TaFG, or HfFG had significantly increased levels of BK, C3a, and C5a, but not those treated with dTaFG13 and dHG-5 (Figure 6b–d).

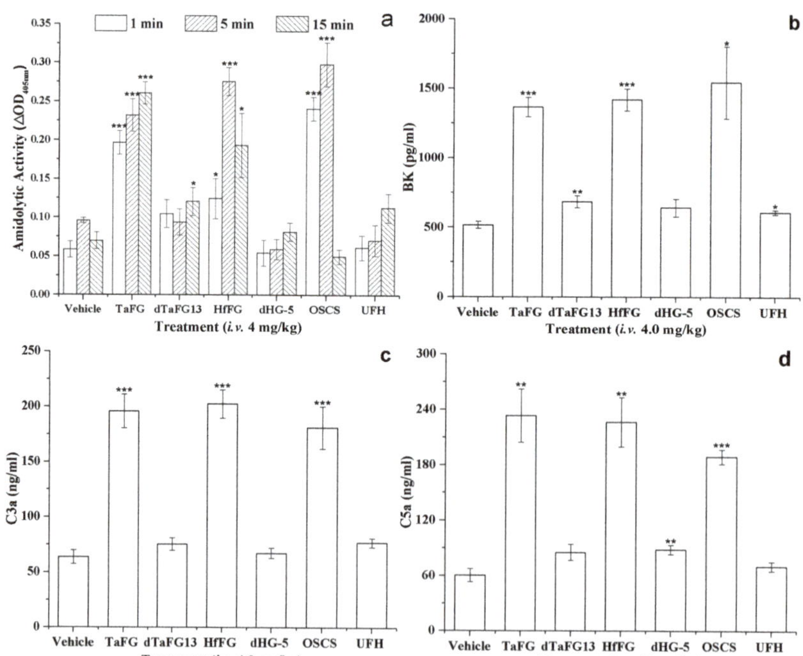

Figure 6. Amidolytic activity (**a**) and levels of BK (**b**), C3a (**c**), and C5a (**d**) in rat plasmas. Amidolytic activity was detected by the chromogenic substrate method and the concentrations of BK, C3a, and C5a were determined by ELISA kits. For each parameter, the value was compared with that of the control (vehicle) group. * $p \leq 0.05$, ** $p \leq 0.01$, and *** $p \leq 0.001$, two-tail *t*-test. Each sample was measured in duplicates and results were expressed as mean ± SEM, n = 3. Abbreviation: BK, bradykinin.

After production from the plasma contact activation–kallikrein/kinin pathway, BK can cause hypotension effect via kinin B2 receptors (B2R) on endothelial cells [24,25]. Similar to OSCS, native FGs are also negatively charged macromolecules that can potently activate the plasma contact system, leading to the production of vasoactive factor BK. We next investigated the involvement of BK in the toxic effect of native FGs using the B2R antagonist HOE140.

2.3. Effects of Inhibitors on the Lethality of Native FGs

HOE140 at 10 μg/kg could completely offset the hypotension induced by 4.0 mg/kg of OSCS or dTaFG13 in rats. However, the hypotension effect of native FGs could not be diminished or abolished by HOE140 (Figure S2). Even rats pretreated with HOE140 at a dose of up to 200 μg/kg showed severe hypotension after TaFG injection. This indicates that the hypotensive effect of native FGs was not mediated by BK, and that their toxic mechanism is different from OSCS. Thus, alternative mechanisms were explored.

Pre-treatment of antiallergic drugs, cholinergic antagonists, and antiasthmatic drugs also cannot prevent the effects of native FGs, whereas we found that pre-treatment of an anticoagulant (heparin or bivalirudin) or antiplatelet (cangrelor) agent can partially mitigate the lethal effect of HfFG (Table 2). In addition, combined the pre-treatment of bivalirudin and cangrelor could completely offset the lethal effect of HfFG. The results indicate that the acute toxic effect of native FGs is mainly caused by thrombosis-mediated embolism, and that their effects of anaphylactoid reaction and cardiovascular failure were secondary pathological changes.

Table 2. Effects of inhibitors on rat mortality induced by FGs (i.v.).

Treatment	Pre-Treatment	Mortality/Total [a]	Average Death Time Post-Dose
HfFG (4.0 mg/kg)	Saline (1 mL/kg)	4/4	5 min
	Heparin (5.0 mg/kg)	1/3	-
	Bivalirudin (2.5 mg/kg)	3/3	15 min
	Cangrelor (0.5 mg/kg)	1/3	-
	Bivalirudin (2.5 mg/kg), Cangrelor (0.5 mg/kg)	0/5	-

[a] The ratio of death numbers to total rats within 60 min after treatment.

Moreover, the direct effects of native FGs on the vessel vascular tension, rat isolated heart, human myocardial cells, and human Ether-a-go-go Related Gene (hERG) potassium channel were also investigated. The results showed that native FGs exhibited no vasodilative activity and had no direct effects on the functions of the rat isolated heart, myocardial cells (cell proliferation, beat rate, and contractility), and hERG channel (Figures S2–S6)

2.4. Effect of FG on Rat Coagulation Function and Platelets

Despite the fact that FG is a potent iXase inhibitor and anticoagulant, high-molecular-weight FGs can also activate FXII and platelets, which may contribute to thrombosis. Previous study showed that, although TaFG had stronger in vitro anticoagulant activity than its depolymerized product, it exhibited weaker in vivo antithrombotic activity in a rat thrombosis model. This may be due to its contradictory activities. Previously, the antithrombotic effect of TaFG was evaluated after subcutaneous injection for 1 h, while the immediate effect of FG after intravenously injection has not been reported.

The anti-iXase IC_{50} values of HfFG and its depolymerized product dHG-5 were 42 ng/mL and 70 ng/mL, respectively [10]. In this work, HfFG and dHG-5 were injected intravenously into rats at 4.0 mg/kg and 6.7 mg/kg, respectively, which were equivalent doses for anti-iXase. Then, the coagulation function was analyzed.

After injection, TaFG and HfFG at 4.0 mg/kg significantly prolonged rat plasma APTT due to their anti-iXase activity, while shortening PT (Table 3). At 15 min, TaFG-treated rats not only had the largest APTT but also had prolonged PT. This indicates that native FG exhibited procoagulant activity in vivo immediately after injection, while coagulopathy occurred later due to the coagulation factor consumption. By comparison, dHG-5 at 6.7 mg/kg prolonged rat plasma APTT and had no effect on PT. According to APTT, TaFG showed stronger anticoagulant activity than HfFG. We also found that TaFG only slightly increased the D-dimer, while HfFG significantly increased the D-dimer level in rat plasmas (Table S2). Combined with the finding of suspected pulmonary embolism in the autopsy of rats treated with native FGs, these results may explain the finding that HfFG (6/6) caused more death than TaFG (2/8).

Table 3. Effects of FGs (i.v.) on the rat coagulation function (mean ± SEM).

Treatment	Dose (i.v.)	Animals	Mortality [a]	Blood Collection Time [b]	APTT (s) [c,d]	PT (s) [d]
TaFG	4.0 mg/kg	8	2	−1 min	15.5 ± 0.3	23.4 ± 0.6
				1 min	295.2 ± 3.8 ***	15.8 ± 0.7 ***
				5 min	208.0 ± 39.9 *	16.7 ± 0.4 ***
				15 min	300.0 ± 0.0 ***	36.5 ± 3.6 **
HfFG	4.0 mg/kg	6	6	−1 min	16.1 ± 0.6	25.1 ± 1.0
				1 min	227.9 ± 10.3 ***	16.2 ± 0.7 ***
				5 min	116.0 ± 6.2 ***	16.1 ± 1.1 ***
				15 min	-	-
dHG-5	6.7 mg/kg	5	0	−1 min	16.2 ± 0.7	26.9 ± 1.1
				1 min	196.8 ± 27.3 **	27.2 ± 1.2
				5 min	73.5 ± 7.7 **	27.4 ± 0.8
				15 min	42.4 ± 4.4 **	26.8 ± 1.3

[a] The death numbers within 15 min after drug treatment. [b] Animals were treated with FG at 0 min (−1 min was before treatment). [c] The APTT beyond detection limit (>300 s) was recordedas 300 s. [d] Compared with the control (before treatment): * $p < 0.05$, ** $p < 0.01$, and *** $p < 0.001$ (two-tail t-test).

After injection, TaFG and HfFG at 4.0 mg/kg dramatically decreased the platelet count by 84.6% and 86.4%, respectively, while dHG-5 at 6.7 mg/kg had no such effect (Table 4). The platelet reduction in blood may be due to the platelet aggregation induced by native FGs, which is consistent with their in vitro activity of platelet activation [5].

Table 4. Effects of FGs (i.v.) on the rat platelet count (mean ± SEM, n = 3).

Treatment	Dose	Platelet Count (10^9/L)		Platelet Count Reduction (%)
		Pretreatment	1 min Post-Dose	
Control	-	363 ± 19	337 ± 8	7.1
TaFG	4.0 mg/kg	331 ± 25	51 ± 9 ***	84.6
HfFG	4.0 mg/kg	250 ± 32	34 ± 8 **	86.4
dHG-5	6.7 mg/kg	327 ± 37	354 ± 28	−8.3

** $p < 0.01$, and *** $p < 0.001$ vs. pretreatment (two-tail t-test).

2.5. Native FGs Induced Rat Pulmonary Embolism

We then further studied the histopathology of rat essential organs including the lung and heart after FG injection. Consistent with their in vivo procoagulant effect, TaFG and HfFG at 4.0 mg/kg caused microvessel embolism and fibrin deposition in rat lungs, while dHG-5 had no effect on lung histology (Figure 7). No microthrombus was found in the heart pathological section in all groups (Table S3). Taken together, the lethal effect of native FG may be mediated by pulmonary embolism due to their activities in the contact system and platelet activation, as confirmed by the remedy effect of pretreated antithrombotic drugs (anticoagulant bivalirudin combined with antiplatelet cangrelor). In addition, after injection of TaFG or HfFG, rats had apnea or respiratory failure prior to cardiovascular dysfunction.

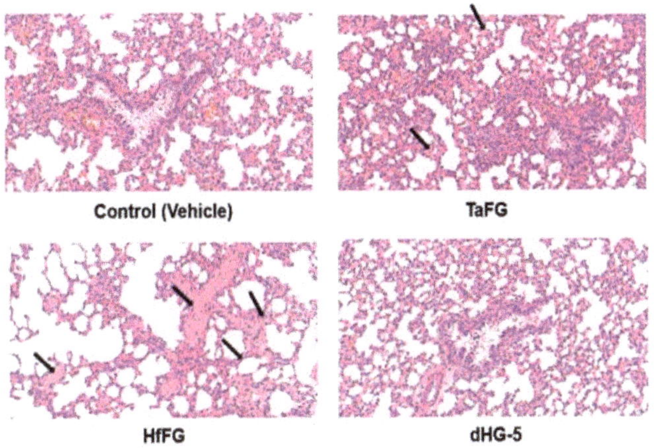

Figure 7. Rat pulmonary microvessel embolism induced by native FGs. Rat were i.v. injected with TaFG (4.0 mg/kg), HfFG (4.0 mg/kg), dHG-5 (6.7 mg/kg), or the vehicle and after 5 min the lung was collected for histopathological assays. Experiments were conducted in triplicates and represented results are shown.

Apart from the activity of procoagulation and inducing the platelet aggregation, the increased production of C3a and C5a by TaFG or HfFG may also contribute to the accumulation and degradation of platelets in the lung, as the involvement of the complement system in lipopolysaccharide-induced anaphylaxis-like shock has been reported [23,26–28].

3. Conclusions

Previous study reported that FGs from sea cucumber have various bioactivities. However, the results of our work indicate that native FGs (TaFG and HfFG) injected intravenously showed strong and acute toxicity in rats. The toxic effect and practical significance of native FGs should be noted when conducting pharmacological studies, especially those involving systemic treatment.

After the injection into rats, the acute toxic effects of TaFG and HfFG involved a dramatic reduction of blood pressure, cardiac dysfunction, respiratory failure, coagulopathy, and decreased platelet count. Although TaFG and HfFG were similar to OSCS in activating the plasma contact system, they showed much more severe toxic effects and the underlying toxic mechanisms were different from OSCS. Pre-treatment of a BK receptor antagonist completely offset the hypotension induced by OSCS, while it had no effect on that induced by TaFG and HfFG. By applying the prophylactic drugs, we found that anticoagulant combined with an antiplatelet can completely prevent the lethality of TaFG and HfFG, while other inhibitors had no obvious preventive effects. This indicates that thrombosis is an initial key factor in the lethal toxicity of these native FGs. Previously, in vitro assays have shown that native FGs can activate platelets and trigger contact activation–coagulation pathway. In this work, these effects were also detected in the rat blood or plasma after treatment. Additionally, TaFG and HfFG caused disseminated pulmonary embolism as revealed by the histopathological analysis.

In summary, the main mechanism of the acute toxicity of native FG was ascribed to the pro-coagulation and platelet activation, which further led to pulmonary embolism and respiratory failure. The side effect of FG depends on the high molecular weight but not the sulfation pattern. After chemical depolymerization, dTaFG13 (13 kDa) and dHG-5 (5 kDa) had no obvious effects on all the tested physiology parameters, which is consistent with their lack of platelet and contact activation activities in vitro. Therefore, depolymerized FG can be safe and effective anticoagulant candidate by selectively targeting the intrinsic coagulation pathway.

4. Materials and Methods

4.1. Drugs and Chemicals

The preparation and physicochemical characteristics of FGs with various molecular weight were described in previous studies [6,7,29]. The compound TaFG and HfFG were prepared from *T. ananas* and *H. fuscopunctata*, respectively. Depolymerized dTaFG13 (13 kD) and dHG-5 (5 kD) were prepared from TaFG and HfFG, respectively. OSCS and heparin (UFH, 197 IU/mg) were purchased from the National Institute for the Control of Pharmaceutical and Biological Products (Beijing, China). CS31(02) (D-Pro-Phe-Arg-pNa, Kallikrein chromogenic substrate) was obtained from Hyphen BioMed (Neuville-sur-Oise, France). HOE140 (B2R antagonist) and the protease inhibitor cocktail were obtained from Sigma-Aldrich. Bivalirudin and cangrelor were obtained from Targetmol. The Rat BK, C3a, and C5a ELISA kits were obtained from Shanghai Guyan Real Co., Ltd. (Shanghai, China). The activated partial thromboplastin time (APTT) and prothrombin time (PT) were obtained from TECO GmbH (Neufahrn, Germany). The Rat D-Dimer (D2D) ELISA Kit was obtained from SAB (Baltimore, MD, USA). All other chemicals were of reagent grade and were obtained commercially.

4.2. Animals and Biological Samples

Male Sprague-Dawley rats (body weight of 250~350 g) were obtained from Hunan SJA Laboratory Animal Co. Ltd. (Changsha, China). Rats were housed in animal rooms with controlled temperatures (18–25 °C) and relative humidity (40~60%), and were allowed to eat and drink ad libitum before experiments. These experiments were reviewed and approved by the Animal Ethics Committee of the Kunming Institute of Botany, Chinese Academy of Sciences (SYXK (Dian) K2018-0005).

Rat blood were collected using a polyethylene catheter from the jugular vein before and after 1, 5, and 15 min following the treatment with FG, OSCS, UFH, or the vehicle (normal saline). Nine volumes of blood was anticoagulated with one volume of 1% EDTA (for the blood used for the clotting assays, 3.8% sodium citrate was used instead). Then, rat plasmas were obtained after centrifugation (1000× *g*, 15 min). Plasma samples were stored at −80 °C or assayed immediately. For the plasma using for BK detection, 10% Protease Inhibitor Cocktail was added.

4.3. Rat Plasma Contact Activation Analysis

Rat blood (0.3 mL) was collected before and after 1 min, 5 min, and 15 min of each treatment. Rat plasma samples were obtained as described in Section 4.2. After dilution with 4 volumes of TS buffer (50 M Tris-HCl, 150 M NaCl, pH 7.4), 100 μL of rat plasma was added into a microwell plate and incubated at 37.0 °C for 60 s. Then, the amidolytic activity of the plasma was assessed by adding 30 μL of 1 mM CS31(02) (chromogenic substrate of kellikrein). The optical density at 405 nM (OD_{405nm}) was recorded at 37.0 °C for 20 min using the Bio-Tek Microplate Reader (ELx 808, Winooski, VT, USA) and change of OD_{405nm} was calculated and expressed as ΔOD_{405nm}.

4.4. Rat Plasma Levels of BK, C3a, and C5a Determination

The concentration of BK, C3a, and C5a in plasma samples collected from rats after 1 min of treatment with FG, OSCS, or normal saline were assayed by means of ELISA (enzyme-linked immunosorbent assay), as specified in the manufacturer's instructions.

4.5. Rat Blood Pressure and Cardiac Function Detection

After rats were anesthetized and fixed, the left jugular vein and right common carotid artery were separated from the tissue and intubated. The inserted catheters were filled with heparinized (125 U/mL) normal saline, the free end of the vein catheter was connected to an injection syringe for drug injection/blood collection, and the artery catheter was connected to a pressure amplifier for blood/ventricular pressure detection. In some experiments, rat electrocardiograms (ECG) were recorded simultaneously by electrodes. Data were acquired by a multi-channel physiological signal-collecting instrument (RM6240, Chengdu, China) with a continuous recording pattern. After a stable period, the compound (1 mL/kg) was intravenously injected.

The systolic blood pressure (SBP), diastolic blood pressure (SDP), and the left ventricular pressure (LVP), were recorded for 10 min before and 30 min after treatment, and the heart rate (HR), mean arterial pressure (AP), pulse pressure (SBP-DBP), and maximal rates of rise of ventricular pressure ($+dP/dt_{max}$) were calculated. The change of AP (ΔAP), HR (ΔHR), or LVP (ΔLVP) were calculated by subtracting the average value of the peak effect after treatment from the average value before treatment (if there was no obvious effect, the value of around 1 min post-dose was used). The change rate, such as ΔAP (%), was calculated as: $\Delta AP/AP(\text{pre-dose}) * 100\%$.

4.6. Rat Respiratory Function Detection

After rats were anesthetized and fixed, the trachea was exposed and intubated. The free end of the catheter was connected to a respiratory flow head which linked to PowerLab (AD instruments, New South Wales, Australia). The respiratory wave was recorded 3 min before and 30 min after treatment. Rat electrocardiograms (ECG) were recorded simultaneously by electrodes connecting to PowerLab.

4.7. Effect of the B2R Antagonist on Rat Hypotension

Anesthetized rats were intravenously injected with 40 μg/kg of HOE140, a highly specific B2R antagonist, 3 min prior to the treatment with FGs or OSCS at 4.0 mg/kg. The record of the rat blood pressure and heart rate were as described in Section 4.5. The dosage of HOE140 was in reference to the literature which stated that 10 μg/ml of HOE140 could fully inhibit the hypotensive effect of 30 μg/kg of BK [16] and validated our pre-experiment.

4.8. Effects of Antithrombotic Drugs on the Toxicity of Native FG

Anesthetized rats were intravenously injected with 1 mL/kg of heparin, bivalirudin, cangrelor bivalirudin combined with cangrelor, or normal saline (control). After 3 min, rats were injected with 4.0 mg/kg of HfFG. The responses of rats after treatment were observed for 1 h and the death time within 1 h was recorded.

4.9. Rat Coagulation Function and Platelet Count Analysis

Rats were anesthetized and the left jugular vein was separated and intubated for drug injection/blood collection. The inserted catheter was filled with normal saline. Rat blood was collected before and after FG treatment, and plasma was obtained as described in Section 4.2. The plasma APTT and PT were detected by a coagulometer (TECO MC-4000, Neufahrn, Germany) using APTT and PT reagents. The plasma D-dimer level was assayed by means of ELISA as specified in the manufacturer's instructions.

4.10. Rat Histopathologic Analysis after Treatment

Anesthetized rats were intravenously injected with FG or saline and after 5 min of treatment, the lung and heart were collected and immersed in 4% paraformaldehyde. After 24 h, the tissue samples were embedded in paraffin, cut into 4 μm thin sections, and stained with hematoxylin-eosine. The samples were examined using the Nikon Eclipse Ci-L microscope and analyzed by 3DHISTECH software (Budapest, Hungary).

4.11. Statistical Analysis

All data are given as the mean ± SEM. The data were analyzed using Student's *t*-test (two-tail) compared with the control group or pretreated values. P values less than 0.05 were considered statistically significant (* $p < 0.05$, ** $p < 0.01$, or *** $p < 0.001$).

Supplementary Materials: The following are available online at https://www.mdpi.com/article/10.3390/md19090487/s1, Table S1: Effects of FGs on rat blood pressure (mean ± SEM); Table S2: Effects of FGs on rat plasma D-dimer (mean ± SEM, n = 3); Table S3: Pathological scores of rat heart and lung after FGs (i.v.) treatment; Figure S1: Effects of FGs on rat left ventricular pressure and the heart rate; Figure S2: Effects of FGs on rat AP and HR in rats with or without HOE140 pre-treatment; Figure S3: Effects of TaFG and HfFG on the pre-contracting rat aortic artery; Figure S4: Effects of HfFG on the function of the rat isolated heart; Figure S5: Effects of native FGs on hERG k^+ current; and Figure S6: Effects of HfFG on the function of human myocardial cells.

Author Contributions: Conceptualization, L.L. and J.Z.; data curation, X.Y.; investigation, L.L.; methodology, L.L., L.Y., and D.L.; resources, N.G., X.J., and J.Z.; software, S.L.; supervision, J.Z.; validation, W.W. and T.Z.; writing—original draft preparation, L.L.; writing—review and editing, L.L. and J.Z. All authors have read and agreed to the published version of the manuscript.

Funding: This work was supported by the National Natural Science Foundation of China [81773737, 82073725, and 81773374], and Yunnan Provincial Science and Technology Department in China [2019ZF011-2].

Institutional Review Board Statement: Animal experiments were approved by the Animal Ethics Committee of the Kunming Institute of Botany, Chinese Academy of Sciences (SYXK (Dian), K2018-0005).

Data Availability Statement: All data is contained within this article and the Supplementary Materials.

Conflicts of Interest: The authors declare no conflict of interest.

References

1. Pomin, V.H. Holothurian Fucosylated Chondroitin Sulfate. *Mar. Drugs* **2014**, *12*, 232–254. [CrossRef]
2. Buyue, Y.; Sheehan, J.P. Fucosylated chondroitin sulfate inhibits plasma thrombin generation via targeting of the factor IXa heparin-binding exosite. *Blood* **2009**, *114*, 3092–3100. [CrossRef]
3. Fonseca, R.J.C.; Santos, G.R.C.; Mourão, P.A.S. Effects of polysaccharides enriched in 2,4-disulfated fucose units on coagulation, thrombosis and bleeding. *Thromb. Haemost.* **2009**, *102*, 829–836. [CrossRef] [PubMed]
4. Sheehan, J.P.; Walke, E.N. Depolymerized holothurian glycosaminoglycan and heparin inhibit the intrinsic tenase complex by a common antithrombin-independent mechanism. *Blood* **2006**, *107*, 3876–3882. [CrossRef]
5. Lin, L.; Zhao, L.; Gao, N.; Yin, R.; Li, S.; Sun, H.; Zhou, L.; Zhao, G.; Purcell, S.W.; Zhao, J. From multi-target anticoagulants to DOACs, and intrinsic coagulation factor inhibitors. *Blood Rev.* **2020**, *39*, 100615. [CrossRef]
6. Gao, N.; Lu, F.; Xiao, C.; Yang, L.; Chen, J.; Zhou, K.; Wen, D.; Li, Z.; Wu, M.; Jiang, J.; et al. β-Eliminative depolymerization of the fucosylated chondroitin sulfate and anticoagulant activities of resulting fragments. *Carbohydr. Polym.* **2015**, *127*, 427–437. [CrossRef]
7. Wu, M.; Wen, D.; Gao, N.; Xiao, C.; Yang, L.; Xu, L.; Lian, W.; Peng, W.; Jiang, J.; Zhao, J. Anticoagulant and antithrombotic evaluation of native fucosylated chondroitin sulfates and their derivatives as selective inhibitors of intrinsic factor Xase. *Eur. J. Med. Chem.* **2015**, *92*, 257–269. [CrossRef] [PubMed]
8. Liu, S.; Zhang, T.; Sun, H.; Lin, L.; Gao, N.; Wang, W.; Li, S.; Zhao, J. Pharmacokinetics and Pharmacodynamics of a Depolymerized Glycosaminoglycan from Holothuria fuscopunctata, a Novel Anticoagulant Candidate, in Rats by Bioanalytical Methods. *Mar. Drugs* **2021**, *19*, 212. [CrossRef]
9. Sun, H.; Gao, N.; Ren, L.; Liu, S.; Lin, L.; Zheng, W.; Zhou, L.; Yin, R.; Zhao, J. The components and activities analysis of a novel anticoagulant candidate dHG-5. *Eur. J. Med. Chem.* **2020**, *207*, 112796. [CrossRef]
10. Zhou, L.; Gao, N.; Sun, H.; Xiao, C.; Yang, L.; Lin, L.; Yin, R.; Li, Z.; Zhang, H.; Ji, X.; et al. Effects of Native Fucosylated Glycosaminoglycan, Its Depolymerized Derivatives on Intrinsic Factor Xase, Coagulation, Thrombosis, and Hemorrhagic Risk. *Thromb. Haemost.* **2020**, *120*, 607–619. [CrossRef]
11. Lin, L.; Xu, L.; Xiao, C.; Zhou, L.; Gao, N.; Wu, M.; Zhao, J. Plasma contact activation by a fucosylated chondroitin sulfate and its structure–activity relationship study. *Glycobiology* **2018**, *28*, 754–764. [CrossRef]
12. Lin, L.; Yang, L.; Chen, J.; Zhou, L.; Li, S.; Gao, N.; Zhao, J. High-molecular-weight fucosylated glycosaminoglycan induces human platelet aggregation depending on alphaIIbbeta3 and platelet secretion. *Platelets* **2020**, *24*, 1–9. [CrossRef]
13. Fonseca, R.J.C.; Oliveira, S.-N.M.C.G.; Pomin, V.H.; Mecawi, A.; Araujo, I.G.; Mourão, P.A.S. Effects of oversulfated and fucosylated chondroitin sulfates on coagulation. *Thromb. Haemost.* **2010**, *103*, 994–1004. [CrossRef]
14. Yan, L.; Wang, D.; Yu, Y.; Zhang, F.; Ye, X.; Linhardt, R.J.; Chen, S. Fucosylated Chondroitin Sulfate 9–18 Oligomers Exhibit Molecular Size-Independent Antithrombotic Activity while Circulating in the Blood. *ACS Chem. Biol.* **2020**, *15*, 2232–2246. [CrossRef] [PubMed]
15. Adam, A.; Montpas, N.; Keire, D.; Désormeaux, A.; Brown, N.J.; Marceau, F.; Westenberger, B. Bradykinin forming capacity of oversulfated chondroitin sulfate contaminated heparin in vitro. *Biomaterials* **2010**, *31*, 5741–5748. [CrossRef] [PubMed]

16. Corbier, A.; Le Berre, N.; Rampe, D.; Meng, H.; Lorenz, M.; Vicat, P.; Potdevin, S.; Doubovetzky, M. Oversulfated Chondroitin Sulfate and OSCS-Contaminated Heparin Cause Dose- and Route-Dependent Hemodynamic Effects in the Rat. *Toxicol. Sci.* **2011**, *121*, 417–427. [CrossRef]
17. Kishimoto, T.K.; Viswanathan, K.; Ganguly, T.; Elankumaran, S.; Smith, S.; Pelzer, K.; Lansing, J.; Sriranganathan, N.; Zhao, G.; Galcheva-Gargova, Z.; et al. Contaminated Heparin Associated with Adverse Clinical Events and Activation of the Contact System. *N. Engl. J. Med.* **2008**, *358*, 2457–2467. [CrossRef]
18. Long, A.T.; Kenne, E.; Jung, R.; Fuchs, T.A.; Renné, T. Contact system revisited: An interface between inflammation, coagulation, and innate immunity. *J. Thromb. Haemost.* **2016**, *14*, 427–437. [CrossRef]
19. Lin, L.; Wu, M.; Zhao, J. The initiation and effects of plasma contact activation: An overview. *Int. J. Hematol.* **2016**, *105*, 235–243. [CrossRef]
20. Wu, M.; Xu, S.; Zhao, J.; Kang, H.; Ding, H. Physicochemical characteristics and anticoagulant activities of low molecular weight fractions by free-radical depolymerization of a fucosylated chondroitin sulphate from sea cucumber Thelenata ananas. *Food Chem.* **2010**, *122*, 716–723. [CrossRef]
21. Discipio, R.G. The activation of the alternative pathway C3 convertase by human plasma kallikrein. *Immunology* **1982**, *45*, 587–595.
22. Wiggins, R.C.; Giclas, P.C.; Henson, P.M. Chemotactic activity generated from the fifth component of complement by plasma kallikrein of the rabbit. *J. Exp. Med.* **1981**, *153*, 1391–1404. [CrossRef]
23. Kawabata, Y.; Yang, S.; Yokochi, T.; Matsushita, M.; Fujita, T.; Shibazaki, M.; Noikura, T.; Endo, Y.; Takada, H. Complement system is involved in anaphylactoid reaction induced by lipopolysaccharides in muramyldipeptide-treated mice. *Shock* **2000**, *14*, 572–577. [CrossRef] [PubMed]
24. Siebeck, M.; Cheronis, J.C.; Fink, E.; Kohl, J.; Spies, B.; Spannagl, M.; Jochum, M.; Fritz, H. Dextran sulfate activates contact system and mediates arterial hypotension via B2 kinin receptors. *J. Appl. Physiol.* **1994**, *77*, 2675–2680. [CrossRef] [PubMed]
25. Yarovaya, G.A.; Neshkova, A.E. Past and present research on the kallikrein-kinin system (on the 90th anniversary of the discovery of the system). *Russ. J. Bioorg. Chem.* **2015**, *41*, 245–259. [CrossRef]
26. Yu, Z.; Saito, H.; Otsuka, H.; Shikama, Y.; Funayama, H.; Sakai, M.; Murai, S.; Nakamura, M.; Yokochi, T.; Takada, H.; et al. Pulmonary platelet accumulation induced by catecholamines: Its involvement in lipopolysaccharide-induced anaphylaxis-like shock. *Int. Immunopharmacol.* **2016**, *43*, 40–52. [CrossRef]
27. Shibazaki, M.; Kawabata, Y.; Yokochi, T.; Nishida, A.; Takada, H.; Endo, Y. Complement-Dependent Accumulation and Degradation of Platelets in the Lung and Liver Induced by Injection of Lipopolysaccharides. *Infect. Immun.* **1999**, *67*, 5186–5191. [CrossRef]
28. Zhao, L.; Ohtaki, Y.; Yamaguchi, K.; Matsushita, M.; Fujita, T.; Yokochi, T.; Takada, H.; Endo, Y. LPS-induced platelet response and rapid shock in mice: Contribution of O-antigen region of LPS and involvement of the lectin pathway of the complement system. *Blood* **2002**, *100*, 3233–3239. [CrossRef] [PubMed]
29. Pan, J.; Qian, Y.; Zhou, X.; Lu, H.; Ramacciotti, E.; Zhang, L. Chemically Oversulfated Glycosaminoglycans Are Potent Modulators of Contact System Activation and Different Cell Signaling Pathways. *J. Biol. Chem.* **2010**, *285*, 22966–22975. [CrossRef]

marine drugs

MDPI

Review

Antiparasitic Effects of Sulfated Polysaccharides from Marine Hydrobionts

Natalya N. Besednova [1,*], Tatyana S. Zaporozhets [1], Boris G. Andryukov [1,2], Sergey P. Kryzhanovsky [3], Svetlana P. Ermakova [4], Tatyana A. Kuznetsova [1], Anastasia N. Voronova [1] and Mikhail Y. Shchelkanov [1,2,5,6]

1 G.P. Somov Research Institute of Epidemiology and Microbiology, Federal Service for Surveillance on Consumer Rights Protection and Human Wellbeing, 690087 Vladivostok, Russia; niiem_vl@mail.ru (T.S.Z.); andrukov_bg@mail.ru (B.G.A.); takuznets@mail.ru (T.A.K.); avoronova92@gmail.com (A.N.V.); adorob@mail.ru (M.Y.S.)
2 School of Biomedicine, Far Eastern Federal University (FEFU), 690091 Vladivostok, Russia
3 Medical Association of the Far Eastern Branch of the Russian Academy of Sciences, 690022 Vladivostok, Russia; priemmodvoran@mail.ru
4 G.B. Elyakov Pacific Institute of Bioorganic Chemistry, Far Eastern Branch of the Russian Academy of Sciences, 690022 Vladivostok, Russia; svetlana_ermakova@hotmail.com
5 National Scientific Center of Marine Biology, Far Eastern Branch of the Russian Academy of Sciences, 690041 Vladivostok, Russia
6 Federal Scientific Center of the East Asia Terrestrial Biodiversity, Far Eastern Branch of the Russian Academy of Sciences, 690022 Vladivostok, Russia
* Correspondence: besednoff_lev@mail.ru; Tel.: +7-(914)792-21-20

Abstract: This review presents materials characterizing sulfated polysaccharides (SPS) of marine hydrobionts (algae and invertebrates) as potential means for the prevention and treatment of protozoa and helminthiasis. The authors have summarized the literature on the pathogenetic targets of protozoa on the host cells and on the antiparasitic potential of polysaccharides from red, brown and green algae as well as certain marine invertebrates. Information about the mechanisms of action of these unique compounds in diseases caused by protozoa has also been summarized. SPS is distinguished by high antiparasitic activity, good solubility and an almost complete absence of toxicity. In the long term, this allows for the consideration of these compounds as effective and attractive candidates on which to base drugs, biologically active food additives and functional food products with antiparasitic activity.

Keywords: sulfated polysaccharides; marine hydrobionts; antiparasitic activity; protozoa; malaria; leishmaniasis; trypanosomiasis; schistosomiasis; cryptosporidiosis; trichomoniasis

Citation: Besednova, N.N.; Zaporozhets, T.S.; Andryukov, B.G.; Kryzhanovsky, S.P.; Ermakova, S.P.; Kuznetsova, T.A.; Voronova, A.N.; Shchelkanov, M.Y. Antiparasitic Effects of Sulfated Polysaccharides from Marine Hydrobionts. *Mar. Drugs* **2021**, *19*, 637. https://doi.org/10.3390/md19110637

Academic Editor: Yuki Fujii

Received: 22 October 2021
Accepted: 10 November 2021
Published: 12 November 2021

Publisher's Note: MDPI stays neutral with regard to jurisdictional claims in published maps and institutional affiliations.

1. Introduction

Protozoal infections (protozoans) caused by parasites belonging to unicellular protozoa remain among the most common human diseases. Protozoa cause severe and often fatal diseases in humans as well as domestic and game animals. About 50 species of protozoa are known to cause disease in humans [1].

Parasites are a very diverse group of organisms that have developed various strategies for infecting hosts and living off them, parasitizing in various organs and tissues including the blood, intestines, central nervous system, liver, lungs, etc. These pathogenic microorganisms can be transmitted to humans through the alimentary route, through arthropod vectors, and sexually. There are three main classes of human pathogenic parasites: protozoa, helminths, and ectoparasites.

Among protozoal infections, malaria [2], leishmaniasis [3,4], trypanosomiasis [5] and intestinal protozoa [6] are of the most significant medical and social importance. Many of the drugs currently used to treat protozoal infections do not meet modern requirements due to toxicity, gradual loss of effectiveness, and the emergence of drug resistance in parasites. Thus, there is an urgent need for new drugs.

Sources of new drugs for the treatment of protozoal infections can be natural products, synthetic molecules, or existing drugs with an extended spectrum of indications. A combination of new methods, including genetic modification of pathogens, bioimaging, molecular docking [7], and robotics has led to the standardization of high-throughput screening platforms for drug discovery [8]. For the treatment of protozoal infections, new effective agents are needed which do not cause at all or cause a minimum of side effects and to which protozoa will not develop resistance in a short time.

Extracts of many seaweeds have pronounced antiparasitic properties [9–12], and have therefore been studied on a wide range of human parasites, such as protozoa (*Plasmodium* spp., *Leishmania* spp., *Trichomonas* spp., etc.) [10,13,14] and helminths (*Neobenedenia* spp.) [15]. Algae extracts can inhibit the binding of parasites to target cells or have a direct toxic effect on protozoa. Thus, R. Moo-Puc et al. [13] showed direct antiprotozoal activity of organic extracts obtained from several algae against the trophozoites of *Trichomonas vaginalis*. At the same time, 44% of the studied seaweed varieties exhibited high or moderate antitrichomonal activity. M.S. Coutinho [16] reported a strong larvicidal effect of extracts of two algae, *Fucus vesiculosus* and *Ulva lactuca* (sea lettuce) against the larval stages of the *Aedes aegypti* mosquito.

However, the authors of this review do not set the goal of a detailed analysis of the results of studying the biological effectiveness of extracts of marine aquatic organisms, since these contain many biologically active substances with different chemical natures. Further in-depth research is required to determine the active component of each. In addition, the chemical composition of the extracts of aquatic organisms depends on the type of algae, the region of extraction, the method of extraction, extragents, and many other factors.

In this regard, sulfated polysaccharides (SPS)—sulfated homo- and heteropolysaccharides which are agonists of receptors of innate and adaptive immunity cells—are of great interest in this regard. They are characterized by low toxicity and the absence of side effects (except in cases of allergic-type reactions when using heparin with an admixture of highly sulfated polysaccharides [15,16]). These biopolymers have various biological actions including antiviral and antibacterial, anti-inflammatory, and immunomodulatory activity [17–19]. SPS encompasses a diverse group of anionic polymers found in various marine organisms from macroalgae to mammals, but have not been found in terrestrial plants [19].

It is known that the most common heteropolysaccharides in the human body are glycosaminoglycans (GAGs), negatively charged long unbranched polymeric polysaccharides consisting of repeating units of disaccharides [20]. The binding of glycosamines to various ligands leads to post-translational modifications that ensure cell migration, proliferation, differentiation, etc. Among GAGs, heparin/heparan sulfates are of particular interest. These substances are present in basement membranes, in the extracellular matrix, and on the surface of cells as part of membranes. They can specifically interact with macromolecules of the extracellular matrix (fibronectin, laminin), enzymes, and a wide class of heparin-binding molecules (growth factors, chemokines) [21]. GAG mimetics, including heparin/heparan sulfates, bind to other macromolecules and interfere with signal transduction pathways in cells, which provides a wide range of biological effects. The natural mimetics of heparan sulfates are the sulfated polysaccharides of seaweed. Fucoidans, carrageenans, and ulvans, isolated from brown, red and green algae, respectively, can mimic the action of endogenous factors and regulate the functions of macroorganism systems through key receptors of cells and enzymes. In this regard, SPS can bind to various receptors on the host cell's surface and compete with viruses, bacteria, and parasites for glycoprotein receptors [20,21].

An essential aspect of the action of SPS is their antiadhesive activity, which largely explains their antibacterial [22], antiviral [23], and as will be shown below, antiparasitic activity.

Fucoidans, SPS derived from brown algae, are the most commonly used in experiments. This is because fucoidans are an essential structural component of algae and account for up to 30% of their dry weight. Furthermore, the pharmacological safety of the polysaccharide is confirmed by absence of toxicity to humans even when consumed in amounts up to 3 g per day [24]. In addition, these biopolymers have a systemic effect, as evidenced by their detection in blood plasma [24].

There are many reports in the literature in which the results of experiments are presented indicating the effectiveness of using SPS from marine aquatic organisms (in most cases, marine macroalgae) as an antiprotozoal agent. In this regard, the purpose of this work is to generalize the available data on the interaction of these compounds with protozoan pathogens and assess the prospects for using them as a basis to create a new generation of antiparasitic drugs.

2. Malaria

Malaria is a life-threatening transmission parasitic disease that is transmitted to humans through the bites of infected female *Anopheles* mosquitoes [25]. This infection continues to be relevant for global health [26]; in 2019, 229 million people fell ill with malaria worldwide, and 409,000 people died. Most malaria cases and deaths, 94%, are reported in the African continent [27]. The most vulnerable contingent for this disease is children under five years of age [28].

The etiological agents of malaria are parasites of the genus *Plasmodium*, whose life cycle occurs in the body of two hosts: a human (intermediate host) and female *Anopheles* mosquitoes (final host). To date, more than 200 Plasmodium species have been officially described, each of which infects a certain range of hosts [29]. In humans, malaria is caused by five species of *Plasmodium*: *P. falciparum* (tropical malaria agent), *P. vivax* (the causative agent of three-day malaria), *P. malariae*, *P. ovale* (the causative agent of malaria oval), and *P. knowlesi*. The first four species are pathogenic to humans, while *P. knowlesi* causes infection in primates and causes zoonotic malaria in Southeast Asia [29]. The most dangerous are *P. falciparum* (mainly distributed in Africa) and *P. vivax* (distributed in the American continent).

All *Plasmodium* species possess high genetic plasticity, allowing parasites to adapt to changes in environmental parameters and rapidly develop resistance to antimalarial therapeutic agents [29]. When an infected female mosquito bites a person, malaria plasmodia at the sporozoite stage enter the bloodstream with insect saliva [26]. Within 10–30 min, they move freely with blood flow and then settle in the liver cells. Active tissue schizonts divide after 6–15 days with the formation of many tissue merozoites, which enter the blood after 1–6 weeks when the infected liver cells are destroyed. The merozoites then penetrate into the erythrocytes, after which the second cycle of erythrocytic schizogony begins.

Penetration of merozoites into erythrocytes is carried out in several stages: (a) initial contact and slight deformation of erythrocytes with the participation of surface antigens of merozoites; (b) severe deformation of erythrocytes with the participation of microneme proteins, rhoptry proteins and the actin–myosin motor of the parasite; (c) opening of pores between the parasite and erythrocytes; (d) internalization [30].

The attachment of merozoites to the membrane of erythrocytes and their invagination into membranes occurs due to the presence of specific receptors on their surface. In the erythrocyte, the merozoites go through four stages of development: i—stage of the ring (trophozoite); ii—stage of amoeboid schizont; iii—stage of morula (fragmentation); iv—formation of erythrocytic merozoites. The erythrocyte membrane is destroyed, and merozoites and their metabolic products are released into the plasma. At this time, the patient begins to suffer an attack of malaria (fever, anemia, and splenomegaly and, in tropical malaria, organ damage). Some of the blood merozoites again penetrate into the erythrocytes and repeat the entire cycle of erythrocytic schizogony. The duration of the erythrocytic schizogony phase is 72 h in *P. malariae* and 48 h in other plasmodia species [31].

Since erythrocytes cannot recycle the parasite and present antigens, the early stages of the erythrocyte ring remain invisible to the immune system until the later stages, trophozoites and schizonts, which modify the erythrocyte membrane in accordance with their needs for membrane transport [32,33]. Thus, the erythrocyte (clinical) stage of schizogony begins [34].

Some of the merozoites turns into gametocytes (immature germ cells). This process is called gametocytogeny. *P. falciparum* gametocytes develop in the deeply located vessels of internal organs. After maturation over a period of 12 days they appear in the peripheral blood, where they can remain viable anywhere from several days to up to 6 weeks. Gametocytes of other plasmodium species develop in peripheral vessels within 2–3 days and, after maturation, die after a few hours.

The most well-characterized is the surface protein of the parasite merozoite-1 (MSP-1), which is of great importance for the successful invasion of erythrocytes [35,36]. It is the largest, and covers the surface of the parasite. After binding to the erythrocyte, it reorients itself so that the apical end of the parasite is aligned with the erythrocyte membrane. This reorientation involves the apical membrane antigen, a transmembrane protein localized at the apical end of the merozoite and binding erythrocytes. The peripheral surface proteins of merozoites, MSP3, MSP6, MSPDBL1, MSPDBL2, and MSP7, bind directly to MSP1 independently of each other to form multiple forms of the MSP1 complex on the parasite's surface. These complexes perform overlapping functions and directly interact with human erythrocytes [35].

Apicomplexan-type parasites, including *Plasmodium* spp., contain three unusual secretory organelles (microneme, rhoptries, and dense granules) required to infect new host cells [37]. Micronemes are located in the apical third of the plasmodium. These organelles secrete several proteins, including the apical membrane protein-1 of *P. falciparum* (PfAMA-1) and proteins of the EBA (erythrocyte binding antigen) family, which bind to receptors on the surface of red blood cells and facilitate the entry of plasmodium into them. It has been suggested that micronemes act in concert with roptria. While micronemes initiate erythrocyte binding, rhoptries secrete proteins to create a PVM, or parasitophorous vacuole membrane, in which the parasite can survive and reproduce [38,39].

Erythrocytes infected with mature forms of plasmodia bind to endothelial cells in capillaries by sequestration, which allows the parasite to multiply and evade destruction in the spleen [40]. Infected erythrocytes can adhere to uninfected ones, forming rosettes that can form clots by binding to other red blood cells mediated by platelets, which can lead to occlusion of the circulatory bed [41].

A significant problem is the ability of the parasite to develop resistance to antimalarial drugs such as chloroquine and sulfadoxine–pyrimethamine. Resistance also develops to artemisinin, part of the malaria elimination program [42,43]. This alarming situation requires the search for and development of new antimalarial agents with an original mechanism of action.

In developing antimalarial drugs, several approaches can be used, ranging from modifying existing agents to developing new agents that act against new targets [44]. Currently, most licensed antimalarial drugs target the intra-erythrocytic stage of the parasite. However, the invasion of merozoites—a pivotal moment in the development of parasites—has been shown by the studies of many authors to have great potential for affecting the surface proteins of plasmodia. Therefore, it is possible to use agents that inhibit the invasion of merozoites in combination with drugs aimed at destroying intraerythrocytic parasites [42,43,45–47].

The invasion of merozoites involves multiple ligand–receptor interactions and multiple pathways of parasitic invasion in host cells [46], which can be blocked by SPS and heparin-like molecules. The surface antigens of the parasite play a vital role in the process of *Plasmodium* invasion.

Studies have shown that SPS, in particular heparin (acidic sulfur-containing GAG) and heparin-like polysaccharides, have an antimalarial effect [47]. Heparin inhibits cytoadhesion and thus reduces parasitemia. On the apical surface of merozoites, various molecules are expressed that interact with receptors on the surface of erythrocytes. Heparin targets a large number of molecules on the parasite's surface that mediate its invasion of red blood cells. In this regard, the authors believe that the risk of developing heparin resistance is very low because, apparently, many molecules at once will not be able to quickly acquire resistance to the polysaccharide. [48].

Heparin binds to proteins of the DBL and RBL families of the parasite; in its presence the interaction between the apical end of merozoites and the surface of erythrocytes is inhibited, and invasion does not occur. In this case, localization is limited only by the apical surface of the parasite, and the polysaccharide has only a slight effect on erythrocytes. A. Leitgeb et al. found that heparin equally inhibits both sialic acid-dependent and sialic acid-independent invasion of *P. falciparum* merozoites [49]. In addition, in the presence of heparin, deformation of the erythrocyte membrane is suppressed. However, heparin was discontinued due to the induction of severe bleeding in patients [45,49].

As an alternative, sevuparin was proposed, a heparin derivative with low anticoagulant activity and high anti-adhesive properties, which destroyed the rosettes (50% at a dose of 250 μg/mL). Sevuparin inhibits the cytoadhesion of plasmodia to epithelial cells, suppresses the invasion of merozoites into erythrocytes, and removes blockage of microcirculation, which alleviates life-threatening symptoms in children [50,51].

Studies carried out on the model of heparin and merozoites of *P. falciparum* have provided an understanding of the mechanism by which other SPS and similar molecules exert antimalarial effects.

SPS from marine plants and invertebrates, which have molecular structures similar to heparin, also exhibited an anticoagulant effect [52,53]. These compounds are an exciting and promising alternative to heparin for antimalarial therapy and contain structures similar to pRBC-binding GAGs [54,55]. For quite a long time [56], the inhibitory effect of fucoidan (SPS of brown alga) on the formation of erythrocyte rosettes by plasmodia has been known; this activity extends to both wild and laboratory strains of the parasite. It has been shown that fucoidans inhibit the invasion of sporozoites into hepatocytes as well as the reinvasion of merozoites into erythrocytes (Figure 1).

Finally, in 2009, Chen et al. [57] presented the results of a study on the antimalarial effect of three fractions of fucoidan from the brown alga *Undaria pinnatifida*.

Experiments were carried out in vitro on *P. falciparum* culture and in vivo on mice infected with *P. berghei*. It was shown that all three fractions of the polysaccharide at a dose of 5 μg/mL were able to significantly inhibit the invasion of merozoites into erythrocytes, and the IC_{50} was similar for both chloroquine-sensitive *P. falciparum* 3D7 strain and K1 strain resistant to this drug. The most active was the fraction with the highest sulfate content. The survival rate of mice in the group of animals infected with *P. berghei* and receiving fucoidan (100 mg/kg) was $37.12 \pm 0.37\%$, while among mice of the control group receiving chloroquine it was (5 mg/kg)—$94.37 \pm 0.94\%$. The protective effect of fucoidan was significantly lower in this case, but compared with the intact control (mice without treatment), the effect did take place. The lifespan of the infected mice was about four days longer than that of the intact animals.

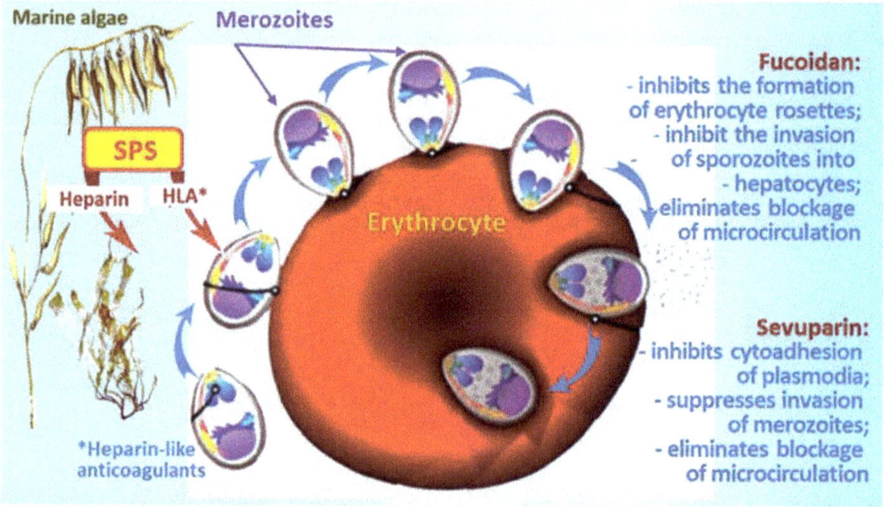

Figure 1. SPS from marine hydrobionts, in particular heparin, heparin-like polysaccharides, and fucoidans as well as the drug sevuparin developed on the same basis, all have antimalarial effects.

Thus, it was shown that fucoidan from *U. pinnatifida* suppresses the invasion of *P. falciparum* merozoites into erythrocytes both in vitro and in vivo, inhibiting the adhesion interaction between merozoites and erythrocytes through the binding of fucoidan sulfate to ligands on the cell surface of host cells, thereby inhibiting parasites. The authors recommended using a dose of fucoidan no higher than 100 mg/kg, as higher doses may have an anticoagulant effect [57]. To search for antimalarial compounds, red algae were also studied.

J. Marques et al. [53] presented the antimalarial and anticoagulant properties of sulfated galactan from the red alga *Botryocladia occidentalis*. The antimalarial activity of this compound is similar to that of heparin. In in vitro cultures, it significantly inhibited the growth of *P. falciparum*. The IC_{50} for galactan was 3.5 μg/mL, IC_{90}—6.8 μg/mL (for comparison: for heparin, the IC_{50} was 4.1 μg/mL, IC_{90}—8.0 μg/mL).

The authors did not establish a correlation between the molecular weight of the polysaccharide and growth inhibition of *P. falciparum* because the two best results were noted for galactan from the red alga *B. occidentalis* (Mw~700 kDa) and polysaccharide from the mollusc *L. grisea* (Mw~30 kDa). Compared with other SPS used by the authors in the experiment, galactan had the most pronounced anticoagulant properties. It has been suggested that the anticoagulant activity and antimalarial properties of galactan are not directly related.

Galactan actively suppressed the penetration of plasmodia into erythrocytes. Microscopic examination 20 h after treating erythrocytes with polysaccharides at a dose of 4 μg/mL showed an apparent decrease in the annular stages, i.e., delayed development of the parasite compared to samples untreated with galactan. Untreated control erythrocytes contained only trophozoites and schizonts. Under the influence of galactan, the rate of invasion of erythrocytes decreased (untreated—3.60 ± 0.10; those treated with galactan—1.10 ± 0.10); however, the maturation rate did not change dramatically (untreated erythrocytes—0.96 ± 0.06; erythrocytes treated with galactan—1.14 ± 0.08). Thus, the main mechanism of the antimalarial action of galactan is inhibition of the invasion of the parasite into erythrocytes [53].

It should be noted that free merozoites are present in the blood for a short period. This allowed the authors to assume that the binding process can also occur inside the erythrocyte. To confirm this position, heparin was added to the culture. Using confocal microscopy, it was shown that heparin penetrated living erythrocytes and bound intra-erythrocyte-developing merozoites [53].

In experiments on mice infected with *P. yoelii* and treated with intravenous galactan, this polysaccharide was shown to significantly reduce parasitemia compared to untreated animals. There were no side effects when the polysaccharide was administered to mice. Antibodies to *Plasmodium* antigens were found.

After 72 days, the mice were infected with plasmodia for the second time. All animals treated with galactan survived without any additional treatment and without disease symptoms until 42 days after the untreated secondary infection (follow-up period). On the positive side, galactan is not obtained from an animal object, i.e., in this case, there is no danger of prion contamination [53].

It is still impossible to talk about the biosynthesis of this promising compound because the mechanism of its enzymatic biosynthesis is unknown. However, an ample supply of algae in nature (found in abundance in Brazil) will make it possible to obtain sulfated galactan in sufficient quantities since its content in dry algae is about 50% [58].

SPS of red algae—galactans including carrageenans—are very promising antimalarial compounds. In [52], the authors investigated the antimalarial properties of a large panel of 50 compounds with inhibitory activity above 20%. Of these, 14 samples had high antimalarial activity at a concentration of 2 μg/mL. In this series, algal SPS were represented by highly sulfated carrageenans, lambda and iota. The authors believe that SPS inhibits the binding of merozoites to erythrocytes by disrupting the interaction of the receptor–ligand with sulfated receptors. It is believed that a variety of merozoite surface proteins mediate these initial contact effects through low-affinity interactions with the erythrocyte surface. Since many of these interactions are likely to be associated with surface sulfate receptors, the ability of heparin-like substances including those from algae to disrupt multiple interactions at several stages of invasion will probably be able to ensure the effectiveness of SPS for all strains of parasites and limit the emergence of drug resistance. This is especially true for SPS of seaweed.

The main mechanism of the pathogenic action of plasmodia is associated with a change in the properties of affected erythrocytes due to the formation of malarial tubercles ("knob structures", "knobs") on their surface, on the top of which there are special "handle" structures formed by the parasitic protein PfEMP1 (*P. falciparum* erythrocyte membrane protein 1). This protein plays the role of the main virulence factor. Moreover, the genome regions on which polypeptides exported by the parasite to the surface of erythrocytes are synthesized are highly polymorphic. The changes caused by exported proteins mediate the transformation of the physical characteristics of erythrocytes and impart adhesion properties to them [59]. Among the exported proteins, the PHIST family, containing 89 proteins, plays a significant role in enhancing the cytoadhesion of the reconstructed host cell [60]. The authors showed that sulfated polysaccharides can deactivate the proteins exported by the parasite, associated with control of the main virulent factor of the parasite *P. falciparum* erythrocyte membrane protein (PfEMP1) [61].

J.M. Mutisya et al. [62] devoted their work using the molecular docking method to the search for agents among SPS that inhibit the PHIST target. Using Sanger sequencing, 86 complete sequences were obtained among which 11 medicinal compounds with antiplasmodial activity were identified. In addition, ten compounds interacted with amino acid residues in the PHISTb (*P. helical* interspersed sub-telomeric b) and RESA (ring-infected erythrocyte surface antigen) domains, demonstrating potential activity against these proteins. At the same time, α-carrageenan from red alga interacted with both the reference and mutant proteins.

As potential sources of antimalarial agents, polysaccharides not only of different species of algae but also invertebrates can be used [63,64]. Thus, fucosylated chondroitin sulfate (FucCS), a unique highly sulfated GAG isolated from some marine organisms, was effective as an inhibitor of *P. falciparum* cytoadhesion to human lung endothelial cells and placental cryosection under static and flow conditions. This compound is low-toxic, rapidly destroyed the rosettes in a dose-dependent fashion, and blocked the development of various phenotypes of parasites, preventing the invasion of merozoites into erythrocytes. The authors proposed FucCS as a candidate for adjunctive therapy in patients with severe malaria. FucCS was also effective in preventing the parasite from adhering to the human placenta, eliminating it. Removal of the sulfated fucose branches from the compound practically eliminated the inhibitory effect, which indicates the central role of these branches in the inhibitory effect of the compound. The effect of FucCS appears to be similar to that of heparin, which inhibits invasion, adhesion, and rosette formation by binding to PfMP-1 regions and the surface protein of MSP-1 merozoites. FucCS can block the interaction of adhesion and invasion proteins with host receptors, which are glycoproteins constitutively expressed in erythrocytes, by binding to conserved regions of these proteins, such as the RII region [64].

FucCS treatment (1 mg/kg/animal per day) slowed down the death of $C_{57}Bl/6$ mice infected with *Plasmodium berghei* ANKA (PbA) in an experimental model of cerebral malaria. The compound had a weaker anticoagulant and antithrombotic effect than heparin, and acted in vivo and in vitro at concentrations lower than those required to trigger its anticoagulant effect. Thus, the authors proposed FucCS as a promising candidate for the adjuvant therapy of severe malaria and the prevention of exacerbations of the disease [63,64].

The later work of M.F. Bastos et al. [65] presented materials on the study of anti-malarial effect and the ability to inhibit cytoadhesion of heparan sulfate from the scallop *Nodipecten nodosus*. The polysaccharide is formed by repeating the disaccharide units of β-D-glucuronic acid 1 → 4 N-acetyl-α-D-glucosamine containing rare sulfate groups at the C2 or C3 positions of glucuronic acid. The polysaccharide blocked invasion up to 91% at the maximum dose (1000 μg/mL). It inhibited *P. falciparum* cytoadhesion to endothelial cells even at the lowest dose tested (1 μg/mL), and reached 86% and 100% inhibition of Pf-iEsCSA and Pf- iEsICAM-1, respectively, at the maximum dose (1000 μg/mL). The polysaccharide dose-dependently destroyed the rosettes at a rate similar to heparin, and showed antithrombotic activity without causing bleeding. Like the previous heparan sulfate, the authors recommended this compound for the adjunctive therapy of severe malaria.

A therapeutic composition consisting of chloroquine (100 nM) and fucoidan (20 μg/mL) has been proposed as a new agent for the prevention and treatment of malaria. The agent prevented the penetration of plasmodium into erythrocytes and did not give side effects, helping to overcome the resistance of the parasite to chloroquine [66].

Placental malaria is characterized by the cytoadhesion of infected erythrocytes and the expression of TNFα in the intervillous tissue, which can interfere with the transport of nutrients to the fetus and negatively affect its development. TNFα stimulates the cytoadhesion of erythrocytes in the capillaries of the placenta by enhancing the ligand bonds of *P. falciparum* Protein-1 (PfMP-1) with chondroitin sulfate A receptors.

The effectiveness of the combined use of SPS and drugs was also demonstrated in the work of Z. Rahmah et al. [67]. They investigated the effects of the seaweed extract of *Eucheuma cottonii* in combination with DHP (dihydroartemisinin-piperaquine) on the degree of parasitemia, cytoadhesion, TNFα level and fetal development in mice infected with *P. berghei*. DHP contains artemisin and chloroquine. *E. cottonii* is a red alga most abundant in Indonesia and containing SPS carrageenan in quantities ranging from 54% to 73%, depending on the place of growth and the type of algae.

The mice in the experiment were divided into four groups: (A)—intact, did not receive treatment with carrageenan; (B)—received DPH; (C)—animals were infected with *P. berghey* and given an extract; (D)—malaria-infected animals which received DHP and extract. Isolation of the placenta and fetus was performed on the 18th day after mating the mice.

Group A showed a high degree of parasitemia, TNFα expression, low fetal weight, and high cytoadhesion in the placenta.

In Group B, DHP therapy reduced parasitemia and cytoadhesion in the placenta. The same was true for the level of TNFα; however, the drug did not prevent low birth weight.

In Group C, there was a significant decrease in parasitemia and cytoadhesion in the placenta, a decrease in TNFα, and no decrease in birth weight.

In Group D, in comparison with group C, all indicators decreased even more. There was no low birth weight, i.e., the combination of the DHP drug with red algae extract containing a large amount of SPS carrageenan was experimentally effective against malaria, reducing cytoadhesion, parasitemia and the level of the proinflammatory cytokine TNFα and eliminating fetal developmental disorders.

Thus, SPS from marine hydrobionts (algae and marine invertebrates) can be a promising basis for creating antimalarial drugs that will have an effect both as independent treatments and as an addition to existing therapy for this protozoal infection.

3. Leishmaniasis

Leishmaniasis is a transmissible disease of humans and animals. The disease occurs in the tropics and subtropics on all continents except Australia [4]. Every year, about one million new cases of infection are registered in the world, up to 75,000 people die from leishmaniasis, and about 400 million people are at risk of getting sick [5].

The disease is caused by 20 species of protozoa of the genus *Leishmania* transmitted by the bite of infected female mosquitoes of the genus *Phlebotomus* (more than 90 species) [68]. Leishmaniasis ranks third among vector-borne diseases after malaria and lymphatic filariasis [69].

In humans and other vertebrates (dogs, rodents, monkeys), leishmanias exist in an immobile stage, or amastigote, in the intestine of the carrier (as opposed to the flagellar stage, or promastigote. When mosquitoes bite infected mammals (a reservoir of about 70 species of animals, including humans) leishmanias enter the intestines of mosquitoes, where they begin to multiply. In the gastrointestinal tract of insects, amastigotes turn into promastigotes, which the female mosquito spits up at the site of the bite on the human skin, while anywhere from 100 to 100,000 amastigotes get into the wound. Here, they multiply in the cytoplasm of the cells of the reticuloendothelial system.

Neutrophils that appear at the entrance gate absorb parasites that do not multiply in these cells. During apoptosis, macrophages absorb neutrophils, where leishmanias turn within 2–5 days into an intracellular morphological form without flagellum, the amastigote. The breeding cycle of amastigotes is 24 h. In the phagolysosomes of macrophages, the parasite replicates and survives due to its resistance to the microbicidal mechanisms of these cells, NO and ROS [70]. The process begins in the skin at the point of penetration of the parasite in the form of a granuloma (leishmanioma).

There are three main forms of leishmaniasis: visceral (VL, black disease, kala-azar, causative agent *Leishmania donovani*), cutaneous (CL, pendinsky ulcer, causative agent *L. tropica*), and mucocutaneous (espundia, causative agent *L. brasiliensis*). In visceral leishmaniasis, the foci of infection are formed in the organs of the reticuloendothelial system [68].

The currently available list of antileishmanial drugs that affect various metabolic pathways of the parasite is limited, and the growing resistance to them is of concern. For the treatment of leishmaniasis, pentavalent antimony compounds, amphotericin B, lipid forms of amphotericin B, miltefosine, and azole preparations are mainly used. However, antimony, like other drugs, has a wide range of side effects and causes high mortality in HIV patients [71,72]. These circumstances have necessitated a search for more harmless means for the treatment and prevention of leishmaniasis.

As presented in the previous section on antimalarial agents, marine macroalgae and invertebrates contain compounds that can control certain pathogens. There is evidence that heparin-binding proteins (HBP) are present on the surface of leishmania and may play a significant role in the parasite's life cycle, determining the success of the binding of the parasite to the host tissue. Thus, L.M.C. Cortes et al. [73] fractionated *L. braziliensis* promastigotes and isolated two heparin-binding proteins present on the promastigote surface. These proteins are involved in the adhesion of these parasites to Lulo cells (this insect cell line is a model for studying the interaction of insects and leishmania as it partially models adhesion events in the intestines of mosquitoes) or to a surface covered with heparin. About 860,000 HBP units are present on the surface of promastigotes. They inhibit the activity of protein kinase C (which will be discussed below) in the host organism by binding to heparin. In addition, the expression of HBP in *L. donovani* is associated with infectious forms of this parasite; HBP predominates in stationary-phase promastigotes, and successive culture passages of these parasites lead to a loss of the ability to bind to heparin.

Since SPSs have a molecular structure similar to heparin, they can bind various ligands, including protein receptors on the surface of the host cell [74].

Thus, to study this antileishmania effect, C.L. Pires et al. [75] used highly purified SPS obtained from extracts of the red algae *Solieria filiformis* (Sf), *Botryocladia occidentalis* (Bo), *Caulerpa racemosa* (Cr), and *Gracilaria caudata* (Gc). The authors evaluated their toxic effects and effects on the growth of *L. amazonensis* amastigotes (Figure 2).

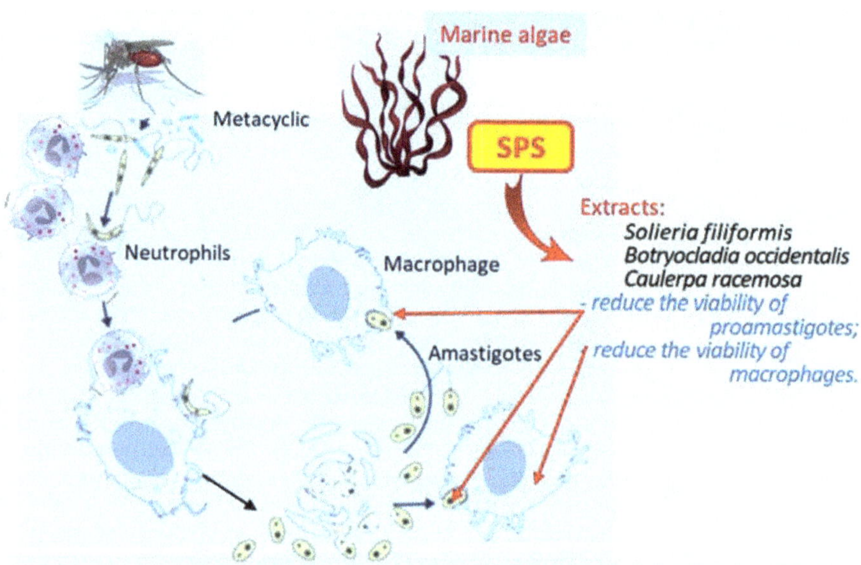

Figure 2. Highly purified SPS obtained from extracts of the red algae *Solieria filiformis*, *Botryocladia occidentalis*, and *Caulerpa racemosa* showed antileishmania activity.

The molecular weight of the polysaccharides was more than 200 kDa (Gc, Sf), 30 kDa (Bo), and 25 kDa (Cr), respectively.

Three of the four SPS were active against *L. amazonensis* (except for Gc), and they dose-dependently reduced the viability of amastigotes. The most active was extract Cr, which significantly reduced the viability of parasites (EC_{50}—34.5 μg/mL). Polysaccharides derived from Bo and Sf showed moderate antileishmania activity, with EC_{50} of 63.7 μg/mL and 137.4 μg/mL, respectively. All polysaccharides reduced the survival of the J774 macrophage cell line, with CC_{50} values of 27.3 μg/mL, 49.3 μg/mL, 73.2 μg/mL and 99.8 μg/mL for Bo, Cr, Gc and Sf, respectively.

However, none of the polysaccharides decreased the viability of peritoneal macrophages. The authors suggested that the cytotoxicity of the compounds was responsible for the destruction of leishmania rather than any direct effect on the parasite. High molecular weight polysaccharides exhibited weak antileishmania activity. The authors proposed SPS as an alternative to heparin for the treatment of leishmaniasis, as they are more effective than alkaloids, terpenoids, and other natural algae products.

A more detailed study of the effectiveness of SPS as an antileishmanial agent is presented in the work of S.A. Minicante et al. [76]. The authors evaluated the effect of polysaccharides of seven species of algae (green, red, and brown) on cell cultures DH82, Vero, L929, MDCK, and U937. Polysaccharides in the studied concentrations (up to 160 μg/mL) did not have a toxic effect on cell cultures and showed an antileishmania effect against *L. infant* (except for two, *Agardhiella subulata* and *Hypnea cornuta*). In addition, a convincing result was obtained using SPS from *U. pinnatifida* and *Sargassum muticum*. At the maximum concentration of polysaccharides (160 μg/mL), the death rate of parasites was 100%, which was confirmed by morphological studies.

At an SPS concentration of 20 μg/mL, inhibition of the growth of leishmania and the presence of an abnormal and round shape of parasites took place. In cultures with a polysaccharide concentration of 80 μg/mL, leishmania cells were aggregated, rounded, and included individuals without flagella. In cultures with an SPS concentration of 160 μg/mL, it was impossible to find whole forms of protozoa, and only apoptotic bodies were determined. Noteworthy is the fact that SPS from *S. muticum* and *U. pinnatifida* at the same concentrations did not act on *Trypanosoma cruzi*, which indicates their specific action against leishmania.

Fucoidan was used orally three times a day for four weeks, starting 15 days post-challenge, at a dose of 200 mg/kg/day in a six-week experiment in BALB/c mice infected with both antimony-sensitive and resistant leishmania strains. The results obtained allowed Kar et al. to establish that this polysaccharide had an inhibitory effect on the multiplication of amastigotes of both strains in macrophages (inhibition > 93% at 50 μg/mL) and also eliminated the parasitic load in the liver and spleen of animals [77].

The participation of NO in the inhibition of intracellular reproduction of amastigotes by fucoidan was established. For this, an iNOS inhibitor (2-amino-5,6-dihydro-6-methyl-4H-1,3-thiazine (AMT)) was used. The in vitro inhibitory effect of fucoidan was markedly reduced after 24 h in the presence of AMT (86% and 77% reductions for AG83 and GE1F8R, respectively). Fucoidan had no direct effect on the viability of promastigotes at doses of 25 and 50 μg/mL. However, slight inhibition (9% and 7% for AG83 and GE1F8R, respectively) was found when the polysaccharide was exposed to a dose of 100 μg/mL for 48 h. The viability of mouse peritoneal macrophages remained unchanged up to 150 μg/mL fucoidan.

In vivo experiments on mice intravenously infected with *L. donovani* promastigotes of strains AG83 and GE1F8R investigated the parasitic load in the liver and spleen. After six weeks all animals treated with fucoidan remained healthy and none of the groups showed a decrease in body weight. The action of fucoidan was dose-dependent and changed over time.

At a lower dose (25 μg/mL), the percentage of parasite death varied between antimony-sensitive (62%) and resistant (45%) strains. However, at a 50 μg/mL concentration for 24 h, inhibition was almost complete (96% and 93% for AG83 and GE1F8R, respectively).

To clarify whether fucoidan provides long-term protection, the cured mice were re-infected two months after the initial infection. The parasite load progressed rapidly in the placebo-treated mice, while the fucoidan-treated mice treated with the polysaccharide at a dose of 200 mg/kg/day were resistant to reinfection, which was observed up to 7 weeks; i.e., fucoidan therapy can form acquired immunity against both susceptible and resistant strains of leishmania. However, the authors emphasized that the mouse model does not reproduce all the features of active human visceral leishmaniasis.

Visceral leishmaniasis is accompanied by impaired proliferation of T-lymphocytes. In this regard, S. Kar et al. [77] conducted a study of the effect of fucoidan on this process. Four weeks after fucoidan treatment, mice infected with AG83 and GE1F8R showed a 12.8- and 11.2-fold increase in spleen T cell proliferation. A comparison of the cytokine expression profile was performed, and it was shown that in both groups of infected animals (AG83 and GE1F8R) the IFNγ, IL-12, and TNFα indices increased after four weeks in AG83 mice, by 7.5, 6.4 and 5.8 times, respectively, and in GE1F8R animals by 6.5, 5.4 and 5.8 times, respectively. IL-10 and TGFβ decreased in AG83 mice treated with fucoidan by

79% and 75%, respectively, and in GE1F8R mice treated with polysaccharide by 72% and 70%, respectively.

Thus, in in vivo situations, fucoidan can provide protection against both antimony-sensitive and resistant strains by direct action on the induction of NO and ROS, stimulation of splenocyte proliferation, and switching the balance of cytokines from the TH2 to the TH1 regime, which creates host protection against leishmania [77].

Leishmanias survive in a hostile environment in macrophages by inhibiting their defense mechanisms, namely, the formation of nitric oxide (NO) and the production of reactive oxygen species (ROS). Parasites use strategies to interfere with a wide range of signalling processes in macrophages, including protein kinase C (PKC), the JAK2/STAT1 cascade, and the MAP kinase pathway [78]. PKC signalling activates NF-kB, a universal transcription factor involved in expressing protective molecules [79,80]. Lipophosphoglycan on the cell surface of *Leishmania* promastigotes is a potent inhibitor of PKC activity in vitro in intact macrophages.

Depletion of PKC creates favourable conditions for the proliferation of *L. donovani* and promotes the survival of the parasite in host cells. At the same time, the expression of atypical isoforms (PKCε and PKCζ) increases, while Ca^{2+}-dependent PKC-β decreases in patients with leishmaniasis. Deactivation of PKCβ correlates with increased IL-10 production and disease progression. Furthermore, PKC signal transduction activates NF-kB, a universal transcription factor involved in expressing protective molecules [79,80]. In addition, there is evidence that *Leishmania* themselves express phosphatases targeting host cell molecules, which also contributes to the intracellular survival of the parasite [81].

Earlier [77], it was found that oral administration of fucoidan in mice at a dose of 200 mg/kg/day for three weeks infected with antimony-sensitive *L. donovani* strain resulted in the relieving of parasitic load in the liver and the inhibition of >93% of amastigotes in macrophages. As mentioned earlier, this healing effect is associated with a Th2 to Th1 switch. In addition, splenocytes from mice treated with fucoidan produced higher levels of NO and ROS.

G. Sharma et al. [70] investigated the cellular mechanisms underlying the antileishmanial effect of fucoidan in macrophage culture in vitro using a model of visceral leishmaniasis. A commercial preparation of fucoidan from *Fucus vesiculosus* contained 97% fucose and trace amounts of galactose, xylose and uronic acid.

When using monoclonal antibodies against IFNγ, TNFα and IL-12, a significant decrease in the protective effect of fucoidan was observed (parasite suppression was 57%, 52% and 63%, respectively) compared to mice receiving fucoidan (parasite inhibition was 99%).

In another series of experiments, mice were infected and then treated with fucoidan and AMT (2-amino-5,6-dihydro-6-methyl-4H-1,3-thiazine). This compound is a known inhibitor of the enzyme NO synthase (NOS) [82]. Mice treated with AMT had a weaker reduction in parasitic load than those treated with fucoidan. The mRNA levels of iNOS (the main enzyme that promotes NO formation) four weeks after infection were increased in infected mice treated with fucoidan (induction 9.15-fold), and remained at the same level after six weeks. Fucoidan mediated the induction of iNOS and proinflammatory cytokines dependent on the activation of p38 and ERK1/2 MAPK, which ultimately led to the activation of NF-kB and the associated antileishmanial response.

Treatment of infected macrophages with fucoidan induced a time-dependent increase in PKC-a, -b1 and b-2 kinase activities. The maximum PKC-a activity was 2.1 times higher 60 min after treatment, and the PKC-b1 and b2 activity were 2.7 and 2.67 times higher than in the untreated control, respectively.

Thus, fucoidan not only protects against leishmanias, it also activates the host's immune response. The results obtained by the reviewed authors indicate the effectiveness of this polysaccharide as a powerful immunoregulator for inclusion in the treatment of visceral leishmaniasis, including when caused by antimony-resistant leishmanias.

As a new therapeutic method for leishmaniasis, M.H.M. Hoseini et al. [83] suggested using chitin. To prove this, the authors injected female BALB/c mice subcutaneously with

Leishmania promastigotes and microparticles of chitin and/or chitosan for two weeks. The animals were sacrificed 12 weeks later, except for one group two weeks after infection. In the groups receiving chitin (0.6 $\pm$ 0.12 mm) and chitosan (1.2 $\pm$ 0.8 mm), the mean lesion sizes were significantly smaller than in the control group (6.2 $\pm$ 1.7 mm).

The parasite load in the lymph nodes of the polysaccharide-treated mice was considerably lower than in the lymph nodes of the control animals. Chitin microparticles induced cell proliferation and increased the production of TNFα and IL-10, that is, they acted as immunomodulators. Moreover, chitin was more effective in experimental leishmaniasis than chitosan. On the other hand, Riezk et al. [84] showed that chitosan and its derivatives at pH 6.5 were approximately 7–20 times more active than at pH 7.5, both against extracellular promastigotes and against intracellular amastigotes of *L. major* and *L. mexicana*. In this case, chitosan with a high molecular weight was more active. This pattern stimulated the production of nitric oxide and reactive oxygen species in both infected and non-infected leishmania macrophages in a time and dose-dependent manner. The rhodamine-labelled polysaccharide was absorbed by pinocytosis and accumulated in the parasitophorous vacuole of macrophages infected with leishmania. Thus, chitosan and chitin require further research as antileishmania drugs.

4. Trypanosomiasis

Trypanosomiasis is a group of transmissible protozoal diseases caused by flagellate protozoa of the genus *Trypanosoma*. The World Health Organization lists these diseases as one of the six most dangerous tropical diseases, distinguishing between American (Chagas disease) and African trypanosomiasis.

There are three stages in the developmental cycle of trypanosomes: (i) the invasive stage, in which the tripomastigote has an elongated shape and the flagellum is an undulating membrane and is therefore very mobile; (ii) the epimastigote exists inside of the carrier; and (iii) the immobile amastigote exists as an intracellular parasite of vertebrates.

The causative agent of American trypanosomiasis, *Trypanosoma cruzi*, is spread by the bite of a triatomine bug and is endemic to the continental region of Latin America. However, in recent years the disease has been found in the United States, Canada, and many European countries [85]. To date, about 6–7 million people around the world are sick with trypanosomiasis and the population at risk of infection is about 65 million.

The invasive stage for the vector, as well as for a vertebrate animal, are metacyclic trypomastigotes, which insects receive when feeding on animals and humans. The faeces of these insects carry the trypomastigotes into wounds or scratches on the skin and cause infection [86]. Moreover, they are also able to invade any germline host cells [86].

The main sites of parasitic trypanosomes are the smooth muscles of the heart and gastrointestinal tract. The life cycle ends when uninfected triatomaceous bugs feed on the blood of infected mammalian hosts and ingest parasites, which grow and differentiate in the insect's gastrointestinal tract and eventually migrate to the hindgut.

The disease can be transmitted through insect vectors, through the placenta from mother to fetus, through blood and tissue transplants, through ingestion of parasitized meat or freshly squeezed fruit juice, or as a result of intra-laboratory infection.

The causative agents of African trypanosomiasis (African sleeping sickness) are *Tripanosoma brucei gambiense* and *Tripanosoma brucei rhodesiense*; the carrier is the tsetse fly. The insect receives metacyclic trypomastigotes from humans or animals (pigs, antelopes, cattle) which act as the host for this parasite. The most anthropophilic insect species is *G. palpalis*. In 2019, 992 cases of African trypanosomiasis were reported. The population at risk of infection is about 65 million.

The minimum invasive dose of trypanosomes is 300–400 parasites. A fly gives off about 400,000 trypanosomes in one bite. One bite of an infected fly is enough for a person to fall ill with sleeping sickness. From about ten days after the bite, a person becomes a source of infestation.

Once inside the host, trypomastigotes migrate and spread throughout the body through the circulatory and lymphatic systems, after which they begin to multiply by binary fission. Ultimately, they disrupt the blood–brain barrier, which leads to infection of the central nervous system and the death of almost 100% of untreated patients [87].

Specific therapy is based on using nifurtimox and benznidazole, both of which are nitroheterocyclic agents that induce the formation of reactive oxygen species. However, these drugs have low efficacy and adverse side effects (impaired liver function, pancreas, skin irritation) [88]. In addition, the annual global cost of treating infected patients is approximately USD 24.7 billion, with 10% of this in the United States and Canada [89]. In connection with this situation, the search for inexpensive, effective and low-toxic compounds that can be used against trypanosomes continues.

As such agents, D. Leal et al. [90] used SPS from brown algae *Lessonia* spp. In addition to alginic acid, brown algae contain sulfated polysaccharides consisting mainly of L-fucopyranose residues linked by $\alpha 1 \rightarrow 3$ and $\alpha 1 \rightarrow 4$ glycosidic bonds. It has already been said above that sulfated fucoidans are of great interest due to their multifaceted high biological activity.

From the extract of *Lessonia* spp. blades, fucoidan was isolated and structurally characterized, then further investigated for trypanocidal activity against epimastigotes, a non-infectious replicative form found in the intestine of insect vectors.

The IC_{50} of fucoidan for epimastigous *T. cruzi* was 250 $\pm$ 3.92 μg/mL. The activity of fucoidan was lower than that of the drug nifurtimox (28.84 $\pm$ 0.84 μg/mL). However, the polysaccharide showed low toxicity in mammalian cells, with a selectivity index of 1.47 towards the *T. cruzi* parasite. S. Kar et al. [77] reported that commercial fucoidan significantly increased ROS and NO levels in *L. donovani*-infected macrophages. The authors showed that fucoidan's possible mechanism of action in this case is associated with an increase in ROS, causing oxidative stress in parasites which lack endogenous antioxidant systems and are therefore susceptible to oxidative stress.

At the same time, if the polysaccharide has antioxidant activity [91] and partially reduces the concentration of free radicals, this can lead to a decrease in the trypanocidal activity of the SPS.

To expand the possibilities for practical application of SPS, W.M. Souza et al. [92] used a sulfated polysaccharide as raw material for the preparation of an anti-trypanosome nanopreparation. To increase SPS activity, the authors synthesized a preparation containing silver nanoparticles and fucoidan obtained from the brown alga *Spatoglossum schroederi*.

These algae synthesize three types of bioactive heterofucans, A, B, and C, of which A has the highest content of fucoidan in comparison with others. Furthermore, it was previously shown in animal studies that the studied polysaccharides were not genotoxic or mutagenic either in vitro or in vivo [93,94].

W.M. Souza et al. [92] obtained seven fractions rich in sulfated fucoidans from the algae extract, which were analyzed by electrophoresis in agarose gel after staining with toluidine blue. Only one band was detected in fractions FO, 5v, and FO6v, which indicated the purity of the obtained polysaccharides. Afterwards, the fucoidans were used to obtain nanoparticles based on silver (Ag). The darkening of the solution marked the completion of the nanoparticle formation process.

The study of the antiparasitic activity of AgFuc, Ag and the FO, 5v fraction showed that at the lowest concentration (25 μg) all of the samples were inactive. Silver had the lowest toxicity against the parasite. FO, 5v affected the epimastigote form of trypanosomes only at the highest concentration used (100 μg). Moreover, its influence was more pronounced after 48 h (59.4 $\pm$ 1.4%). The highest time-dependent effect was obtained with AgFuc nanoparticles (after 24 h, the inhibition rate was 58.9 $\pm$ 1.2%; after 48 h, 67.3 $\pm$ 2.1%). Sample FO, 5v did not kill the parasites. Instead, AgFuc significantly increased the number of necrotic and apoptotic cells (Figure 3).

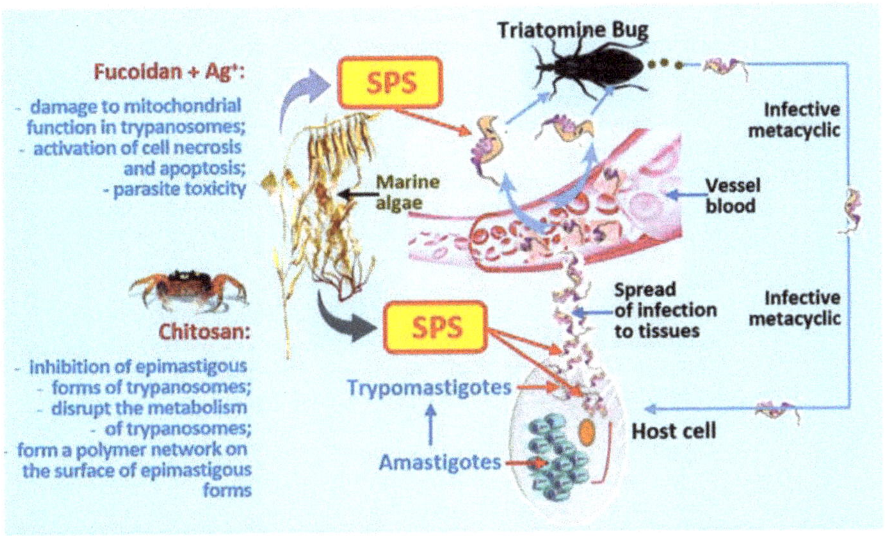

Figure 3. Sulfated biopolymers from marine hydrobiont fucoidans (brown algae) in combination with silver nanoparticles, as well as chitosan from crustaceans, have an antitrypanosomal effect.

Rhodamine is a marker of mitochondrial function. Application of FO, 5v reduced the number of rhodamine-stained parasites. AgFuc caused damage to the function of mitochondria in parasites, resulting in their death. According to the authors, parasites may have receptors that recognize polysaccharides, particularly fucoidans, facilitating the penetration of nanoparticles into the parasite's cytoplasm. Inside the parasite, they decompose and the silver component causes the formation of ROS, damaging mitochondria and causing the death of the parasite. These results are very interesting and promising for creating antitrypanosomal drugs; however, they require further in-depth studies.

Cell labelling with annexin V-FITC and P1 helps differentiate apoptotic, necrotic and viable cells. AgFuc significantly increased the number of cells stained with P1, indicating a higher level of necrosis than control samples. There were also many cells stained with annexin (19.7%), which indicated the apoptotic effect of the nanopreparation.

Another promising study concerns the use of chitosan oligosaccharides [95]. We present data on chitosan among the results obtained using SPS since the mechanisms of antiparasitic action of these compounds are similar.

The commercial preparation of chitin is obtained from the shells of marine arthropods. Every year in the world, seafood processing generates from 6 to 8 million tons of shell waste of which, most is thrown away [96]. Chitosan is a copolymer of N-acetyl-D-glucosamine (GlcNAc) and D-glucosamine (GlcN) obtained by partial deacetylation of chitin. Chitosan is biocompatible, nontoxic and biodegradable, and is therefore widely used in many fields [97]. In addition, the cationic nature of chitosan leads to enhanced bioadhesion on the membranes of microorganisms and fungi [98].

DaSilva et al. [95] used this property of chitosan against trypanosomes. However, its high molecular weight, degree of deacetylation, and, most importantly, poor solubility remains major obstacles to its widespread use in medicine. Therefore, the authors obtained small soluble chitooligosaccharides (COS) by enzymatic hydrolysis of chitosan using chitosanase from *B. toyonensis*.

Under the microscope, trypanosomes looked like elongated cells with a single thin flagellum and a smooth surface. After 72 h of incubation with COS, a network was observed on the surface of epimastigote forms of *T. cruzi* which adhered to the parasites, forming aggregates. The use of energy-dispersive X-ray spectroscopy (EDS) made it possible to establish that the grid consisted of organic material. The authors were unable to determine whether the network was COS or an extravasate of the internal contents of the parasites. Still, it appeared to be a combination of an oligomer and an extravasate. Thus, studies have shown that COS can form a polymer network that interacts with the parasite, preventing movement or possibly the exchange of nutrients by the parasite through membrane receptors by disrupting its metabolism. The oligomers also caused a time-dependent inhibition of the epimastigote forms of trypanosomes.

Thus, COS can be used to control *T. cruzi*, possibly in combination with drugs or in drug delivery systems, to enhance their action or overcome drug resistance.

5. Schistosomiasis

Schistosomes, or blood flukes, are the causative agents of schistosomiasis. In humans, the three species are most often found are *Shistosoma haematobium*, the causative agent of genitourinary schistosomiasis, and *S. japonicum* and *S. mansoni*, the causative agents of intestinal schistosomiasis [68]. It is estimated that at least 236.6 million people required treatment in 2019, of whom only 105.4 million received treatment. Helminthiasis occurs primarily in tropical and subtropical developing countries [99].

The main pathological manifestations of Japanese schistosomiasis are granuloma and subsequent liver fibrosis caused by the retention of worm eggs in this organ, which leads to fatal irreversible damage in the liver and intestines of the host [100,101]. At the same time, a dominant CD4 + Th2 immune response mediated by IL-4 and IL-13 develops, leading to the development of granulomas and fibrosis. During the development of liver fibrosis, macrophages of the M2 phenotype act as a key cell population (30%). These cells play a critical role in the immune response, as they contribute to the TH2 response and are required to suppress the Th1 response.

It is known that fucoidans have anti-inflammatory effects, inhibiting LPS-induced inflammation in macrophages and blocking the signalling pathway of the nuclear factor NF-kB, TLR4 [102]. A commercial preparation of fucoidan from the brown alga *Sargassum hemiphyllum* reduced the expression of cytokines (TIMP-1, CXCL-1, MCP-1, and MIP-2), reducing the intensity of the development of lung fibrosis in mice [103].

X. Bai et al. [104] used fucoidan obtained from the brown alga *F. vesiculosus* as an anti-inflammatory and hepatoprotective agent in experimental schistosomiasis (500 mg/kg twice a day for 40 days). When using fucoidan to treat mice infected with schistosomes, a significant decrease in the size of liver granulomas and the degree of fibrosis was observed during the infectious process. The levels of mRNA and anti-inflammatory cytokines (IL-4 and IL-13) increased, as did infiltration of the liver and spleen tissue with Treg cells as well as the levels of IL-10 and TGFβ mRNA.

Among the splenocytes stimulated in vitro by fucoidan, the number of Treg cells increased along with increased expression of chemokine receptors CCR4 and CXCR5 on Treg-cells. In addition, an increase in the levels of IL-4 and IL-13 mRNA was observed in macrophages.

Thus, the use of fucoidan in schistosomiasis reduces pathological changes in the liver and prevents the progression of the process caused by *S. japonica*, which may be a new potential strategy for the treatment of patients with pathology caused by this helminthiasis in the future.

6. Cryptosporidiosis

Cryptosporidiosis is a parasitic disease of the intestinal tract of mammals, including humans, caused by the protozoa *Cryptosporidium parvum*. *Cryptosporidia* are capable of multiplying not only in cells but also extracellularly [105]. Cryptosporidiosis is spread by food, often through contaminated water. The agent causes severe watery diarrhea in humans and other mammals [106], and is especially dangerous for immunocompromised individuals. Cryptosporidiosis is a major problem for animal farming, as these protozoa are one of the most common enteropathogens in young farm animals and poultry, resulting in large economic losses [107]. In Europe, about 4.7 million disease cases are registered annually, with about 500 deaths per year [108].

The only drug approved by the Food and Drug Administration (FDA, Silver Spring, MD, USA) for the treatment of immunocompromised people with cryptosporidiosis is nitazoxanide, the effectiveness of which is low; therefore, as with other parasitic diseases, a search is underway for effective new drugs for the treatment and prevention of this disease.

Infection with cryptosporidia occurs when the host animal swallows the oocysts of the pathogen, after which they secrete sporozoites as they pass through the stomach and duodenum. Sporozoites attach to intestinal epithelial cells, envelop the host cell membrane, and develop into trophozoites.

Sporozoites, however, cannot directly interact with intestinal epithelial cells. Instead, they contain a glycocalyx, a filamentous layer of branched carbohydrates [109]. This acts as a protective barrier [110] and includes a high level of transmembrane mucin glycoproteins of the proteoglycan type [111]. Proteoglycans are macromolecules located on the cell surface or in the extracellular matrix in which one or more glycosamine chains are covalently linked to a membrane or secreted protein. Various pathogens use these structures to invade host cells [112–114].

A. Inomata et al. [115] suggested that *C. parvum* interacts with GAG on host cells, and some polysaccharides can inhibit the parasite's attachment to cells [116]. The authors evaluated the anticryptosporidial effects of five sulfated polysaccharides on parasite invasion and found that they all have an inhibitory effect on this process. In this case, the most powerful effect was exerted by heparin at a dose of 1 μg/mL. Among the studied SPS was fucoidan, which had a dose-dependent inhibitory effect, reducing the invasion of parasites by about 50% at a concentration of 100 μg/mL.

The results of this study also indicated that polysaccharides (in particular, this was shown in the fucoidan model) compete with some factor(s) on the surface of the parasite and that this is are involved in the invasion of HCT-8 cells. It has also been shown that cell surface heparan sulfate plays a role in *C. parvum* invasion in vitro.

The authors did not aim to study the antiparasitic action of fucoidan in detail. However, they showed the mechanism of action of SPS when interacting with parasites and cells, as well as the need to continue research in this direction in order to create new effective drugs against cryptosporidiosis.

H. Maruyama et al. [116] investigated the effect of native (obtained from the sporophylls of *U. pinnatifida*) and desulfated fucoidans on the adhesion of cryptosporidia to the cell culture of human intestinal epithelium (Intestinal 407), as well as the infectious process in newborn mice. The adhesion of *C. parvum* to intestinal cells was significantly reduced when the cells were treated with a low dose of fucoidan (1%, 50 μg/mL). The number of *C. parvum* oocysts in mice receiving fucoidan decreased by almost five times compared with control animals that did not receive the polysaccharide. Thus, the authors have convincingly proven that fucoidan effectively inhibits the growth of *C. parvum* in mice and also prevents the adhesion of parasites to the epithelial cells of the human intestine. At the same time, fucoidan can inhibit cryptosporidium due to direct binding to functional mediators derived from *C. parvum* in the intestinal epithelial cells of newborn mice. However, the desulfated fucoidan had little effect on the growth of parasites.

Chitosan has promising potential in the prevention and treatment of cryptosporidiosis. To assess the effect of two chitosans (Chitosan NAG and Chitosan Mix) on oocyst secretion in newborn CD-1 mice, *C. parvum* oocysts were orally inoculated, treated with chitosans and compared with those of untreated animals [117]. Paromomycin, a classic drug used in veterinary medicine, was used as a control. Treatment with both chitosans and paromomycin significantly reduced the release of parasites in infected animals that received chitosan (−56%, −34.5%, 58%, respectively).

In the in vitro experiments, after 24 h incubation at 37 °C a significant decrease in the viability of cryptosporidium oocysts was observed (>95%). In addition, paromomycin, chitosan NAG, and chitosan Mix all dose-dependently inhibited the reproduction of C.

parvum in the NCT-8 and Caco-2 cell lines. The authors considered these results as the first evidence of the effectiveness of chitosan as an anticryptosporidial compound.

7. Trichomoniasis

Trichomonas vaginalis is a flagellated protozoan that causes trichomoniasis, a sexually transmitted disease in humans. Infection is possible through non-sterile gynecological instruments and gloves. The parasite lives on the surface of the epithelium of the urogenital tract of women and men. The parasite replicates by binary fission of the nucleus. The trophozoite form is the only stage of this protozoan [118].

The disease is widespread. According to the WHO, about 170 million people are infected with vaginal trichomonas. In developed countries, the prevalence of this pathology is 2–10%; in developing countries, it reaches 40%. 5-nitroimidazole is used to treat the disease [119]. Oral administration of this medication remains the recommended treatment regimen for trichomoniasis. However, the adverse side effects on the gastrointestinal tract, allergenic profile, and increasing drug resistance of parasites necessitate a search for new ways to combat trichomonas [120].

As with other parasitic diseases, scientists are searching for new effective compounds in order to create drugs against this infection, including among biologically active substances from marine hydrobionts, particularly algae. Thus, C.B.S. Telles et al. [121] analyzed the activity of five heterofucans from the brown alga *S. filipendula*, which have strong antiproliferative and antioxidant effects. Two out of the five polysaccharides were inactive towards *Trichomonas trophozoites*

Three fucans showed activity after 24 h of vaginal treatment in mice. The polysaccharides exerted a cytotoxic effect on protozoa, almost completely suppressing their viability. As shown in this work, the degree of sulfation of fucans is an essential factor in anti-*T. vaginalis* activity. The authors placed great hopes on algal polysaccharides; however, they also drew attention to the fact that further serious research is needed in order to clarify the complete structure of these polysaccharides, the configuration of glycosidic bonds, their position, and the number of sulfate groups and branch points.

8. Conclusions

As follows from the above materials, SPS from marine hydrobionts are promising antiparasitic compounds. Parasites' development of resistance to traditional drugs leads to the loss of the effectiveness of existing drugs as therapy. In addition, the toxicity and adverse side effects of many drugs used to treat parasitic infections stimulates the search for alternative remedies. Active screening of biologically active natural compounds, including among the metabolites of marine aquatic organisms on which basis effective drugs can be obtained, has led to the development of new strategies in the therapy and prevention of parasitic diseases.

SPS from marine algae and invertebrates are practically nontoxic, and only in rare cases slightly toxic in the body. Polysaccharides such as, for example, fucoidan, have long been widely used by the population of many countries and are produced in the form of food additives [17,18].

However, the mechanisms of the antiparasitic action of these unique compounds at the cellular and molecular levels are far from being fully understood. It can be considered that at present there is an accumulation of knowledge, and attempts are being made to explain many processes, as the interaction of each type of parasite with different SPS is specific and each requires a special approach.

Numerous studies have shown that both algae extracts and the metabolites of these aquatic organisms, including SPS, have not only powerful antiparasitic, but also antioxidant, anti-inflammatory, immunomodulatory and antitoxic potential [18,122,123], which cannot but enhance their action against parasitic invasion. Therefore, it is necessary to study polysaccharides with a characterized structure in order to determine how such important structural parameters as monosaccharide composition, type of glycoside bond, molecular weight, the content of sulfate groups, and uronic acids affect the antiparasitic activity of SPS. Such materials have already been obtained when studying the effectiveness of SPS in bacterial and viral infections [124] and, to a lesser extent, in parasitic invasions [53,57,125]. This also seems to occur in the interaction of SPE with parasites of various taxonomic groups [53,57]. Thus, it is necessary to expand research on compounds with a high antiparasitic effect in order to determine the relationship between their structure and biological activity.

Of great interest is the study of the nature of the direct action of SPS on protozoa using contemporary methods, as is the case in the study of the interaction between bacteria and polysaccharides. These compounds bind to the surface of the bacteria, causing damage to the membrane or leakage of nutrients from the microorganism. This is confirmed by the discovery of nucleic acids [126] and proteins [127] released after the treatment of bacteria with SPS. It is possible to carry out the same research with unicellular and multicellular parasites. It is also possible that SPS traps and binds nutrients in the environment, leading to a loss of bioavailable of nutrients for parasites and causing a decrease in viability or even death in protozoa.

A careful approach in determining the antiparasitic action of SPS requires an adequate extrapolation of the data [128] obtained in vitro to the field of application in vivo, given that most studies of marine polysaccharides at present are conducted outside the body. The possibility of a discrepancy between the concentrations of drugs used for in vitro experiments to affecting the body's cells and protozoans and the actual conditions of the human body should be borne in mind.

Questions about the targets for the SPS of marine hydrobionts on each species of protozoa, as well as about whether parasites can develop resistance to these compounds, also require in-depth study.

Despite the abundance of unresolved issues, SPS from marine hydrobionts, which combine high antiparasitic potential with antitoxic, anti-inflammatory, immunomodulatory and antioxidant properties, are a promising basis for creating new drugs, food supplements, and functional food products to fight parasitic infections. It is also necessary to conduct in-depth studies of the possibility of the combined use of antiparasitic drugs and SPS to reduce the toxicity of antiparasitic drugs and eliminate their undesirable side effects. The multivalent action of SPS has been realized at the cellular and molecular level. It is a significant factor in increasing the antiparasitic therapeutic potential of biologically active substances in aquatic organisms. SPS, unlike many drugs, has a substantial number of targets for the implementation of their action, and their antiparasitic effect is a combination of several possible mechanisms leading to the death of parasites.

Difficulties in the development of drugs based on SPS remain, largely due to the complexity of their standardization as they need to be standardized in terms of such physicochemical parameters as molecular weight, monosaccharide composition and degree of sulfation, and structure of side chains as well as the type or combination of types of bonds between fucose and other residual monosaccharides that make up the SPS. Obtaining chemically pure, structurally characterized and homogeneous samples with low molecular weight or oligomeric fractions with polydispersity indices close to unity from native polysaccharides is a difficult task. One of the approaches to solving it consists in the use of polysaccharide-degrading enzymes such as fucoidan hydrolases (fucoidanase), alginate lyase, and carrageenans sulfatases [129,130].

Despite this, the combination of new methods including genetic modification of pathogens, molecular docking, bioimaging, and the use of physicochemical methods for research, which led to the standardization of high-throughput screening platforms in drug development, will undoubtedly allow these strategies to be applied to the creation of antiparasitic drugs based on biologically active substances from marine hydrobionts.

Author Contributions: N.N.B.—idea and writing plan, concept, methodology, approval of the final review; B.G.A.—approval of the final version, drawing illustrations, and translation of the text; T.S.Z. and T.A.K.—methodology, conceptualization, validation; S.P.K.—editing and validation of the manuscript, collection and analysis of literature data; M.Y.S.—analysis and interpretation of data, preparation of the original layout; S.P.E.—collection and analysis of literature, preparation of a draft manuscript; A.N.V.—writing a conclusion, proofreading of the manuscript. All authors have read and agreed to the published version of the manuscript.

Funding: The study was supported by the "Far East" Integrated Program of Basic Research, Far Eastern Branch, Russian, project no. 18-03-059, supported the study 2019–2021.

Conflicts of Interest: The authors declare no conflict of interest.

References

1. Fletcher, S.; Caprarelli, G.; Merif, J.; Andresen, D.; van Hal, S.; Stark, D.; Ellis, J. Epidemiology and geographical distribution of enteric protozoan infections in Sydney, Australia. *J. Public Health Res.* **2014**, *3*, 298. [CrossRef]
2. Henry, N.B.; Sermé, S.S.; Siciliano, G.; Sombié, S.; Diarra, A.; Sagnon, N.; Traoré, A.S.; Sirima, S.B.; Soulama, I.; Alano, P. Biology of *Plasmodium falciparum* gametocyte sex ratio and implications in malaria parasite transmission. *Malar. J.* **2019**, *18*, 70. [CrossRef] [PubMed]
3. Szempruch, A.J.; Dennison, L.; Kieft, R.; Harrington, J.M.; Hajduk, S.L. Sending a message: Extracellular vesicles of pathogenic protozoan parasites. *Nat. Rev. Genet.* **2016**, *14*, 669–675. [CrossRef] [PubMed]
4. Okwor, I.; Uzonna, J. Social and economic burden of human leishmaniasis. *Am. J. Trop. Med. Hyg.* **2016**, *94*, 489–493. [CrossRef] [PubMed]
5. Strasen, J.; Williams, T.; Ertl, G.; Zoller, T.; Stich, A.; Ritter, O. Epidemiology of Chagas disease in Europe: Many calculations, little knowledge. *Clin. Res. Cardiol.* **2014**, *103*, 1–10. [CrossRef]
6. Kim, S.-B.; Paulmurugan, R. Bioluminescent imaging systems for assay developments. *Anal. Sci.* **2021**, *37*, 233–247. [CrossRef]
7. Álvarez-Bardón, M.; Pérez-Pertejo, Y.; Ordóñez, C.; Sepúlveda-Crespo, D.; Carballeira, N.M.; Tekwani, B.L.; Murugesan, S.; Martinez-Valladares, M.; García-Estrada, C.; Reguera, R.M.; et al. Screening marine natural products for new drug leads against trypanosomatids and malaria. *Mar. Drugs* **2020**, *18*, 187. [CrossRef] [PubMed]
8. Stein, É.M.; Machado, L.P.; Roffato, H.K.; Miyasato, P.A.; Nakano, E.; Colepicolo, P.; Andreguetti, D.X. Antischistosomal activity from Brazilian marine algae. *Rev. Bras. Farm.* **2015**, *25*, 663–667. [CrossRef]
9. Vonthron-Sénécheau, C.; Kaiser, M.; Devambez, I.; Vastel, A.; Mussio, I.; Rusig, A.-M. Antiprotozoal activities of organic extracts from french marine seaweeds. *Mar. Drugs* **2011**, *9*, 922–933. [CrossRef] [PubMed]
10. Powers, J.L.; Zhang, X.; Kim, C.Y.; Abugri, D.A.; Witola, W.H. Activity of green algae extracts against *Toxoplasma gondii*. *Med. Aromat. Plants* **2017**, *6*, 3. [CrossRef]
11. Torres, F.A.; Passalacqua, T.G.; Velásquez, A.M.A.; de Souza, R.A.; Colepicolo, P.; Graminha, M.A. New drugs with antiprotozoal activity from marine algae: A review. *Rev. Bras. Farm.* **2014**, *24*, 265–276. [CrossRef]
12. Moo-Puc, R.; Robledo, D.; Freile-Pelegrin, Y. Evaluation of selected tropical seaweeds for in vitro anti-trichomonal activity. *J. Ethnopharmacol.* **2008**, *120*, 92–97. [CrossRef]
13. Yamthe, L.R.T.; Appiah-Opong, R.; Fokou, P.V.T.; Tsabang, N.; Boyom, F.F.; Nyarko, A.K.; Wilson, M.D. Marine algae as source of novel antileishmanial drugs: A review. *Mar. Drugs* **2017**, *15*, 323. [CrossRef] [PubMed]
14. Hutson, K.S.; Mata, L.; Paul, N.A.; de Nys, R. Seaweed extracts as a natural control against the monogenean ectoparasite, *Neobenedenia* sp., infecting farmed barramundi (*Lates calcarifer*). *Int. J. Parasitol.* **2012**, *42*, 1135–1141. [CrossRef] [PubMed]
15. Kishimoto, T.K.; Viswanathan, K.; Ganguly, T.; Elankumaran, S.; Smith, S.; Pelzer, K.; Lansing, J.; Sriranganathan, N.; Zhao, G.; Galcheva-Gargova, Z.; et al. Contaminated heparin associated with adverse clinical events and activation of the contact system. *N. Engl. J. Med.* **2008**, *358*, 2457–2467. [CrossRef] [PubMed]
16. Guerrini, M.; Beccati, D.; Shriver, Z.; Naggi, A.M.; Viswanathan, K.; Bisio, A.; Capila, I.; Lansing, J.C.; Guglieri, S.; Fraser, B.; et al. Oversulfated chondroitin sulfate is a contaminant in heparin associated with adverse clinical events. *Nat. Biotechnol.* **2008**, *26*, 669–675. [CrossRef] [PubMed]
17. Arlov, Ø.; Rütsche, D.; Korayem, M.A.; Öztürk, E.; Zenobi-Wong, M. Engineered sulfated polysaccharides for biomedical applications. *Adv. Funct. Mater.* **2021**, *31*, 2010732. [CrossRef]
18. Zeng, K.; Groth, T.; Zhang, K. Recent advances in artificially sulfated polysaccharides for applications in cell growth and differentiation, drug delivery, and tissue engineering. *ChemBioChem* **2019**, *20*, 737–746. [CrossRef]
19. Gandhi, N.S.; Mancera, R.L. The structure of glycosaminoglycans and their interactions with proteins. *Chem. Biol. Drug Des.* **2008**, *72*, 455–482. [CrossRef] [PubMed]
20. Zaporozhets, T.; Besednova, N. Prospects for the therapeutic application of sulfated polysaccharides of brown algae in diseases of the cardiovascular system: Review. *Pharm. Biol.* **2016**, *54*, 3126–3135. [CrossRef] [PubMed]
21. Meneghetti, M.C.Z.; Hughes, A.; Rudd, T.; Nader, H.B.; Powell, A.K.; Yates, E.A.; Lima, M.A. Heparan sulfate and heparin interactions with proteins. *J. R. Soc. Interface* **2015**, *12*, 20150589. [CrossRef] [PubMed]

22. Liu, Y.; Liu, W.; Wang, Y.; Ma, Y.; Huang, L.; Zou, C.; Li, D.; Cao, M.-J.; Liu, G.-M. Inhibitory effect of depolymerized sulfated galactans from marine red algae on the growth and adhesion of diarrheagenic *Escherichia coli*. *Mar. Drugs* **2019**, *17*, 694. [CrossRef] [PubMed]
23. Kwon, P.S.; Oh, H.; Kwon, S.-J.; Jin, W.; Zhang, F.; Fraser, K.; Hong, J.J.; Linhardt, R.J.; Dordick, J.S. Sulfated polysaccharides effectively inhibit SARS-CoV-2 In Vitro. *Cell Discov.* **2020**, *6*, 50. [CrossRef] [PubMed]
24. Irhimeh, M.R.; Fitton, J.H.; Lowenthal, R.M.; Kongtawelert, P. A quantitative method to detect fucoidan in human plasma using a novel antibody. *Methods Find. Exp. Clin. Pharmacol.* **2005**, *27*, 705–710. [CrossRef] [PubMed]
25. Varo, R.; Chaccour, C.; Bassat, Q. Update on malaria. *Medicina Clínica* **2020**, *155*, 395–402. [CrossRef] [PubMed]
26. Graumans, W.; Jacobs, E.; Bousema, T.; Sinnis, P. When is a plasmodium-infected mosquito an infectious mosquito? *Trends Parasitol.* **2020**, *36*, 705–716. [CrossRef] [PubMed]
27. Center for Disease Control and Prevention. Malaria. Available online: https://www.cdc.gov/parasites/malaria/ (accessed on 1 October 2021).
28. Goerdeler, F.; Seeberger, P.H.; Moscovitz, O. Unveiling the Sugary Secrets of Plasmodium Parasites. *Front Microbiol.* **2021**, *12*, 712538. [CrossRef]
29. Sato, S. Plasmodium—A brief introduction to the parasites causing human malaria and their basic biology. *J. Physiol. Anthr.* **2021**, *40*, 1. [CrossRef]
30. Weiss, G.E.; Gilson, P.R.; Taechalertpaisarn, T.; Tham, W.-H.; de Jong, N.; Harvey, K.L.; Fowkes, F.; Barlow, P.N.; Rayner, J.C.; Wright, G.; et al. Revealing the sequence and resulting cellular morphology of receptor-ligand interactions during *Plasmodium falciparum* invasion of erythrocytes. *PLoS Pathog.* **2015**, *11*, e1004670. [CrossRef]
31. Venugopal, K.; Hentzschel, F.; Valkiūnas, G.; Marti, M. Plasmodium asexual growth and sexual development in the haematopoietic niche of the host. *Nat. Rev. Genet.* **2020**, *18*, 177–189. [CrossRef] [PubMed]
32. Maier, A.G.; Cooke, B.M.; Cowman, A.F.; Tilley, L. Malaria parasite proteins that remodel the host erythrocyte. *Nat. Rev. Genet.* **2009**, *7*, 341–354. [CrossRef] [PubMed]
33. Counihan, N.A.; Modak, J.K.; de Koning-Ward, T.F. How malaria parasites acquire nutrients from their host. *Front. Cell Dev. Biol.* **2021**, *9*, 649184. [CrossRef] [PubMed]
34. Marques, J.; Valle-Delgado, J.J.; Urban, P.; Baró, E.; Prohens, R.; Mayor, A.; Cisteró, P.; Delves, M.; Sinden, R.E.; Grandfils, C.; et al. Adaptation of targeted nanocarriers to changing requirements in antimalarial drug delivery. *Nanomed. Nanotechnol. Biol. Med.* **2017**, *13*, 515–525. [CrossRef] [PubMed]
35. Lin, C.S.; Uboldi, A.D.; Epp, C.; Bujard, H.; Tsuboi, T.; Czabotar, P.; Cowman, A.F. Multiple *Plasmodium falciparum* merozoite surface protein 1 complexes mediate merozoite binding to human erythrocytes. *J. Biol. Chem.* **2016**, *291*, 7703–7715. [CrossRef] [PubMed]
36. Dijkman, P.M.; Marzluf, T.; Zhang, Y.; Chang, S.-Y.S.; Helm, D.; Lanzer, M.; Bujard, H.; Kudryashev, M. Structure of the merozoite surface protein 1 from *Plasmodium falciparum*. *Sci. Adv.* **2021**, *7*, eabg0465. [CrossRef] [PubMed]
37. Counihan, N.A.; Kalanon, M.; Coppel, R.; de Koning-Ward, T. Plasmodium rhoptry proteins: Why order is important. *Trends Parasitol.* **2013**, *29*, 228–236. [CrossRef] [PubMed]
38. Lingelbach, K.; Joiner, K. The parasitophorous vacuole membrane surrounding *Plasmodium* and *Toxoplasma*: An unusual compartment in infected cells. *J. Cell Sci.* **1998**, *111*, 1467–1475. [CrossRef] [PubMed]
39. Cao, J.; Kaneko, O.; Thongkukiatkul, A.; Tachibana, M.; Otsuki, H.; Gao, Q.; Tsuboi, T.; Torii, M. Rhoptry neck protein RON2 forms a complex with microneme protein AMA1 in *Plasmodium falciparum* merozoites. *Parasitol. Int.* **2009**, *58*, 29–35. [CrossRef] [PubMed]
40. Miller, L.H.; Ackerman, H.C.; Su, X.; Wellems, T.E. Malaria biology and disease pathogenesis: Insights for new treatments. *Nat. Med.* **2013**, *19*, 156–167. [CrossRef] [PubMed]
41. Lee, W.-C.; Russel, B.; Renia, L. Sticking for a cause: The falciparum malaria parasites cytoadherence paradigm. *Front. Immunol.* **2019**, *10*, 1444. [CrossRef]
42. Takala-Harrison, S.; Jacob, C.G.; Arze, C.; Cummings, M.P.; Silva, J.C.; Dondorp, A.M.; Fukuda, M.M.; Hien, T.T.; Mayxay, M.; Noedl, H.; et al. Independent emergence of artemisinin resistance mutations among *Plasmodium falciparum* in southeast Asia. *J. Infect. Dis.* **2015**, *211*, 670–679. [CrossRef] [PubMed]
43. Shibeshi, M.A.; Kifle, Z.D.; Atnafie, S.A. Antimalarial drug resistance and novel targets for antimalarial drug discovery. *Infect. Drug Resist.* **2020**, *13*, 4047–4060. [CrossRef]
44. Belete, T.M. Recent progress in the development of new antimalarial drugs with novel targets. *Drug Des. Dev. Ther.* **2020**, *14*, 3875–3889. [CrossRef]
45. Dans, M.G.; Weiss, G.E.; Wilson, D.; Sleebs, B.E.; Crabb, B.S.; de Koning-Ward, T.F.; Gilson, P.R. Screening the medicines for malaria venture pathogen box for invasion and egress inhibitors of the blood stage of *Plasmodium falciparum* reveals several inhibitory compounds. *Int. J. Parasitol.* **2020**, *50*, 235–252. [CrossRef] [PubMed]
46. Beeson, J.G.; Drew, D.R.; Boyle, M.; Feng, G.; Fowkes, F.; Richards, J.S. Merozoite surface proteins in red blood cell invasion, immunity and vaccines against malaria. *FEMS Microbiol. Rev.* **2016**, *40*, 343–372. [CrossRef] [PubMed]
47. Lantero, E.; Aláez-Versón, C.; Romero, P.; Sierra, T.; Fernàndez-Busquets, X. Repurposing heparin as antimalarial: Evaluation of multiple modifications toward In Vivo application. *Pharmaceutics* **2020**, *12*, 825. [CrossRef]

48. Kobayashi, K.; Takano, R.; Takemae, H.; Sugi, T.; Ishiwa, A.; Gong, H.; Recuenco, F.C.; Iwanaga, T.; Horimoto, T.; Akashi, H.; et al. Analyses of interactions between heparin and the apical surface proteins of *Plasmodium falciparum*. *Sci. Rep.* **2013**, *3*, 3178. [CrossRef] [PubMed]
49. Leitgeb, A.M.; Nde, P.; Cho-Ngwa, F.; Titanji, V.; Samje, M.; Blomqvist, K.; Wahlgren, M. Low anticoagulant heparin disrupts *Plasmodium falciparum* rosettes in fresh clinical isolates. *Am. J. Trop. Med. Hyg.* **2011**, *84*, 390–396. [CrossRef] [PubMed]
50. Saiwaew, S.; Sritabal, J.; Piaraksa, N.; Keayarsa, S.; Ruengweerayut, R.; Utaisin, C.; Sila, P.; Niramis, R.; Udomsangpetch, R.; Charunwatthana, P.; et al. Effects of sevuparin on rosette formation and cytoadherence of *Plasmodium falciparum* infected erythrocytes. *PLoS ONE* **2017**, *12*, e0172718. [CrossRef] [PubMed]
51. McQuaid, F.; Rowe, J.A. Rosetting revisited: A critical look at the evidence for host erythrocyte receptors in *Plasmodium falciparum* rosetting. *Parasitology* **2020**, *147*, 1–11. [CrossRef]
52. Boyle, M.J.; Skidmore, M.; Dickerman, B.; Cooper, L.; Devlin, A.; Yates, E.; Horrocks, P.; Freeman, C.; Chai, W.; Beeson, J.G. Identification of heparin modifications and polysaccharide inhibitors of *Plasmodium falciparum* merozoite invasion that have potential for novel drug development. *Antimicrob. Agents Chemother.* **2017**, *61*, 00709–00717. [CrossRef] [PubMed]
53. Marques, J.; Vilanova, E.; Mourão, P.A.S.; Fernàndez-Busquets, X. Marine organism sulfated polysaccharides exhibiting significant antimalarial activity and inhibition of red blood cell invasion by *Plasmodium*. *Sci. Rep.* **2016**, *6*, 24368. [CrossRef]
54. Mourão, P.A.S.; Pereira, M.S.; Pavão, M.S.G.; Mulloy, B.; Tollefsen, D.M.; Mowinckel, M.-C.; Abildgaard, U. Structure and anticoagulant activity of a fucosylated chondroitin sulfate from echinoderm. *J. Biol. Chem.* **1996**, *271*, 23973–23984. [CrossRef]
55. Chen, S.; Hu, Y.; Ye, X.; Li, G.; Yu, G.; Xue, C.; Chai, W. Sequence determination and anticoagulant and antithrombotic activities of a novel sulfated fucan isolated from the sea cucumber *Isostichopus badionotus*. *Biochim. Biophys. Acta Gen. Subj.* **2012**, *1820*, 989–1000. [CrossRef] [PubMed]
56. Rowe, A.; Berendt, A.; Marsh, K.; Newbold, C. *Plasmodium falciparum*: A family of sulfated glycoconjugates disrupts erythrocyte rosettes. *Exp. Parasitol.* **1994**, *79*, 506–516. [CrossRef]
57. Chen, J.-H.; Lim, J.D.; Sohn, E.-H.; Choi, Y.-S.; Han, E.-T. Growth-Inhibitory effect of a fucoidan from brown seaweed *Undaria pinnatifida* on *Plasmodium parasites*. *Parasitol. Res.* **2008**, *104*, 245–250. [CrossRef] [PubMed]
58. Melo, F.R.; Pereira, M.S.; Foguel, D.; Mourão, P.A.S. Antithrombin-Mediated anticoagulant activity of sulfated polysaccharides: Different mechanisms for heparin and sulfated galactans. *J. Biol. Chem.* **2004**, *279*, 20824–20835. [CrossRef] [PubMed]
59. Boddey, J.A.; O'Neill, M.T.; Lopaticki, S.; Carvalho, T.; Hodder, A.N.; Nebl, T.; Wawra, S.; Van West, P.; Ebrahimzadeh, Z.; Richard, D.; et al. Export of malaria proteins requires co-translational processing of the PEXEL motif independent of phosphatidylinositol-3-phosphate binding. *Nat. Commun.* **2016**, *7*, 10470. [CrossRef] [PubMed]
60. Warncke, J.D.; Vakonakis, I.; Beck, H.-P. Plasmodium Helical Interspersed Subtelomeric (PHIST) proteins, at the center of host cell remodeling. *Microbiol. Mol. Biol. Rev.* **2016**, *80*, 905–927. [CrossRef] [PubMed]
61. Tarr, S.J.; Moon, R.W.; Hardege, I.; Osborne, A.R. A conserved domain targets exported PHISTb family proteins to the periphery of Plasmodium infected erythrocytes. *Mol. Biochem. Parasitol.* **2014**, *196*, 29–40. [CrossRef] [PubMed]
62. Mutisya, J.M.; Mobegi, V.A.; Kinyua, J.K.; Kivecu, M.N.; Okoth, R.O.; Chemwor, G.C.; Mwakio, E.W.; Cheruiyot, A.C.; Yeda, R.A.; Okello, C.O.; et al. Characterization of sulfated polysaccharide activity against virulent *Plasmodium falciparum* PHISTb/RLP1 protein. *F1000Research* **2020**, *9*, 1268. [CrossRef]
63. Bastos, M.F.; Albrecht, L.; Kozlowski, E.O.; Lopes, S.C.P.; Blanco, Y.C.; Carlos, B.C.; Castiñeiras, C.; Vicente, C.P.; Werneck, C.C.; Wunderlich, G.; et al. Fucosylated chondroitin sulfate inhibits *Plasmodium falciparum* cytoadhesion and merozoite invasion. *Antimicrob. Agents Chemother.* **2014**, *58*, 1862–1871. [CrossRef] [PubMed]
64. Burns, A.L.; Dans, M.G.; Balbin, J.M.; de Koning-Ward, T.F.; Gilson, P.R.; Beeson, J.G.; Boyle, M.J.; Wilson, D.W. Targeting malaria parasite invasion of red blood cells as an antimalarial strategy. *FEMS Microbiol Rev.* **2019**, *43*, 223–238. [CrossRef] [PubMed]
65. Bastos, M.F.; Albrecht, L.; Gomes, A.M.; Lopes, S.; Vicente, C.P.; De Almeida, R.P.; Cassiano, G.C.; Fonseca, R.J.C.; Werneck, C.C.; Pavão, M.S.; et al. A new heparan sulfate from the mollusk Nodipecten nodosus inhibits merozoite invasion and disrupts rosetting and cytoadherence of *Plasmodium falciparum*. *Memórias Inst. Oswaldo Cruz* **2019**, *114*, e190088. [CrossRef] [PubMed]
66. Eun, Y.; Tae, H. Pharmaceutical Composition for Preventing or Treating Malaria, Containing Fucoidan as Active. International Patent Application No. WO2021060862, 24 September 2020.
67. Rahmah, Z.; Indriana, N.; Astari, L.F.; Sutikno, A.M. The combined effect of extract seaweed and DHP on placental malaria. *Adv. Soc. Sci. Educ. Humanit. Res.* **2020**, *59*, 452–458. [CrossRef]
68. Atun, R.A.; Bennett, S.; Duran, A. When do Vertical (Stand-Alone) Programmes Have a Place in Health Systems? WHO POLICY Brief. 2008. Available online: https://www.who.int/management/district/services/WhenDoVerticalProgrammesPlaceHealthSystems.pdf (accessed on 12 May 2021).
69. Rahi, M.; Chaturvedi, R.; Das, P.; Sharma, A. India can consider integration of three eliminable disease control programmes on malaria, lymphatic filariasis, and visceral leishmaniasis. *PLoS Pathog.* **2021**, *17*, e1009492. [CrossRef]
70. Sharma, G.; Kar, S.; Ball, W.B.; Ghosh, K.; Das, P.K. The curative effect of fucoidan on visceral leishmaniasis is mediated by activation of MAP kinases through specific protein kinase C isoforms. *Cell. Mol. Immunol.* **2014**, *11*, 263–274. [CrossRef]
71. Tasdemir, D.; Kaiser, M.; Brun, R.; Yardley, V.; Schmidt, T.J.; Tosun, F.; Rüedi, P. Antitrypanosomal and antileishmanial activities of flavonoids and their analogues: In Vitro, In Vivo, Structure-activity relationship, and quantitative structure-activity relationship studies. *Antimicrob. Agents Chemother.* **2006**, *50*, 1352–1364. [CrossRef]

72. Berbert, T.R.N.; De Mello, T.F.P.; Nassif, P.W.; Mota, C.A.; Silveira, A.V.; Duarte, G.C.; Demarchi, I.G.; Aristides, S.M.A.; Lonardoni, M.V.C.; Teixeira, J.J.V.; et al. Pentavalent antimonials combined with other therapeutic alternatives for the treatment of cutaneous and mucocutaneous leishmaniasis: A systematic review. *Dermatol. Res. Pract.* **2018**, *2018*, 9014726. [CrossRef] [PubMed]
73. De Castro Côrtes, L.M.; de Souza Pereira, M.C.; da Silva, F.S.; Pereira, B.A.S.; de Oliveira Junior, F.O.; de Araújo Soares, R.O.; Brazil, R.P.; Toma, L.; Vicente, C.M.; Nader, H.B.; et al. Participation of heparin binding proteins from the surface of Leishmania (Viannia) braziliensis promastigotes in the adhesion of parasites to *Lutzomyia longipalpis* cells (Lulo) In Vitro. *Parasites Vectors* **2012**, *5*, 142. [CrossRef] [PubMed]
74. Maciej-Hulme, M.L.; Skidmore, M.; Price, H.P. The role of heparan sulfate in host macrophage infection by *Leishmania species*. *Biochem. Soc. Trans.* **2018**, *46*, 789–796. [CrossRef] [PubMed]
75. Pires, C.L.; Rodrigues, S.D.; Bristot, D.; Gaeta, H.H.; Toyama, D.d.; Farias, W.R.L.; Toyama, M.H. Evaluation of macroalgae sulfated polysaccharides on the *Leishmania amazoensis* promastigote. *Mar. Drugs.* **2013**, *11*, 934–943. [CrossRef]
76. Minicante, S.A.; Michelet, S.; Bruno, F.; Castelli, G.; Vitale, F.; Sfriso, A.; Morabito, M.; Genovese, G. Bioactivity of phycocolloids against the mediterranean protozoan *Leishmania infantum*: An inceptive study. *Sustainability* **2016**, *8*, 1131. [CrossRef]
77. Kar, S.; Sharma, G.; Das, P.K. Fucoidan cures infection with both antimony-susceptible and -resistant strains of *Leishmania donovani* through Th1 response and macrophage-derived oxidants. *J. Antimicrob. Chemother.* **2011**, *66*, 618–625. [CrossRef] [PubMed]
78. Shadab, M.; Ali, N. Evasion of host defence by *Leishmania donovani*: Subversion of signaling pathways. *Mol. Biol. Int.* **2011**, *2011*, 1–10. [CrossRef]
79. Olivier, M.; Brownsey, R.W.; Reiner, N.E. Defective stimulus-response coupling in human monocytes infected with Leishmania donovani is associated with altered activation and translocation of protein kinase C. *Proc. Natl. Acad. Sci. USA* **1992**, *89*, 7481–7485. [CrossRef] [PubMed]
80. Solano-Gálvez, S.-G.; Álvarez-Hernández, D.-A.; Gutiérrez-Kobeh, L.; Vázquez-López, R. Leishmania: Manipulation of signaling pathways to inhibit host cell apoptosis. *Ther. Adv. Infect. Dis.* **2021**, *8*, 20499361211014977. [CrossRef] [PubMed]
81. Soulat, D.; Bogdan, C. Function of macrophage and parasite phosphatases in leishmaniasis. *Front. Immunol.* **2017**, *8*, 1838. [CrossRef] [PubMed]
82. De Martini, C.C.; De Andrade, J.T.; De Almeida, S.K.M.; Silva, K.L.O.; Eugenio, F.D.R.; Dos Santos, P.S.P.; De Lima, V.M.F. Cellular apoptosis and nitric oxide production in PBMC and spleen from dogs with visceral leishmaniasis. *Comp. Immunol. Microbiol. Infect. Dis.* **2018**, *57*, 1–7. [CrossRef] [PubMed]
83. Hoseini, M.H.M.; Moradi, M.; Alimohammadian, M.H.; Shahgoli, V.K.; Darabi, H.; Rostami, A. Immunotherapeutic effects of chitin in comparison with chitosan against Leishmania major infection. *Parasitol. Int.* **2016**, *65*, 99–104. [CrossRef] [PubMed]
84. Riezk, A.; Raynes, J.G.; Yardley, V.; Murdan, S.; Croft, S.L. Activity of chitosan and its derivatives against *Leishmania* major and *Leishmania mexicana* In Vitro. *Antimicrob. Agents Chemother.* **2020**, *64*, 01772-19. [CrossRef]
85. Antinori, S.; Galimberti, L.; Bianco, R.; Grande, R.; Galli, M.; Corbellino, M. Chagas disease in Europe: A review for the internist in the globalized world. *Eur. J. Intern. Med.* **2017**, *43*, 6–15. [CrossRef]
86. Rassi, A.; Marin-Neto, J.A. Chagas Disease. *Negl. Trop. Dis.* **2015**, *375*, 45–71. [CrossRef]
87. Varikuti, S.; Jha, B.K.; Volpedo, G.; Ryan, N.; Halsey, G.; Hamza, O.M.; McGwire, B.S.; Satoskar, A.R. Host-directed drug therapies for neglected tropical diseases caused by protozoan parasites. *Front. Microbiol.* **2018**, *9*, 2655. [CrossRef] [PubMed]
88. Stillwaggon, E.; Perez-Zetune, V.; Bialek, S.R.; Montgomery, S.P. Congenital chagas disease in the United States: Cost savings through maternal screening. *Am. J. Trop. Med. Hyg.* **2018**, *98*, 1733–1742. [CrossRef] [PubMed]
89. Jackson, Y.; Alirol, E.; Getaz, L.; Wolff, H.; Combescure, C.; Chappuis, F. Tolerance and safety of nifurtimox in patients with chronic chagas disease. *Clin. Infect. Dis.* **2010**, *51*, e69–e75. [CrossRef] [PubMed]
90. Leal, D.; Mansilla, A.; Matsuhiro, B.; Moncada-Basualto, M.; Lapier, M.; Maya, J.; Olea-Azar, C.; De Borggraeve, W. Chemical structure and biological properties of sulfated fucan from the sequential extraction of sub Antarctic *Lessonia* sp. (*Phaeophyceae*). *Carbohydr. Polym.* **2018**, *199*, 304–313. [CrossRef] [PubMed]
91. Wang, L.; Jayawardena, T.U.; Yang, H.-W.; Lee, H.G.; Kang, M.-C.; Sanjeewa, K.K.A.; Oh, J.Y.; Jeon, Y.-J. Isolation, characterization, and antioxidant activity evaluation of a fucoidan from an enzymatic digest of the edible seaweed, *Hizikia fusiforme*. *Antioxidants* **2020**, *9*, 363. [CrossRef] [PubMed]
92. Sousa, W.M.; Silva, R.O.; Bezerra, F.F.; Bingana, R.D.; Barros, F.C.N.; Costa, L.E.C.; Sombra, V.G.; Soares, P.M.G.; Feitosa, J.P.A.; de Paula, R.C.M.; et al. Sulfated polysaccharide fraction from marine algae Solieria filiformis: Structural characterization, gastroprotective and antioxidant effects. *Carbohydr. Polym.* **2016**, *152*, 140–148. [CrossRef] [PubMed]
93. Usoltseva, R.V.; Anastyuk, S.D.; Ishina, I.; Isakov, V.V.; Zvyagintseva, T.N.; Thinh, P.D.; Zadorozhny, P.A.; Dmitrenok, P.S.; Ermakova, S.P. Structural characteristics and anticancer activity in vitro of fucoidan from brown alga Padina boryana. *Carbohydr. Polym.* **2018**, *184*, 260–268. [CrossRef] [PubMed]
94. Barroso, E.M.A.; Costa, L.S.; Medeiros, V.P.; Cordeiro, S.L.; Costa, M.S.S.P.; Franco, C.R.C.; Nader, H.B.; Leite, E.L.; Rocha, H.A.O. A non-anticoagulant heterofucan has antithrombotic activityin vivo. *Planta Medica* **2008**, *74*, 712–718. [CrossRef] [PubMed]
95. Silva, N.; Araújo, N.; Daniele-Silva, A.; Oliveira, J.; Medeiros, J.; Araújo, R.; Ferreira, L.; Rocha, H.; Silva-Junior, A.; Silva, M.; et al. Antimicrobial activity of chitosan oligosaccharides with special attention to antiparasitic potential. *Mar. Drugs* **2021**, *19*, 110. [CrossRef] [PubMed]

96. Yan, N.; Chen, X. Don't waste seafood waste: Turning cast-off shells into nitrogen-rich chemicals would benefit economics and the environment. *Nature* **2015**, *524*, 155–157. [CrossRef] [PubMed]
97. Šimat, V.; ElAbed, N.; Kulawik, P.; Ceylan, Z.; Jamroz, E.; Yazgan, H.; Čagalj, M.; Regenstein, J.M.; Özogul, F. Recent advances in marine-based nutraceuticals and their health benefits. *Mar. Drugs* **2020**, *18*, 627. [CrossRef] [PubMed]
98. Ke, C.-L.; Deng, F.-S.; Chuang, C.-Y.; Lin, C.-H. Antimicrobial actions and applications of chitosan. *Polymers* **2021**, *13*, 904. [CrossRef] [PubMed]
99. McManus, D.P.; Dunne, D.; Sacko, M.; Utzinger, J.; Vennervald, B.J.; Zhou, X.-N. Schistosomiasis. *Nat. Rev. Dis. Prim.* **2018**, *4*, 13. [CrossRef] [PubMed]
100. Chuah, C.; Jones, M.; Burke, M.; McManus, D.P.; Gobert, G.N. Cellular and chemokine-mediated regulation in schistosome-induced hepatic pathology. *Trends Parasitol.* **2014**, *30*, 141–150. [CrossRef] [PubMed]
101. Colley, D.G.; Bustinduy, A.L.; Secor, W.E.; King, C.H. Human schistosomiasis. *Lancet* **2014**, *383*, 2253–2264. [CrossRef]
102. Sanjeewa, K.K.A.; Jayawardena, T.U.; Kim, H.-S.; Kim, S.-Y.; Fernando, I.S.; Wang, L.; Abetunga, D.; Kim, W.-S.; Lee, D.-S.; Jeon, Y.-J. Fucoidan isolated from *Padina commersonii* inhibit LPS-induced inflammation in macrophages blocking TLR/NF-κB signal pathway. *Carbohydr. Polym.* **2019**, *224*, 115195. [CrossRef]
103. Yu, H.-H.; Ko, E.C.; Chang, C.-L.; Yuan, K.S.-P.; Wu, A.T.H.; Shan, Y.-S.; Wu, S.-Y. Fucoidan inhibits radiation-induced pneumonitis and lung fibrosis by reducing inflammatory cytokine expression in lung tissues. *Mar. Drugs* **2018**, *16*, 392. [CrossRef] [PubMed]
104. Bai, X.; Li, M.; Wang, X.; Chang, H.; Ni, Y.; Li, C.; He, K.; Wang, H.; Yang, Y.; Tian, T.; et al. Therapeutic potential of fucoidan in the reduction of hepatic pathology in murine schistosomiasis japonica. *Parasites Vectors* **2020**, *13*, 451. [CrossRef]
105. Clode, P.L.; Koh, W.H.; Thompson, R.A. Life without a host cell: What is cryptosporidium? *Trends Parasitol.* **2015**, *31*, 614–624. [CrossRef] [PubMed]
106. Gunasekera, S.; Zahedi, A.; O'Dea, M.; King, B.; Monis, P.; Thierry, B.; Carr, J.M.; Ryan, U. Organoids and bioengineered intestinal models: Potential solutions to the cryptosporidium culturing dilemma. *Microorganisms* **2020**, *8*, 715. [CrossRef]
107. Mammeri, M.; Chevillot, A.; Chenafi, I.; Thomas, M.; Julien, C.; Vallée, I.; Polack, B.; Follet, J.; Adjou, K.T. Molecular characterization of *Cryptosporidium* isolates from diarrheal dairy calves in France. *Veter. Parasitol. Reg. Stud. Rep.* **2019**, *18*, 100323. [CrossRef]
108. Li, X.; Nguyen, T.; Xiao, C.; Levy, A.; Akagi, Y.; Silkie, S.; Atwill, E.R. Prevalence and genotypes of cryptosporidium in wildlife populations co-located in a protected watershed in the pacific northwest, 2013 to 2016. *Microorganisms* **2020**, *8*, 914. [CrossRef]
109. Merga, Y.; Campbell, B.J.; Rhodes, J.M. mucosal barrier, bacteria and inflammatory bowel disease: Possibilities for therapy. *Dig. Dis.* **2014**, *32*, 475–483. [CrossRef] [PubMed]
110. Kato, K.; Ishiwa, A. The role of carbohydrates in infection strategies of enteric pathogens. *Trop. Med. Health* **2015**, *43*, 41–52. [CrossRef] [PubMed]
111. McGuckin, M.A.; Lindén, S.K.; Sutton, P.; Florin, T.H. Mucin dynamics and enteric pathogens. *Nat. Rev. Genet.* **2011**, *9*, 265–278. [CrossRef] [PubMed]
112. Ortega-Barria, E.; Boothroyd, J.C. A toxoplasma lectin-like activity specific for sulfated polysaccharides is involved in host cell infection. *J. Biol. Chem.* **1999**, *274*, 1267–1276. [CrossRef]
113. Ishiwa, A.; Kobayashi, K.; Takemae, H.; Sugi, T.; Gong, H.; Recuenco, F.C.; Murakoshi, F.; Inomata, A.; Horimoto, T.; Kato, K. Effects of dextran sulfates on the acute infection and growth stages of *Toxoplasma gondii*. *Parasitol. Res.* **2013**, *112*, 4169–4176. [CrossRef] [PubMed]
114. Recuenco, F.C.; Kobayashi, K.; Ishiwa, A.; Enomoto-Rogers, Y.; Fundador, N.G.V.; Sugi, T.; Takemae, H.; Iwanaga, T.; Murakoshi, F.; Gong, H.; et al. Gellan sulfate inhibits *Plasmodium falciparum* growth and invasion of red blood cells In Vitro. *Sci. Rep.* **2014**, *4*, 4723. [CrossRef] [PubMed]
115. Inomata, A.; Murakoshi, F.; Ishiwa, A.; Takano, R.; Takemae, H.; Sugi, T.; Recuenco, F.; Horimoto, T.; Kato, K. Heparin interacts with elongation factor 1α of *Cryptosporidium parvum* and inhibits invasion. *Sci. Rep.* **2015**, *5*, 11599. [CrossRef] [PubMed]
116. Maruyama, H.; Tanaka, M.; Hashimoto, M.; Inoue, M.; Sasahara, T. The suppressive effect of *Mekabu fucoidan* on an attachment of *Cryptosporidium parvum* oocysts to the intestinal epithelial cells in neonatal mice. *Life Sci.* **2007**, *80*, 775–781. [CrossRef] [PubMed]
117. Mammeri, M.; Chevillot, A.; Thomas, M.; Polack, B.; Julien, C.; Marden, J.-P.; Auclair, E.; Vallee, I.; Adjou, K.T. Efficacy of chitosan, a natural polysaccharide, against *Cryptosporidium parvum* In Vitro and In Vivo in neonatal mice. *Exp. Parasitol.* **2018**, *194*, 1–8. [CrossRef] [PubMed]
118. Mérou, N.; Lecadet, C.; Pouvreau, S.; Arzul, I. An eDNA/eRNA-based approach to investigate the life cycle of non-cultivable shellfish micro-parasites: The case of *Bonamia ostreae*, a parasite of the European flat oyster *Ostrea edulis*. *Microb. Biotechnol.* **2020**, *13*, 1807–1818. [CrossRef] [PubMed]
119. Paulish-Miller, T.E.; Augostini, P.; Schuyler, J.A.; Smith, W.L.; Mordechai, E.; Adelson, M.E.; Gygax, S.E.; Secor, W.E.; Hilbert, D.W. *Trichomonas vaginalis* metronidazole resistance is associated with single nucleotide polymorphisms in the nitroreductase genes ntr4Tv and ntr6Tv. *Antimicrob. Agents Chemother.* **2014**, *58*, 2938–2943. [CrossRef] [PubMed]
120. Bouchemal, K.; Bories, C.; Loiseau, P. Strategies for prevention and treatment of *Trichomonas vaginalis* infections. *Clin. Microbiol. Rev.* **2017**, *30*, 811–825. [CrossRef]
121. Telles, C.B.S.; Mendes-Aguilar, C.; Fidelis, G.P.; Frasson, A.P.; Pereira, W.O.; Scortecci, K.C.; Camara, R.B.G.; Nobre, L.T.D.B.; Costa, L.S.; Tasca, T.; et al. Immunomodulatory effects and antimicrobial activity of heterofucans from *Sargassum filipendula*. *J. Appl. Phycol.* **2018**, *30*, 569–578. [CrossRef]

122. Asker, M.S.; Kady, E.M.; Mahmoud, M.G. New trends of the polysaccharides as a drug. *World J. Agric. Soil Sci.* **2019**, *3*, 114–119. [CrossRef]
123. Sanjeewa, K.; Jeon, Y.-J. Fucoidans as scientifically and commercially important algal polysaccharides. *Mar. Drugs* **2021**, *19*, 284. [CrossRef] [PubMed]
124. Andrew, M.; Jayaraman, G. Marine sulfated polysaccharides as potential antiviral drug candidates to treat Corona Virus disease (COVID-19). *Carbohydr. Res.* **2021**, *505*, 108326. [CrossRef] [PubMed]
125. Azzouz, N.; Kamena, F.; Laurino, P.; Kikkeri, R.; Mercier, C.; Cesbron-Delauw, M.-F.; Dubremetz, J.-F.; De Cola, L.; Seeberger, P.H. *Toxoplasma gondii* secretory proteins bind to sulfated heparin structures. *Glycobiology* **2013**, *23*, 106–120. [CrossRef] [PubMed]
126. Liu, M.; Liu, Y.; Cao, M.-J.; Liu, G.-M.; Chen, Q.; Sun, L.; Chen, H. Antibacterial activity and mechanisms of depolymerized fucoidans isolated from *Laminaria japonica*. *Carbohydr. Polym.* **2017**, *172*, 294–305. [CrossRef]
127. Palanisamy, S.; Vinosha, M.; Rajasekar, P.; Anjali, R.; Sathiyaraj, G.; Marudhupandi, T.; Selvam, S.; Prabhu, N.M.; You, S. Antibacterial efficacy of a fucoidan fraction (Fu-F2) extracted from *Sargassum polycystum*. *Int. J. Biol. Macromol.* **2019**, *125*, 485–495. [CrossRef] [PubMed]
128. Choi, G.-W.; Lee, Y.-B.; Cho, H.-Y. Interpretation of non-clinical data for prediction of human pharmacokinetic parameters: In Vitro-In Vivo extrapolation and allometric scaling. *Pharmaceutics* **2019**, *11*, 168. [CrossRef] [PubMed]
129. Silchenko, A.S.; Rasin, A.B.; Kusaykin, M.I.; Malyarenko, O.S.; Shevchenko, N.M.; Zueva, A.O.; Kalinovsky, A.I.; Zvyagintseva, T.N.; Ermakova, S.P. Modification of native fucoidan from *Fucus evanescens* by recombinant fucoidanase from marine bacteria Formosa algae. *Carbohydr. Polym.* **2018**, *193*, 189–195. [CrossRef]
130. Belik, A.; Silchenko, A.; Malyarenko, O.; Rasin, A.; Kiseleva, M.; Kusaykin, M.; Ermakova, S. Two new alginate lyases of PL7 and PL6 families from polysaccharide-degrading bacterium *Formosa algae* KMM 3553T: Structure, properties, and products analysis. *Mar. Drugs* **2020**, *18*, 130. [CrossRef] [PubMed]

marine drugs

MDPI

Article

A Novel C1q Domain-Containing Protein Isolated from the Mollusk *Modiolus kurilensis* Recognizing Glycans Enriched with Acidic Galactans and Mannans

Andrei V. Grinchenko [1], Alex von Kriegsheim [2], Nikita A. Shved [1,3], Anna E. Egorova [3], Diana V. Ilyaskina [3], Tatiana D. Karp [3], Nikolay V. Goncharov [1,3], Irina Y. Petrova [1] and Vadim V. Kumeiko [1,3,*]

1 A.V. Zhirmunsky National Scientific Center of Marine Biology, Far Eastern Branch, Russian Academy of Sciences, 690041 Vladivostok, Russia; grishagrin@mail.ru (A.V.G.); nikitawayfarer@yandex.ru (N.A.S.); goncharovnv.GN@gmail.com (N.V.G.); iupet@mail.ru (I.Y.P.)

2 Institute of Genetics and Cancer, The University of Edinburgh, Edinburgh EH4 2XU, UK; Alex.VonKriegsheim@ed.ac.uk

3 Institute of Life Sciences and Biomedicine, Far Eastern Federal University, 690922 Vladivostok, Russia; bioanna1995@gmail.com (A.E.E.); Ilyaskinadiana0506@gmail.com (D.V.I.); tachellabio@gmail.com (T.D.K.)

* Correspondence: vkumeiko@yandex.ru; Tel.: +7-902-555-1821

Abstract: C1q domain-containing (C1qDC) proteins are a group of biopolymers involved in immune response as pattern recognition receptors (PRRs) in a lectin-like manner. A new protein MkC1qDC from the hemolymph plasma of *Modiolus kurilensis* bivalve mollusk widespread in the Northwest Pacific was purified. The isolation procedure included ammonium sulfate precipitation followed by affinity chromatography on pectin-Sepharose. The full-length MkC1qDC sequence was assembled using de novo mass-spectrometry peptide sequencing complemented with N-terminal Edman's degradation, and included 176 amino acid residues with molecular mass of 19 kDa displaying high homology to bivalve C1qDC proteins. MkC1qDC demonstrated antibacterial properties against Gram-negative and Gram-positive strains. MkC1qDC binds to a number of saccharides in Ca^{2+}-dependent manner which characterized by structural meta-similarity in acidic group enrichment of galactose and mannose derivatives incorporated in diversified molecular species of glycans. Alginate, κ-carrageenan, fucoidan, and pectin were found to be highly effective inhibitors of MkC1qDC activity. Yeast mannan, lipopolysaccharide (LPS), peptidoglycan (PGN) and mucin showed an inhibitory effect at concentrations three orders of magnitude greater than for the most effective saccharides. MkC1qDC localized to the mussel hemal system and interstitial compartment. Intriguingly, MkC1qDC was found to suppress proliferation of human adenocarcinoma HeLa cells in a dose-dependent manner, indicating to the biomedical potential of MkC1qDC protein.

Keywords: bivalve mollusk; C1q domain-containing; lectin-like; pattern recognition receptor; polysaccharides; interstitial compartment

Citation: Grinchenko, A.V.; von Kriegsheim, A.; Shved, N.A.; Egorova, A.E.; Ilyaskina, D.V.; Karp, T.D.; Goncharov, N.V.; Petrova, I.Y.; Kumeiko, V.V. A Novel C1q Domain-Containing Protein Isolated from the Mollusk *Modiolus kurilensis* Recognizing Glycans Enriched with Acidic Galactans and Mannans. *Mar. Drugs* **2021**, *19*, 668. https://doi.org/10.3390/md19120668

Academic Editors: Yasuhiro Ozeki, Yuki Fujii and Marco Gerdol

Received: 28 October 2021
Accepted: 24 November 2021
Published: 26 November 2021

1. Introduction

The specific interaction of proteins with carbohydrate underlies a variety of cell-to-cell communications and cooperation with extracellular matrix (ECM). Cell surface forms unique glycosylation patterns, which play a significant role in the immune defense providing the self and non-self-discrimination abilities. Meanwhile, the most known group of carbohydrate binding proteins are lectins, which demonstrated diversified affinity to a variety of carbohydrate determinants. Initially all carbohydrate binding proteins were recognized as lectins. However, extensive genomic research distinguished separate groups of proteins recognizing glycans of pathogen-associated molecular patterns (PAMPs) [1]. Among these biopolymers, the C1qDC proteins form very interesting family widespread in different groups of invertebrates that probably predicts an origin of vertebrates complement system involved in the immune response.

In the vertebrates, the C1q protein is a key component of the complement system, triggering the classical pathway of its activation. In invertebrates, C1q domains and corresponding proteins show a wide range of ligands, including a variety of PAMPs. Interestingly, the largest number of genes encoding C1qDC proteins was found in bivalve species genomes: 337 in *Crassostrea gigas* [2], 296 in *Pinctada fucata* [3], 445 in *Modiolus philippinarum* [4], 1182 in *Ruditapes philippinarum* [5], 554 in *Saccostrea glomerata* [6] and 476 in *Crassostrea virginica* [7]. This unusually large number of genes probably arose due to multiple duplications of genomic fragments. Studies suggest that this process occurred independently in different species [5,7]. Presumably, the abundance of C1qDC proteins allows covering the protective needs of the bivalve against various pathogens due to the potential structural diversity of PAMPs. Many bivalve C1qDC proteins are soluble, secreting PRRs that agglutinate and opsonize foreign agents by PAMPs recognizing [8–12], but several studies also have shown that they are involved in embryogenesis [9,10,13], shell formation [14–16] and interaction with predators [14]. Until recent years, some bivalve C1qDC proteins are classified either as lectins [8,17–19] or as lectin-like proteins, which emphasizes their probable origin as lectins with subsequent diversification [2,7,20].

Currently, lectins as biomedical and biotechnological tools are a very active area of research [21–25]. In recent years, great attention has been paid to carbohydrate binding proteins derived from marine organisms. Bivalves have the most extensive repertoires of lectins, which allows them to thrive in an environment saturated with pathogens [26]. In addition to antimicrobial properties, bivalve lectins show promising antitumor and antiviral activity [22,24]. Bivalve C1qCD proteins are a large group of carbohydrate-recognizing molecules which are interesting due to functional and property similarities with lectins and can be a new object of biotechnological usage.

In this work, a novel C1qDC bivalve protein from *M. kurilensis* (MkC1qDC) was identified and an effective protocol for its isolation was developed. Immunohistochemical detection showed intracellular localization of target protein only in hemocytes and MkC1qDC presence to the hemal system, ECM and interstitial components. The physicochemical and functional properties of MkC1qDC was characterized, including the carbohydrate specificity and antimicrobial activity. Furthermore, MkC1qDC demonstrated the inhibition of HeLa proliferation in a dose-dependent manner, suggesting biomedical potential of this protein.

2. Results

2.1. MkC1qDC Purification and Electrophoretic Properties

Cell-free hemolymph (plasma) of *M. kurilensis* is characterized by the highest hemagglutination (HA) activity towards human erythrocytes equally for all groups of the AB0 system displaying 1/64–1/256 titers against 6×10^7 cells per mL. Analysis of the carbohydrate specificity of *M. kurilensis* plasma agglutinins by hemagglutination inhibition (HAI) assay showed that mucins type II and type III, mannan, N-acetyl-D-galactosamine, N-acetyl-D-glucosamine, sialic acid, D-(–)-ribose, as well as D-glucuronic and D-galacturonic acids, were inhibitory. Fraction 0–15% of ammonium sulfate precipitation of plasma proteins had no agglutinating activity. Fractions 15−30%, 30−45%, 45−60%, 60−75% and 75−85% had agglutinating activity, but only 60−75% and 75−85% was inhibited by uronic acids and citrus pectin containing a polygalacturonic acid structure. The inhibition activity was Ca^{2+}-dependent as 30 mM Na_2EDTA disposed of it. Thus, in the presented final scheme for the isolation of the target protein (Figure 1a), ammonium sulfate precipitation was reduced to two fractions: 0–60% and 60–85%. Many of the major plasma proteins (Figure 1b, Lane 1), such as 41 kDa, 36 kDa, 29 kDa and some others, was found only in the 0–60% fraction (Figure 1b, Lane 2), while in the fraction 60–85% there were significantly fewer bands, but as a result of concentration, a band with a mass of 19 kDa, corresponding to the target MkC1qDC protein, appears (Figure 1b, Lane 3). After chromatography purification of the 60−85% fraction using pectin-immobilized affinity column (Figure 1c) the band disappeared in the flow-through (Figure 1b, Lane 4), but was recovered in the

Ca^{2+}-binding eluate (Figure 1b, Lane 5). Typically, the protein eluted as a narrow peak upon changing the eluate solution, as noticed by an increase in conductivity (Figure 1c). On average, we were able to purify 2-4 mg of MkC1qDC from 1000 mL of *M. kurilensis* hemolymph plasma.

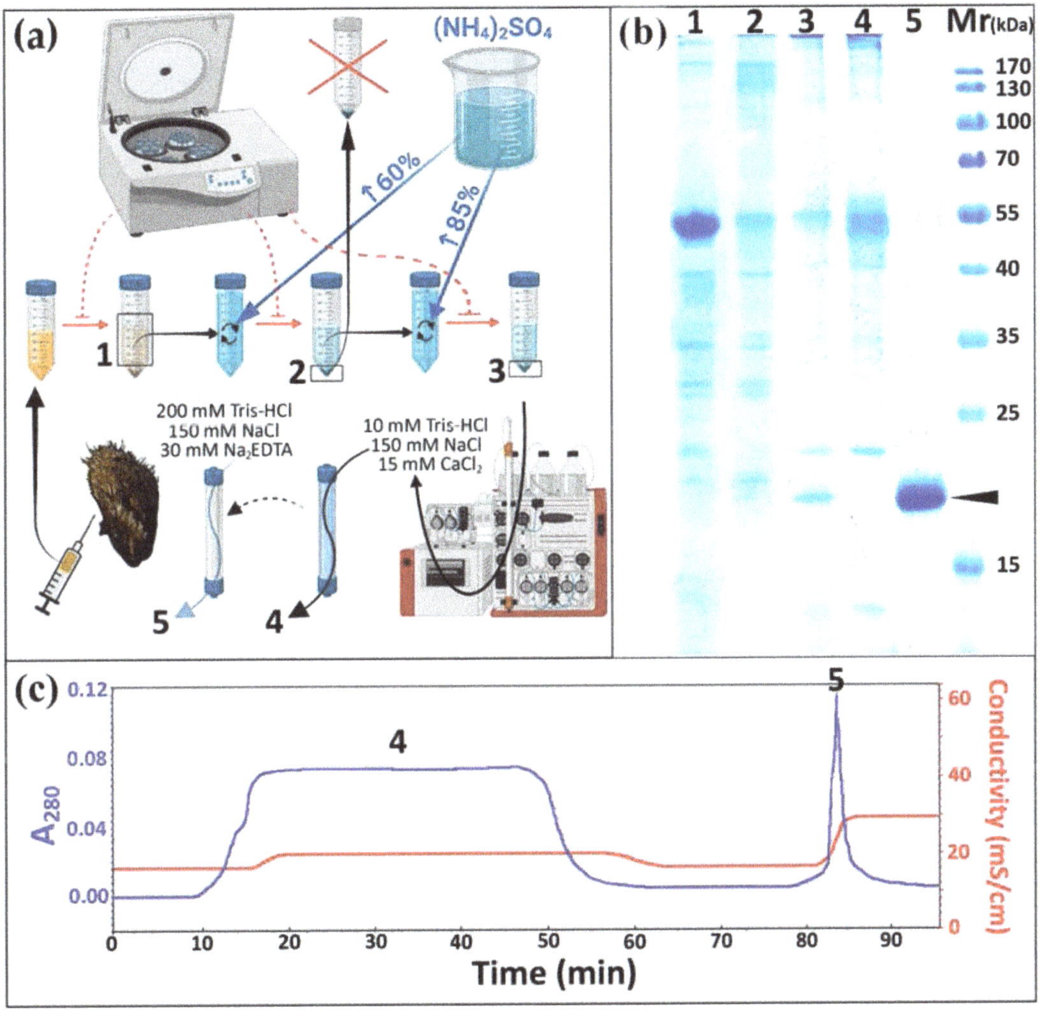

Figure 1. The scheme of MkC1qDC isolation (**a**); sodium dodecyl sulfate polyacrylamide gel electrophoresis (SDS-PAGE) of samples obtained at different stages of MkC1qDC isolation (**b**); MkC1qDC band labeled; elution profile of affinity chromatography on pectin-Sepharose CL-4B (**c**). 1—cell-free hemolymph; 2—sample after precipitation by ammonium sulfate 0–60% saturation; 3—second precipitation by ammonium sulfate before affinity chromatography; 4—flow-through of affinity chromatography; 5—eluted fraction; Mr—molecular weight standards.

2.2. Sequencing Analysis

We initially sequenced the protein by Edman sequencing. MkC1qDC samples were purified independently at five different time points, and we were able to sequence 13 to 41 residues (Figure 2a). However, we were unable to find any significant matches against any of the public databases when using the Basic Local Alignment Search Tool (BLAST, https://blast.ncbi.nlm.nih.gov/Blast.cgi, accessed on 26 May 2021).

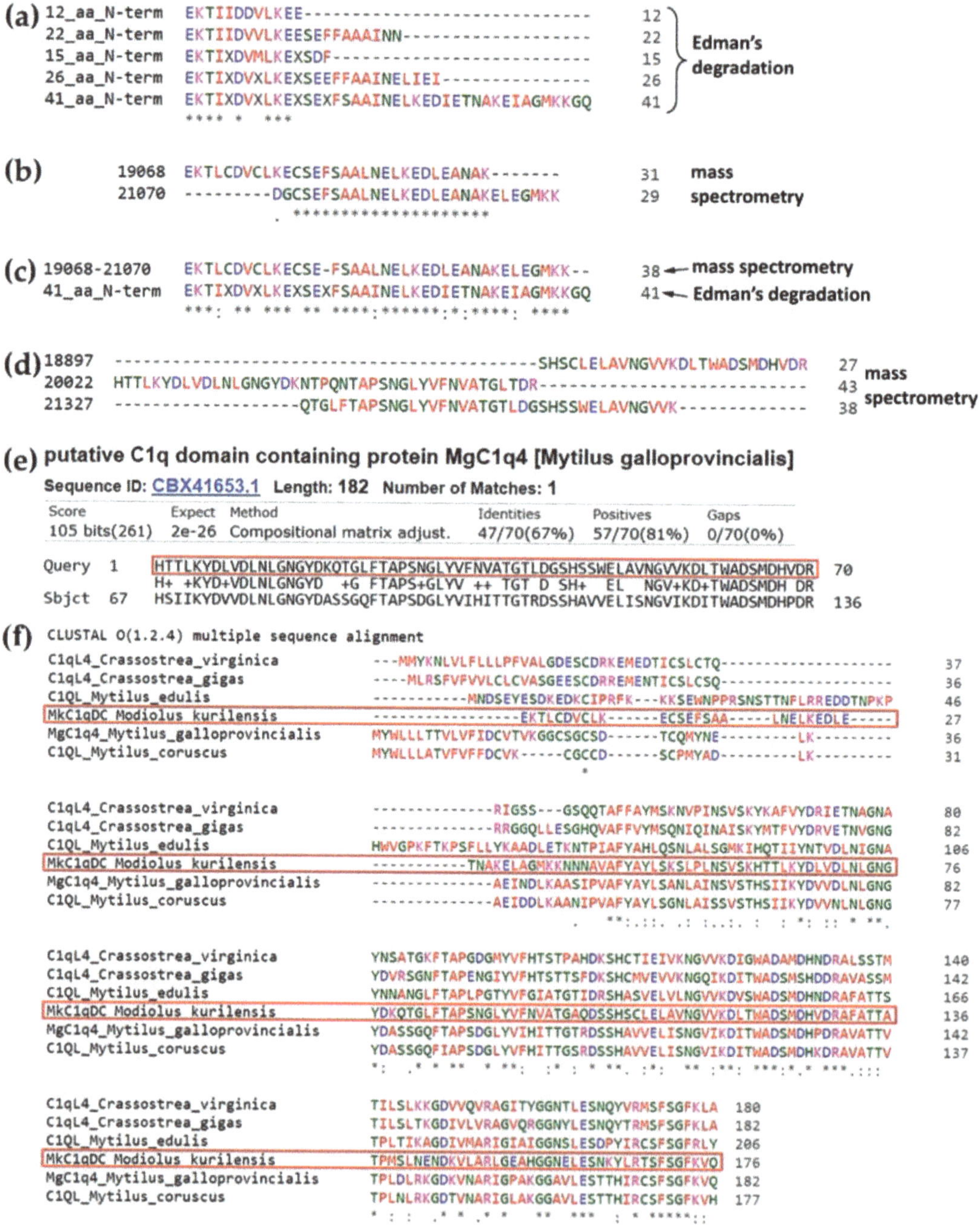

Figure 2. Amino acids sequencing of MkC1qDC: N-terminal peptides from Edman's degradation (**a**); alignment of two fragments obtained by mass-spectrometric de novo sequencing (**b**) identical to Edman's degradation results (**c**); alignment of three de novo peptides (**d**) that had high quality of matches with C1qDC or C1q-like proteins from other Bivalvia (**e**); alignment of full-length MkC1qDC sequence assembly with nearest homologues (**f**).

To complement the data, we applied de novo mass spectrometry peptide sequencing to determine the amino acid sequence of the protein. The purified protein was digested with trypsin and several peptides of up to 50 residues with an Average Local Confidence >80% were identified. Analysis of these fragments allowed us to assemble a sequence that matched to data obtained by Edman sequencing (Figure 2b,c). In addition, a 70 amino acids sequence was generated by alignment of three overlapping peptides (Figure 2d). Analysis of this fragment by BLAST found a match with C1qDC or C1q-like proteins of Bivalvia with high statistical significance and up to 67% identity for CBX41653.1 (Figure 2e and Figure S1). The full-length MkC1qDC sequence (Figure S2) was subsequently assembled using the longer stretches obtained by using the de Bruijn assembler ALPS [27] and CBX41653.1 as a reference. Analysis of assembled MkC1qDC sequence by BLAST detected a putative conserved C1q superfamily domain and high homology to other Bivalvia C1qDC proteins: CBX41653.1; CAC5364865.1; CAG2251157.1; XP_022294274.1; XP_034307311.1 (Figure 2f). The homologous proteins had one C1q domain at the C-terminus, a short unique N-terminus and were in the range of 177–206 residues. Full-length MkC1qDC has 176 amino acids in length, a predicted molecular weight of 19181 Da and a pI 5.2.

2.3. Physicochemical Properties

To further explore the properties of MkC1qDC we used 2D electrophoresis. Purified MkC1qDC was found to be a single polypeptide with a molecular mass of 19 kDa mass (Figure 3a) and a pI of 5.2, both of which match the theoretical pI and molecular weight.

Purified MkC1qDC had high agglutinating activity after incubation for 1 h at temperatures of 0–40 °C, which slightly decreased at 50 °C, and were completely lost at 60 °C or higher (Figure 3b). At the same time, the maximum HA activity was detected after 1 h of incubation in the pH range of 7–8 (Figure 3c). Increasing the pH to nine led to a slight decrease in activity and no activity was detected above pH 10. Acidic solutions up to pH 3 had only a weak effect on HA activity, indicating the wide range of pH stability.

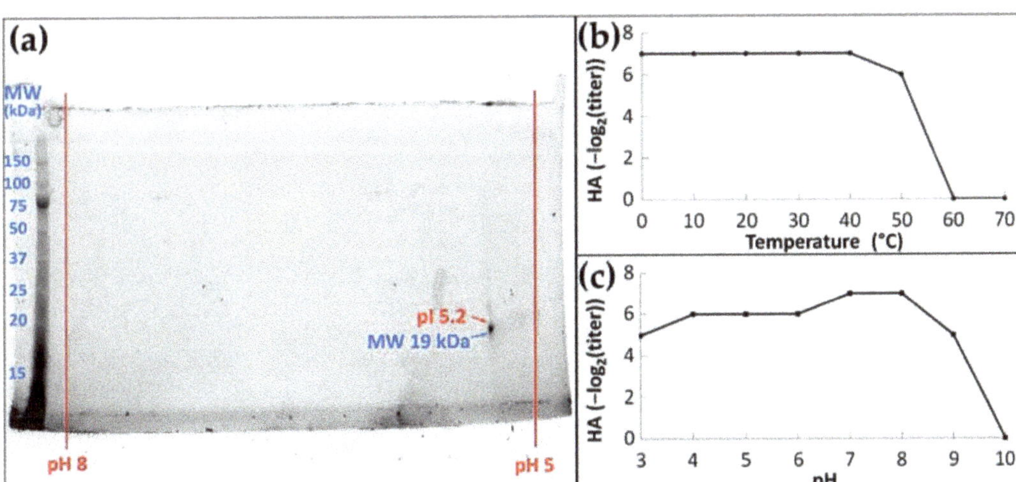

Figure 3. Basic physicochemical properties of the MkC1qDC: 2D electrophoresis of purified MkC1qDC (**a**); the activity after 1 h of incubation at different temperatures (**b**) or pH (**c**). HA: hemagglutination.

2.4. Carbohydrate Specificity

The highest MkC1qDC agglutination activity was against human erythrocytes with no prevalence with respect to blood group. Therefore, all succeeding experiments was performed only with human group 0 erythrocytes.

Among 23 monosaccharides tested by the HAI assay, only six were inhibitory (Table 1). L-gulose, sialic (N-acetylneuraminic), D-galacturonic and D-glucuronic acids showed the highest inhibitory effect among monosaccharides with half maximal inhibitory concentration (IC50) in the range of 1.16–1.46 mg/mL. D-galactose and 2-deoxy-D-galactose were effective in the range of 2.46–5.4 mg/mL. Furthermore, disaccharides such as 2α-mannobiose and D-lactose were noted as inhibitory molecules displaying IC50s starting from 5.13 mg/mL. Mannan from *Saccharomyces cerevisiae*, LPS from *Escherichia coli*, PDG from *Staphylococcus aureus* and mucin type II from porcine stomach were moderately inhibitory at concentrations three orders of magnitude greater than for the most effective saccharides (0.10–0.49 mg/mL). Alginate, κ-carrageenans, fucoidan and citrus pectin were highly effective inhibitors characterized by the lowest IC50 (less than 0.002 mg/mL).

Table 1. Carbohydrates used in the hemagglutination inhibition with MkC1qDC.

Carbohydrates		IC50s
Polysaccharides	alginate	<0.30 μg/mL
	κ-carrageenan	<0.66 μg/mL
	fucoidan	<0.62 μg/mL
	pectin	1.6 μg/mL
	mannan	101 μg/mL
	LPS	125 μg/mL
	PDG	250 μg/mL
	mucin type II	493 μg/mL
	hyaluronic acid	—
	chitosan	—
	agarose	—
	dextran	—
Oligosaccharides	D-lactose	15 mM, 5.13 mg/mL
	2α- mannobiose	30 mM, 10.26 mg/mL
	N-acetyl-D-lactosamine	—
	D-melibiose	—
	D-maltose	—
	D-raffinose	—
Monosaccharides	N-acetylneuraminic (sialic) acid	3.75 mM, 1.16 mg/mL
	D-galacturonic acid	7.5 mM, 1.46 mg/mL
	D-glucuronic acid	7.5 mM, 1.46 mg/mL
	L-gulose	7.5 mM, 1.35 mg/mL
	2-deoxy-D-galactose	15 mM, 2.46 mg/mL
	D-galactose	30 mM, 5.40 mg/mL
	D-mannose	—
	D-glucose	—
	D-fucose	—
	L-fucose	—
	N-acetyl-D-galactosamine	—
	N-acetyl-D-glucosamine	—
	N-acetyl-D-mannosamine	—
	D-glucosamine	—
	α-methyl-D-glucopyranose	—

Table 1. *Cont.*

Carbohydrates		IC50s
Monosaccharides	L-rhamnose	—
	D-ribose	—
	myo-inositol	—
	DL-arabinose	—
	D-xylose	—
	L-sorbose	—
	methyl-β-xylopyranose	—
	D-glucurono-6,3-lactone	—

2.5. Bacterial Agglutination and Antimicrobial Activity

MkC1qDC had antibacterial activity against both Gram-positive (*Bacillus subtilis*, *S. aureus*) and Gram-negative (*Vibrio* sp., *Ruegeria* sp., *E. coli*, *Pseudoalteromonas* sp.) bacteria (Figure 4). At the same time, agglutination occurred most effectively in the cases of *Pseudoalteromonas* sp. and *B. subtilis*, which in the presence of MkC1qDC formed the largest conglomerates, in contrast to *Vibrio* sp., which were united into small sparse groups of 4–15 cells. In addition, MkC1qDC showed bacteriostatic properties against most of the above strains as significant growth suppression started after 4–12 h ($p < 0.05$). The only exception was *Vibrio* sp. The decrease in the density of the *Vibrio* sp. cultures in the presence of MkC1qDC occurred only after 26 h of cultivation ($p < 0.05$) when the control was in a stationary phase (Figure 4).

2.6. Antibody Production and Immunohistochemical Localization of MkC1qDC in Mussel Tissues

To further characterize MkC1qDC polyclonal antibodies were raised in rabbits. The antibody reacted strongly with purified MkC1qDC as well as a polypeptide with the same molecular weight in the plasma of *M. kurilensis* (Figure 5a,b). The upper ~40 kDa band appeared to be an incompletely dissociated dimer, as evidenced by its molecular weight, an increase in this fraction during storage and the disappearance of the component when 2-mercaptoethanol is replaced by dithiothreitol. Finally, antisera and purified immunoglobulin G fraction were tested in indirect enzyme-linked immunosorbent assay (ELISA) and Western blotting, and proved to be suitable for an investigation of tissue-specific MkC1qDC localization taking 1/500 dilution as a working titer.

The intracellular localization of MkC1qDC was most obvious in hemocytes (Figure 5c). A subset of the large cells had stained granules that could occupy almost the entire cytoplasm. In contrast, most of the agranulocytes and all small juvenile hemocytes were not stained. Control samples with pre-immune rabbit serum treatment showed no staining (Figure 5d).

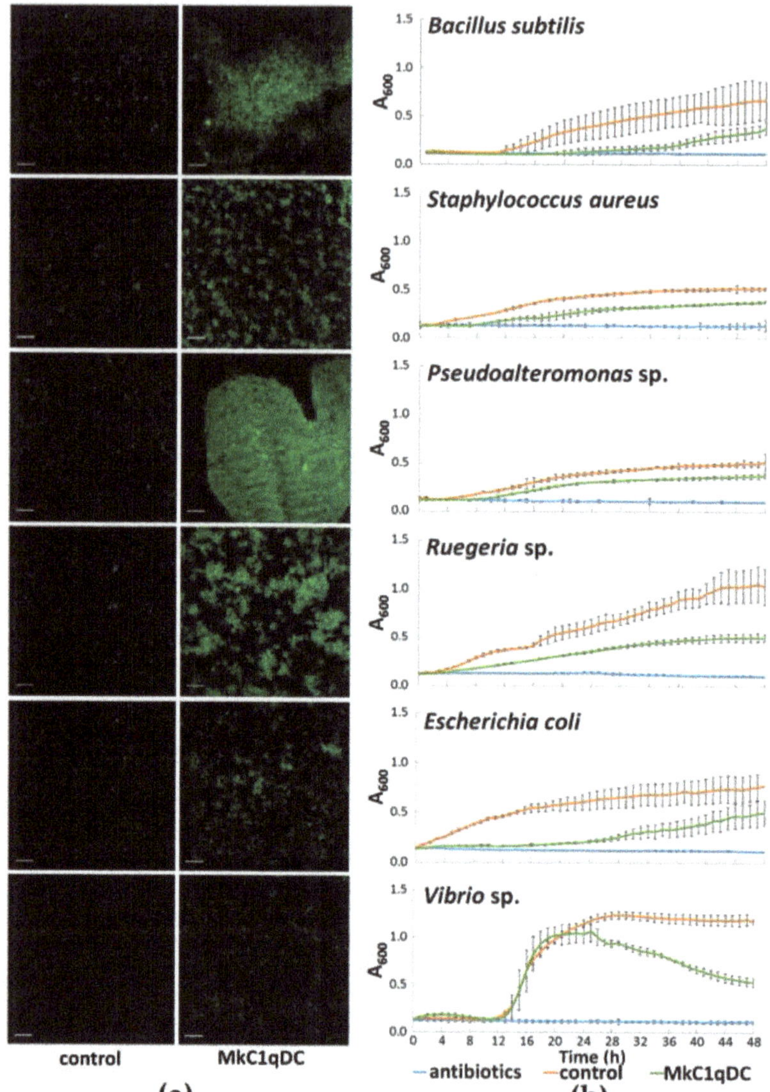

Figure 4. Antibacterial properties of MkC1qDC (0.1 mg/mL) against paraformaldehyde fixed FITC-labeled bacteria (**a**), and live bacterial cultures (**b**). The data on the graphs are presented as a mean ± 95% confidence interval. Scale bars: 10 μm.

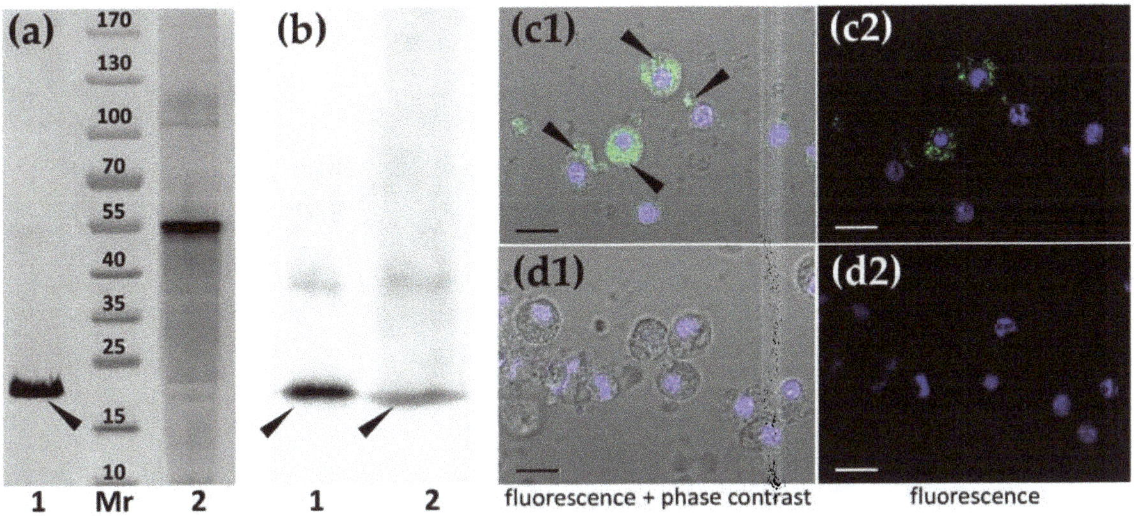

Figure 5. Detection of MkC1qDC (narrow labeled) in the components of M. kurilensis hemolymph: in plasma (**a**,**b**) and in hemocytes (**c**,**d**). (**a**) Electrophoresis; (**b**) Western blotting (1—purified MkC1qDC; 2—plasma; Mr—molecular weight standards); (**c1**,**c2**) treatment with MkC1qDC antibodies; (**d1**,**d2**) control with pre-immune rabbit serum. Nuclei are colored in blue by DAPI. Scale bars: 10 μm.

In paraffin sections MkC1qDC was found in structures associated with the hemal system (Figure 6). The protein was mainly in hemocytes and the walls of hemal sinuses and vessels, as well as in the ECM of connective tissue and interstitial space. Gills were one of the organs expressing the highest levels of the protein. Intense immunofluorescent staining was well defined to the walls of bronchial sinuses inside the ctenidia. Moreover, hemocytes that were present in large numbers inside the sinus, including the abfrontal narrowing, were brightly stained and stood out, particularly compared to other organs (Figure 6a). Mantle, another organ with massive contact with the water environment, had some differences with respect to MkC1qDC localization. As in the previous case, staining was detected along the walls of the hemal sinuses and in the hemocytes located inside them. In addition, intensely stained hemocytes were found in the epithelial wall of the papillae. However, fibrous components were stained most evidently in the marginal zone of the mantle (Figure 6b). The distribution pattern of MkC1qDC in the internal epithelium of different parts of the intestine was identical: the walls of hemal sinuses and hemocytes located inside them or intercalated in between cells of the internal epithelium were stained most intensely (Figure 6c). The digestive gland of *M. kurilensis*, which is formed by the digestive tubules surrounded by the interstitium, hemocytes, as well as the walls of the interstitial space and fibers of relatively large fibrous formations, were positive for MkC1qDC (Figure 6d). The staining for MkC1qDC was weak in the kidney tubules, formed by nephrocytes with concretions and large vacuoles, while the relatively narrow interstitial space between the tubules, filled with fibrous tissue with hemocytes, was stained more intensely. In addition, the walls of concretions, which contain collagen, had slightly more intense staining compared to the control (Figure 6e). In gonads, which are mainly filled with gametes in different stages of development, intense fluorescence was confined to the interstitial space formed by the connective tissue, in which the walls and hemocytes were best stained (Figure 6f). In general, the intensity of the signal in the pericardium was intense, which might be explained by the abundance of hemal canals and hemocytes located within. Additionally, the pericardial epithelium was characterized by slightly more intense fluorescence (Figure 6g). Analysis of the posterior adductor muscle also showed the presence of MkC1qDC at the edge of the hemal vessels and in the hemocytes located inside them with a slight color of the tissue (Figure 6h).

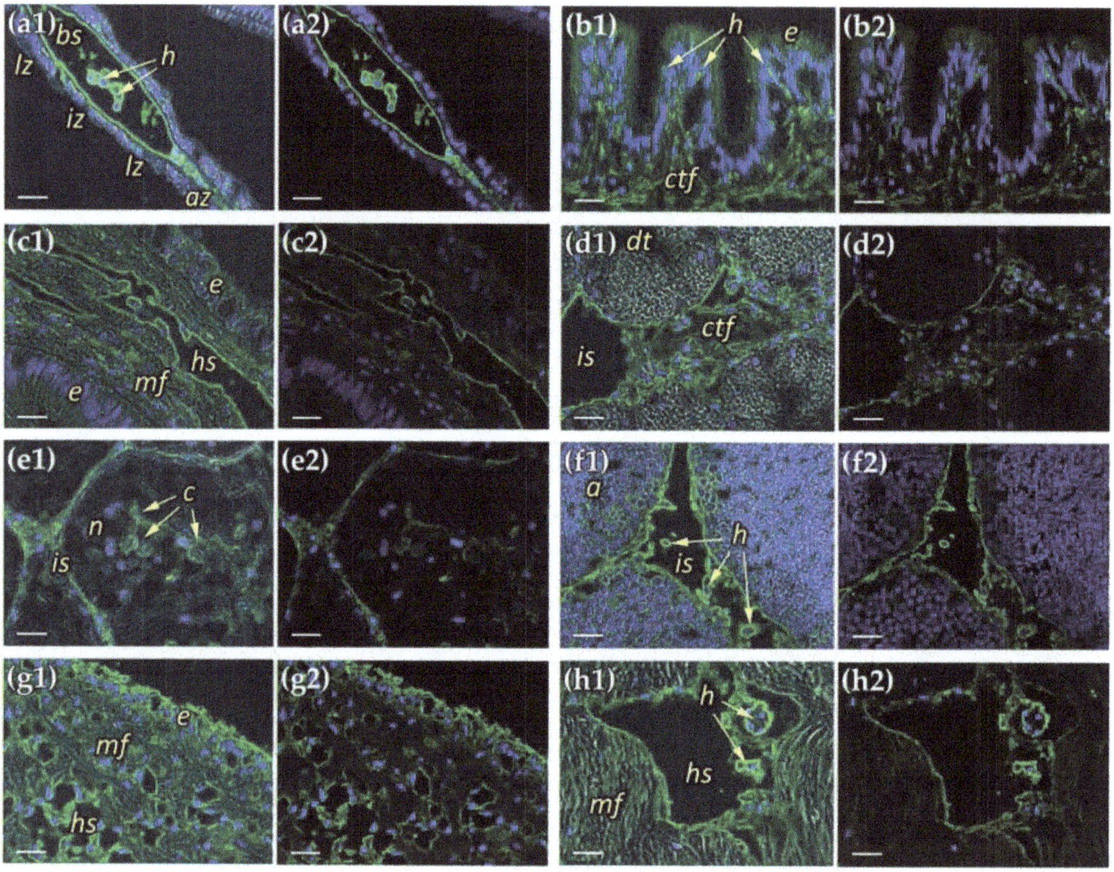

Figure 6. The MkC1qDC localization in organs of *M. kurilensis*: (**a1**,**a2**) gills; (**b1**,**b2**) mantle; (**c1**,**c2**) intestine; (**d1**,**d2**) digestive gland; (**e1**,**e2**) kidney; (**f1**,**f2**) gonad; (**g1**,**g2**) pericardium; (**h1**,**h2**) muscle; (**a1**–**h1**) fluorescence with phase contrast; (**a2**–**h2**) fluorescence only. Abfrontal zone: *az*; acinus: *a*; hemocytes: *h*; hemal sinuses/vessels: *hs*; intermediate zone: *iz*; bronchial sinus: *bs*; interstitial space: *is*; concretions: *c*; lateral zone: *lz*; muscle fibers: *mf*; nephrocytes: *n*; digestive tubule: *dt*; connective tissue fibers: *ctf*; epithelium: *e*. Scale bars: 20 µm.

2.7. Antiproliferative Activity on HeLa Cell Line

Finally, we tested if MkC1qDC may have antiproliferative activity in a mammalian context. To determine this, we monitored HeLa proliferation using high-content, time-lapse microscopy. We used a machine-learning algorithm to segment individual cells and determined cell proliferation in the presence or absence of MkC1qDC. Significant differences between control and treated groups were detected after 4 h (4 µg/mL of MkC1qDC) or 16 h (1 and 2 µg/mL) (Figure 7). Further analysis showed that MkC1qDCa elicited a dose-dependent antiproliferative effect. At the highest concentration, we detected next to no cell proliferation and most cells displayed morphology akin to dead cells.

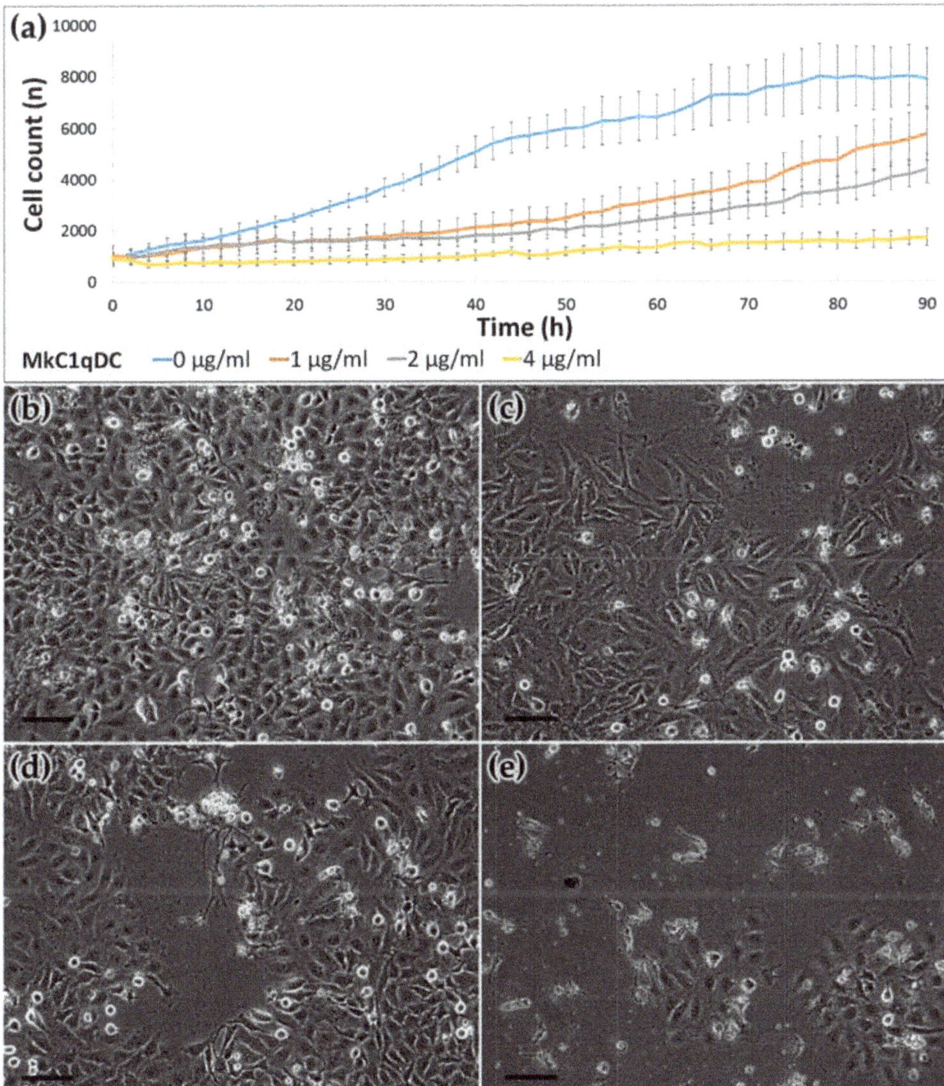

Figure 7. The MkC1qDC effect on HeLa cell line proliferation. Time-lapse cell monitoring and dynamic count based on Cell iQ machine learning during 90 h MkC1qDC treatment (**a**); the HeLa cell cultures after 90 h cultivation with MkC1qDC concentrations 0 µg/mL (control, **b**), 1 µg/mL (**c**), 2 µg/mL (**d**) and 4 µg/mL (**e**). Scale bars: 50 µm.

3. Discussion

Using a HAI technique to study the carbohydrate specificity of the *M. kurilensis* plasma, we observed a strong inhibitory effect on D-glucuronic acids. Earlier, in a closely related species *Modiolus modiolus*, the same analysis did not reveal the same specificity [28]. Lectins modiolin H and E isolated from *M. modiolus* had a different molecular weight [28,29] from lectin-like MkC1qDC observed in this study (Figures 1b and 3a). Obtaining N-terminal sequence of 41 amino acids, any significant homology matches were not identified, which indicates the uniqueness of the protein's N-terminus. All obtained data indicate that a new protein with unique carbohydrate-recognizing properties has been isolated. From mass spectrometry and de novo assembling we observed the full-length MkC1qDC sequence with high similarity to MgC1q4 (Figure 2d–f). We find out that MkC1qDC has similar structure to the group of Bivalvia sialic acid-binding lectins determined as C1qDC [30]. Moreover, molecular weight and pI of MkC1qDC were very similar to that of other globular head C1qDC bivalve proteins isolated from *Chlamys farreri* (CfC1qDC) and *Argopecten irradians* (AiC1qDC-1) [8,9].

C1qDC proteins have been found to be immune factors in many cases. Various pathogens and their components induce an increase in C1qDC proteins expression, as well as C1qDC proteins demonstrating the ability to bind PDG, LPS, poly I:C, mannan, β-1,3-glucan, yeast glucan and live bacteria, generally indicating their role as PRRs [9–13,18–20,31–38]. Agglutinating activity of MkC1qDC was observed against Gram-positive and Gram-negative bacteria, but with slightly more pronounced selectivity for certain strains, such as *Pseudoalteromonas* sp. and *B. subtilis* (Figure 4). Agglutinating activity similar to that discussed in this study was observed for recombinant proteins from *C. farreri* (CfC1qDC) and *A. irradians* (AiC1qDC-1, AiC1qDC-2) scallops [8,9,11]. Interestingly, HA correlates with the content of the 19 kDa polypeptide in the *M. kurilensis* plasma [39], which probably means the importance of MkC1qDC as a soluble PRR with agglutinating function.

It is currently known that C1qDC proteins in the bivalve can be expressed in in all organs [8,16,17,20,37], especially in hepatopancreas [9,11,19,33], hemocytes [12,13,40], mantle [12,18] and gills [19]. However, transcription in hemocytes invariably increased upon immune stimulation. This fact indicates the clearly inducible nature of C1qDC synthesis and their involvement in the immune response with hemolymph. MkC1qDC was found in all organs of the mollusk; however, the only structures with intracellular localization were hemocytes (Figure 5), which are ubiquitous in the elements of the hemal system or sometimes migrate in the epithelium of the mantle and the intestine. In all other cases, MkC1qDC was associated with extracellular structures, mainly with the walls of the hemal system and interstitial space, as well as with ECM fibers (Figure 6). Organs with the highest expression are the gills, the edge of the mantle and the pericardium—these are tissues with the highest exposure to pathogens. Thus, it can be assumed that the main cells synthesizing MkC1qDC are hemocytes, and due to good solubility, it circulates throughout the body and further can be fixed on the components of the extracellular matrix, performing immune or other possible additional functions.

The role of C1qDC proteins as PRRs implies their specificity for various PAMPs; this study also showed the high sensitivity of MkC1qDC towards mannan, LPS and PDG (Table 1). However, the specificity toward alginate, κ-carrageenan, fucoidan and pectin was significantly higher, making MkC1qDC protein characterized by most pronounced specificity to glycans enriched with acid galactans and mannans. These polysaccharides are mainly components of algal cell walls, which explains some of their structural similarity. Besides their obvious feeding role, microalgae can be invasive in bivalve and initiate some pathogenicity processes. In particular, earlier this was shown for *M. kurilensis* and *Coccomyxa parasitica* [41]. At the same time, for Chlorophyta the content of alginate and pectin in the cell walls is known [42]. Regarding selectivity to monosaccharides, the affinity of bivalve C1qDC proteins for sialic acid was only shown [17,18,30], while in the current study the highest activity was detected for sialic acid, but also for other acetylated monosaccharides such as uronic acids, as well as for D-galactose, 2-deoxy-D-galactose, L-gulose and disaccharides D-lactose and 2α-mannobiose. At the same time, many lectins showed the wide spectrum of carbohydrate specificity, especially for oligosaccharide with different structures [43,44]. It was also demonstrated that point amino acid substitutions in the carbohydrate-recognition domain of C-type lectin can lead to a significant change in affinity from mannose to methyl L-fucosides [45] or galactose [46] as well as a ricin-B chain-like galactose-binding protein which was obtained from sialic acid-binding lectin [47].

The range of carbohydrate specificity determines the potential in the use of lectins and lectin-like proteins in biotechnology and biomedical science. During tumor cells transformation, a change in the carbohydrates types of their surface occurs. Therefore, the lectins and lectin-like proteins that are able to recognize particular carbohydrate patterns also show antitumor activity. Impotence of cell surface sialylation in tumorigenesis and metastasis was known for a long time [48,49]. The sialic acid-binding lectin from the mollusk *Haliotis discus discus*, which is also a representative of the C1qDC proteins [30], possesses pronounced antitumor properties [50–52]. Other important saccharides for tumorigenic cell surface alteration are mannosides [53]. Hence, the mannose binding lectin was already used in designing a selective drug delivery system [54,55]. Additionally, oncodiagnostic applications of different lectins with affinity to β-branched galactosides, mannosides, as well as to sialylation and fucosylation glycoconjugates, was shown [56,57]. In the current study, MkC1qDC, which showed sialic acid binding lectin-like activity and the highest affinity to acidic galactans and mannans, significantly inhibited the growth of HeLa at the concentration 1 mg/mL, while at 4 mg/mL caused death of most of the cells, resulting in the absence of the significant growth of cell line even after 90 h of cultivation (Figure 7). This result is only a preliminary indicator of the biomedical potential of the new protein, and further analysis should include the study of both normal and cancer cell lines in order to identify the mechanisms of the antiproliferative activity of MkC1qDC.

Thus, this study first described a novel C1qDC protein, which has immune functions and is associated with the hemal system and interstitial compartment of the marine mollusk *M. kurilensis*. The PRR properties such as antibacterial activity and interactions with PAMPs (mannan, LPS, PDG) were observed. At the same time, agglutination activity and growth inhibition of bacteria had various effects on different strains, although MkC1qDC properties were not associated with their Gram classification. The investigated range of carbohydrate affinity of C1qDC protein has a unique character that, together with its ability to inhibit the growth of the HeLa cells, indicates possible implementation of MkC1qDC protein in biotechnology and biomedicine. Taking into account that the MkC1qDC protein is capable of recognizing acidic galactans and mannans, it may be of interest for creating systems for targeted delivery of therapeutic agents to cancer cells, and can also be included in diagnostic kits for phenotyping the cell surface carbohydrate profile. This is of interest for further research and development involving both native and recombinant MkC1qDC.

4. Materials and Methods

4.1. Purification and Electrophoretic Properties of the MkC1qDC

The MkC1qDC was purified from cell-free hemolymph (plasma) of *M. kurilensis*. Healthy adult mollusks with shells over 70 mm in length were selected from Peter the Great Bay, Sea of Japan (42.892078° N, 132.737502° E and 43.200788° N, 131.914084° E). Hemolymph was isolated immediately from the hemal sinus of the dorsal adductor muscle. Cells were removed by centrifugation for 15 min at 4 °C and 300× *g*. Plasma from different individuals was mixed, clarified by centrifugation for 30 min at 4000× *g* and 4 °C and stored at −80 °C until the next stage.

The isolation of MkC1qDC was carried out in a few steps. Firstly, ammonium sulfate precipitation [58] of *M. kurilensis* plasma proteins with 0−15%, 15−30%, 30−45%, 45−60%, 60−75% and 75−85% salt saturation at 0 °C was produced. Protein coagulates were centrifuged for 30 min at 4000× *g* and 0 °C to precipitate. The obtained precipitates were dissolved in Tris buffer saline (TBS: 10 mM Tris-HCl, 150 mM NaCl, pH 7.5), dialyzed against a large volume of TBS three times at 4 °C, clarified by centrifugation (30 min, 10,000× *g*, 4 °C) and added $CaCl_2$ up to 15 mM just before using. The target activity was determined by hemagglutination assays described below, and included inhibition by uronic (D-glucuronic, D-galacturonic) acids and adding chelants (EDTA, and EGTA) for Ca^{2+}-dependence determination.

Citrus pectin (Copenhagen Pectin A/S) covalently linked to Sepharose CL-4B (Sigma-Aldrich, St. Louis, MO, USA) by divinyl sulfone (Sigma-Aldrich, St. Louis, MO, USA) was applied for affinity chromatography on the final purification step. Loading of prepared fractions with target activity and column washing from unbound components was performed in TBS with 15 mM $CaCl_2$. The elution was carried out with a large buffer solution containing a chelant (200 mM Tris-HCl, 150 mM NaCl, 30 mM Na_2EDTA, pH 7.5). All stages were performed with the flow 0.2 mL/min. The eluate was dialyzed against TBS at 4 °C and divided into aliquots containing 100 µg of protein in cryovials, which were frozen and stored in liquid nitrogen at −196 °C.

Protein concentration was determined by the Bradford method [59], using BSA as standard, and measured by NanoPhotometer P360 (Implen) at a wavelength of 595 nm.

To analyze the quality of protein isolation at each stage, SDS-PAGE based on 12% polyacrylamide gel stained with Coomassie Brilliant Blue G-250 solution was produced [60].

The isoelectric point (pI) and composition of the purified MkC1qDC were determined by 2D electrophoresis using ReadyStrip™ IPG strips pH of 4−7 (Bio-Rad, Hercules, CA, USA) in the first dimension and a gradient of 4−20% polyacrylamide gel in the second dimension.

4.2. Amino Acid Sequencing

N-terminal sequencing was performed according to Edman's degradation method [61] by a Procise 492 Protein Sequencer (Applied Biosystems, Waltham, MA, USA), using the manufacturer's protocol.

For mass spectrometry, MkC1qDC was dissolved in 6 M guanidinium-HCl, 100 mM Tris pH 8, 10 mM chloroacetamide, 5 mM Tris (2-carboxyethyl) phosphine hydrochloride, heated at 95 °C for 5 min. The sample was diluted with water 10-fold and 1/100 trypsin/substrate was added. Digestion was performed at 37 °C for 4 h. The digest was acidified to 1% trifluoroacetic acid final concentration, spun at 15,000× *g* for 10 min and desalted on homemade C18 stage tips as described previously [62]. Peptides were analyzed on a Fusion Lumos mass spectrometer (Thermo Fisher Scientific, Milan, Italy) coupled to an RSLnano uHPLC pump (Thermo Fisher Scientific, Milan, Italy). Peptides were separated on a 15 cm C18 Aurora column (IonOptiks, Victoria, Australia) using a 40 min gradient from 5−35% acetonitrile with 0.05% acetic acid. The instrument was operating in a high/high mode with a resolution of 120k/35k (MS and MS/MS). De novo sequencing was performed using the PEAKS software suite (Bioinformatics Solutions) with 5 ppm/0.01 Da mass accuracy (MS and MS/MS). Cysteine carbamylation fixed and Methionine oxidation (Variable) were selected as modification; trypsin was selected as protease with 2 missed cleavages. The list of peptides was trimmed using a cut-off of 80% Average Local Confidence Score. These high-confidence peptides were assembled into longer stretches using ALPS, a de Bruijn assembler, using a k-value of 7 as described [27]. This resulted in the assembly of 11 protein sequences. Initial BLAST searching against a non-redundant Bivalvia sequence database revealed a significant homology to CBX41653.1. Therefore, it was used as a template to stitch the peptides together by aligning them to CBX41653.1 using Clustal Omega [63]. The theoretical molecular weight and pI of assembled MkC1qDC were determined by ExPASy (http://web.expasy.org/compute_pi/, accessed on 26 May 2021).

4.3. Hemagglutination, Carbohydrate Specificity Assay, pH and Temperature Effects

Hemagglutination (HA) was performed as described previously [64], using fixed human (0, A, B, AB), rabbit, rat and sheep erythrocytes. In the case of purified MkC1qDC the protein concentration of 0.1 mg/mL was used.

To investigate the carbohydrate specificity of MkC1qDC, a hemagglutination inhibition (HAI) test with the presence of one of 41 different saccharides was performed, 12 of which were polymers, 6 were oligomers and others were monomers (Table 1). Each carbohydrate was used in a series concentration of two-fold dilutions from 30 to 0.0029 mM (for polymers, concentration was calculated based on average monomers weight), and the inhibitory effect was determined by a decrease in the lectin HA activity (titer decline for one 2-fold dilution).

Isotonic (150 mM NaCl) buffer solutions were used to determine the pH dependence: glycine-HCl (pH 3), sodium acetate-acetic acid (pH 4, pH 5), sodium cacodylate-HCl (pH 6, pH 7), Tris-HCl (pH 8), glycine-NaOH (pH 9) and carbonate-bicarbonate (pH 10) at a concentration of 20 mM. HA activity was checked after 1 h of incubation at room temperature in TBS/Ca^{2+} with high buffering capacity (50 mM Tris-HCl, 150 mM NaCl, 15 mM $CaCl_2$, pH 7.5).

To study the thermal lability, MkC1qDC samples were incubated for 1 h at temperatures of 0, 10, 20, 30, 40, 50, 60 and 70 °C and then were brought to room temperature before HA reaction.

4.4. Bacterial Agglutination and Bacteriostatic Assays

Gram-negative (*Vibrio* sp., *Ruegeria* sp., *E. coli*, *Pseudoalteromonas* sp.) and Gram-positive (*S. aureus*, *B. subtilis*) bacteria from marine organisms were used for analysis. Strains grew for 2-4 days to log-phase at room temperature in the liquid medium based on sterile-filtered sea water [65].

Agglutination assay was performed as described previously [66] with some modifications. After that, growing bacteria were collected by centrifugation (3000× *g*, 20 min at 4 °C), washed three times with HEPES buffer saline (HBS: 0.1 M HEPES-NaOH, 1.5 M NaCl, pH 7.4) and fixed by 4% PFA solution in HBS. Next, bacterial cells were stained with a fluorescent dye Fluorescein isothiocyanate (FITC, Thermo Fisher Scientific, Milan, Italy) with a final concentration of 0.1 mg/mL, vortexed at 800 rpm for 1 h, and washed 3 times in HBS. The stained suspensions with a final optical density (A_{600}) of 1 were mixed in HBS with 50 mM $CaCl_2$ and 0.1 mg/mL MkC1qDC, or without it as control (3 replicates for each condition). The mixtures were vortexed at 100 rpm for 1 h at room temperature. Visualization was performed using laser scanning microscope FV1200MPE-FV12M-5XX-3XX (Olympus).

For the evaluation of MkC1qDC antimicrobial activity, bacteria were diluted in poor broth medium (PBM: 1.5% peptone, 1.5% NaCl, pH 7.2) to A_{600} = 0.05 and 50 µL placed into each well of the 96-well flat-bottom microplate. Then 50 µL of 0.2 mg/mL MkC1qDC solution in PBM was added in experimental wells, or 50 µL PBM as a positive growth control, or 50 µL antibiotic solution (penicillin 2000 units/mL and streptomycin 2 mg/mL) in PBM as negative control (6 replicates for each condition). The results of bacterial growth inhibition were estimated at 600 nm using the Cytation 5 Cell Imaging Multi-Mode Reader (BioTek). The inhibition efficiency was evaluated by the delay time in the bacteria growth in the presence of lectin vs the positive control.

4.5. Preparation and Validation of Polyclonal Antibodies

The commonly used immunization protocols [67] with some modifications were employed to generate the rabbit polyclonal antibody to MkC1qDC. Briefly, the first intramuscular injection of purified MkC1qDC was performed with complete Freund's adjuvant; then two consecutive injections of the immunogen with incomplete Freund's adjuvant were given at intervals of 20 and 60 days, respectively; the boost was carried out in 14 days by a subcutaneous injection of MkC1qDC in sterile TBS. Each injection contained 250 µg of purified MkC1qDC. Blood was taken 10-14 days after each injection from the ear vein, and the isolated serum was stored at –80 °C.

The immunoglobulin fraction was purified by standard protocol [68] using ammonium sulfate precipitation and Sephadex G-25 (GE Healthcare, Chicago, IL, USA) gel filtration for desalting.

Primary antibody activity was measured by the commonly used protocol of indirect ELISA [69] with horseradish peroxidase-conjugated secondary antibody and o-phenylenediamine as substrate. Results were detected by iMark Microplate Absorbance Reader (Bio-Rad, Hercules, CA, USA) at 492 nm.

In addition, antibody validation using Western blotting was performed by Mini Trans-Blot®Module for Mini-PROTEAN®Tetra Cell (Bio-Rad, Hercules, CA, USA). After electrophoresis, samples were transferred onto the PVDF membrane and placed for 2 h in a 3% BSA solution in TBS with 0.05% Tween-20 (TBST). The purified primary antibody was used at 1/2500, 1/5000 and 1/10000 dilutions (calculated based on ELISA result) in 0.5% BSA prepared on TBST, followed by incubation at 23 °C for 2 h. Then it was incubated with mouse antirabbit secondary antibody conjugated with horseradish peroxidase (Thermo Fisher Scientific, Milan, Italy) at 1/20000 dilution in TBST, and incubated for 1 h at 23 °C. After that, a substrate for chemiluminescent staining Pierce ECL Plus Western Blotting Substrate (Thermo Fisher Scientific, Milan, Italy) was applied to the membrane according to the industrial protocol. The result was recorded on a ChemiDoc Touch Imaging System (Bio-Rad, Hercules, CA, USA). Analysis was performed using native *M. kurilensis* plasma and purified MkC1qDC as samples.

4.6. Immunohistochemistry and Protein Localization

Tissue fragments of the intestine, mantle, muscle, pericardium, gills, gonads, digestive gland and kidney were excised and immediately fixed for 2 h in paraformaldehyde (PFA) solution in phosphate buffered saline (PBS: 10 mM Na_2HPO_4-KH_2PO_4, 137 mM NaCl, 2.7 mM KCl, pH 7.4), washed 10 times in PBS and dehydrated through a progressive series of ethanol, infiltrated by xylene and embedded in paraffin. Tissue sections with a thickness of 10 μm were obtained using an HM-360 rotary microtome (MICROM International GmbH, Germany). The paraffin sections were dewaxed and washed 3 times in PBS with 0.05% Tween-20 (PBST) with further staining without drying.

Suspension of live hemocytes was obtained as previously described [64], diluted in RPMI-1640 culture medium (PanEco, Tokyo, Japan) and placed in an 8-well chamber (Ibidi GmbH, Gräfelfing, Germany) for 30 min at 10 °C for their adhesion. Then the cells were fixed for 10 min at 23 °C by adding a 16% PFA in artificial sea water (ASW: 460 mM NaCl, 9.4 mM KCl, 48.3 mM $MgCl_2$, 6 mM $NaHCO_3$, 10.8 mM $CaCl_2$, 10 mM HEPES, pH 7.5) directly to the culture medium to a final concentration of 1.5%. Then the liquid was taken from the wells, 300 μL of cold methanol was added to them and left at 4 °C for 10 min, after which the chamber was left for longer storage at –20 °C. Immediately before staining, hemocyte preparations were washed 3 times with PBST solution.

Further, in order to permeabilize the membranes, PBS with 0.5% Triton X-100 was applied to the sections and hemocytes for 10 min at 23 °C. The primary antibodies in PBS with 0.5% BSA at a dilution of 1/500 were incubated with preparations for 2 h at 23 °C; secondary goat antibodies conjugated with Alexa Fluor-488 (Thermo Fisher Scientific, Milan, Italy) with the same dilution were incubated for 1 h at 23 °C; nuclei were stained by DAPI (4,6-diamino-2-phenylindole, dihydrochloride; Invitrogen) with a concentration of 1 μg/mL in PBS for 5 min at 23 °C. Between all steps, samples were washed 3 times for 10 min in PBST and, after the last step of staining, embedded in the water-soluble medium Mowiol 4-88 (Sigma-Aldrich, St. Louis, MO, USA). The described procedure with pre-immune rabbit serum as primary antibody was used as a control. The samples were analyzed using FluoView FV1200MPE-FV12M-5XX-3XX laser scanning microscope (Olympus).

4.7. Human Cell Culture and Proliferation Assay

Cell proliferation assay was performed using a high-content imaging system Cell-iQ MLF (CM Technologies). Human adenocarcinoma cell line HeLa (ATCC) was plated in tissue-culture grade 24-well plates at a density of 15,000 cells/cm^2. The cells were cultured at 37 °C and 5% CO_2 in Dulbecco's Modified Eagle's medium (DMEM, Gibco) supplemented with 10% FBS. The cells were cultured for their adhesion and growth stabilization for 18 h, then treated with 1, 2 and 4 μg/mL of MkC1qDC and monitored for 90 h applying time-lapse phase-contrast imaging. Untreated wells were utilized as a control sample group. All series were cultivated and analyzed in six replicates. Cell proliferation data are presented as growth curves indicating total cell count produced automatically based on machine learning.

4.8. Experimental Design and Statistical Rationales

Proteomics experiments were conducted on one purified protein sample. For the analysis of MkC1qDC effect on bacteria and HeLa growing, the Mann–Whitney U test and Kruskal–Wallis H test were used. A p-value less than 0.05 was considered to be statistically significant. Experimental results are presented as a mean ±95% confidence interval.

Supplementary Materials: The following are available online at https://www.mdpi.com/article/10.3390/md19120668/s1, Figure S1: Mass spectrometry data of peptides using for alignment with N-terminus peptides obtained by Edman's degradation (a, b) and for construction 70 amino acids sequence (c–e) with high homology to Bivalvia C1qDC proteins (see Figure 2), Figure S2: The full-length MkC1qDC amino acids sequence in fasta format.

Author Contributions: Conceptualization, V.V.K. and A.V.G.; methodology, V.V.K., A.V.G. and I.Y.P.; validation, A.V.G. and N.A.S.; formal analysis, A.V.G., A.v.K.; investigation, A.V.G., A.v.K., N.A.S., A.E.E., D.V.I., T.D.K. and N.V.G.; data curation, A.v.K.; writing—original draft preparation, A.V.G.; writing—review and editing, V.V.K. and A.v.K.; visualization, A.V.G.; supervision, V.V.K. and A.V.G.; project administration, V.V.K.; funding acquisition, V.V.K. All authors have read and agreed to the published version of the manuscript.

Funding: Protein isolation, characterization of its localization and physicochemical properties were supported by the Ministry of Science and Higher Education of the Russian Federation (project # 0657-2020-0004). Protein structural investigations and anticancer activity were supported by Russian Science Foundation (RSF) grant # 20-15-00378.

Institutional Review Board Statement: The study was conducted according to the guidelines of the Declaration of Helsinki and approved by the Ethics Committee of the Far Eastern Federal University, School of Biomedicine (Ref. #7/12-19-2017).

Data Availability Statement: The mass spectrometry proteomics data have been deposited to the ProteomeXchange Consortium with the dataset identifier PXD027507.

Acknowledgments: The authors are grateful for the opportunity of bivalves collection, hemolymph isolation and initial fractionation that were carried out at the Vostok Marine Biological Station (121082600034-5) and the CKP «Primorsky aquarium», A.V. Zhirmunsky National Scientific Center of Marine Biology FEB RAS (Vladivostok, Russia).

Conflicts of Interest: The authors declare no conflict of interest.

References

1. Wang, W.; Song, X.; Wang, L.; Song, L. Pathogen-Derived Carbohydrate Recognition in Molluscs Immune Defense. *Int. J. Mol. Sci.* **2018**, *19*, 721. [CrossRef]
2. Gerdol, M.; Venier, P.; Pallavicini, A. The genome of the Pacific oyster Crassostrea gigas brings new insights on the massive expansion of the C1q gene family in Bivalvia. *Dev. Comp. Immunol.* **2015**, *49*, 59–71. [CrossRef]
3. Takeuchi, T.; Koyanagi, R.; Gyoja, F.; Kanda, M.; Hisata, K.; Fujie, M.; Goto, H.; Yamasaki, S.; Nagai, K.; Morino, Y.; et al. Bivalve-specific gene expansion in the pearl oyster genome: Implications of adaptation to a sessile lifestyle. *Zool. Lett.* **2016**, *2*, 3. [CrossRef]
4. Sun, J.; Zhang, Y.; Xu, T.; Zhang, Y.; Mu, H.; Zhang, Y.; Lan, Y.; Fields, C.J.; Hui, J.H.L.; Zhang, W.; et al. Adaptation to deep-sea chemosynthetic environments as revealed by mussel genomes. *Nat. Ecol. Evol.* **2017**, *1*, 121. [CrossRef] [PubMed]
5. Mun, S.; Kim, Y.-J.; Markkandan, K.; Shin, W.; Oh, S.; Woo, J.; Yoo, J.; An, H.; Han, K. The Whole-Genome and Transcriptome of the Manila Clam (Ruditapes philippinarum). *Genome Biol. Evol.* **2017**, *9*, 1487–1498. [CrossRef] [PubMed]
6. Powell, D.; Subramanian, S.; Suwansa-ard, S.; Zhao, M.; O'Connor, W.; Raftos, D.; Elizur, A. The genome of the oyster *Saccostrea* offers insight into the environmental resilience of bivalves. *DNA Res.* **2018**, *25*, 655–665. [CrossRef] [PubMed]
7. Gerdol, M.; Greco, S.; Pallavicini, A. Extensive Tandem Duplication Events Drive the Expansion of the C1q-Domain-Containing Gene Family in Bivalves. *Mar. Drugs* **2019**, *17*, 583. [CrossRef] [PubMed]
8. Zhang, H.; Song, L.; Li, C.; Zhao, J.; Wang, H.; Qiu, L.; Ni, D.; Zhang, Y. A novel C1q-domain-containing protein from Zhikong scallop Chlamys farreri with lipopolysaccharide binding activity. *Fish Shellfish Immunol.* **2008**, *25*, 281–289. [CrossRef] [PubMed]
9. Kong, P.; Zhang, H.; Wang, L.; Zhou, Z.; Yang, J.; Zhang, Y.; Qiu, L.; Wang, L.; Song, L. AiC1qDC-1, a novel gC1q-domain-containing protein from bay scallop Argopecten irradians with fungi agglutinating activity. *Dev. Comp. Immunol.* **2010**, *34*, 837–846. [CrossRef]
10. Wang, L.; Wang, L.; Zhang, H.; Zhou, Z.; Siva, V.S.; Song, L. A C1q Domain Containing Protein from Scallop Chlamys farreri Serving as Pattern Recognition Receptor with Heat-Aggregated IgG Binding Activity. *PLoS ONE* **2012**, *7*, e43289. [CrossRef]

11. Wang, L.; Wang, L.; Kong, P.; Yang, J.; Zhang, H.; Wang, M.; Zhou, Z.; Qiu, L.; Song, L. A novel C1qDC protein acting as pattern recognition receptor in scallop Argopecten irradians. *Fish Shellfish Immunol.* **2012**, *33*, 427–435. [CrossRef]
12. Jiang, S.; Li, H.; Zhang, D.; Zhang, H.; Wang, L.; Sun, J.; Song, L. A C1q domain containing protein from Crassostrea gigas serves as pattern recognition receptor and opsonin with high binding affinity to LPS. *Fish Shellfish Immunol.* **2015**, *45*, 583–591. [CrossRef] [PubMed]
13. Gestal, C.; Pallavicini, A.; Venier, P.; Novoa, B.; Figueras, A. MgC1q, a novel C1q-domain-containing protein involved in the immune response of Mytilus galloprovincialis. *Dev. Comp. Immunol.* **2010**, *34*, 926–934. [CrossRef] [PubMed]
14. Zimmer, R.K.; Ferrier, G.A.; Kim, S.J.; Ogorzalek Loo, R.R.; Zimmer, C.A.; Loo, J.A. Keystone predation and molecules of keystone significance. *Ecology* **2017**, *98*, 1710–1721. [CrossRef] [PubMed]
15. Xie, B.; He, Q.; Hao, R.; Zheng, Z.; Du, X. Molecular and functional analysis of PmC1qDC in nacre formation of Pinctada fucata martensii. *Fish Shellfish Immunol.* **2020**, *106*, 621–627. [CrossRef]
16. Xiong, X.; Li, C.; Zheng, Z.; Du, X. Novel globular C1q domain-containing protein (PmC1qDC-1) participates in shell formation and responses to pathogen-associated molecular patterns stimulation in Pinctada fucata martensii. *Sci. Rep.* **2021**, *11*, 1105. [CrossRef] [PubMed]
17. He, X.; Zhang, Y.; Yu, F.; Yu, Z. A novel sialic acid binding lectin with anti-bacterial activity from the Hong Kong oyster (Crassostrea hongkongensis). *Fish Shellfish Immunol.* **2011**, *31*, 1247–1250. [CrossRef]
18. Li, C.; Yu, S.; Zhao, J.; Su, X.; Li, T. Cloning and characterization of a sialic acid binding lectins (SABL) from Manila clam Venerupis philippinarum. *Fish Shellfish Immunol.* **2011**, *30*, 1202–1206. [CrossRef]
19. Yang, J.; Wei, X.; Liu, X.; Xu, J.; Yang, D.; Yang, J.; Fang, J.; Hu, X. Cloning and transcriptional analysis of two sialic acid-binding lectins (SABLs) from razor clam Solen grandis. *Fish Shellfish Immunol.* **2012**, *32*, 578–585. [CrossRef]
20. Gerdol, M.; Manfrin, C.; De Moro, G.; Figueras, A.; Novoa, B.; Venier, P.; Pallavicini, A. The C1q domain containing proteins of the Mediterranean mussel Mytilus galloprovincialis: A widespread and diverse family of immune-related molecules. *Dev. Comp. Immunol.* **2011**, *35*, 635–643. [CrossRef]
21. Kilpatrick, D. Animal lectins: A historical introduction and overview. *Biochim. Biophys. Acta-Gen. Subj.* **2002**, *1572*, 187–197. [CrossRef]
22. Cheung, R.C.F.; Wong, J.H.; Pan, W.; Chan, Y.S.; Yin, C.; Dan, X.; Ng, T.B. Marine lectins and their medicinal applications. *Appl. Microbiol. Biotechnol.* **2015**, *99*, 3755–3773. [CrossRef]
23. Mayer, S.; Raulf, M.-K.; Lepenies, B. C-type lectins: Their network and roles in pathogen recognition and immunity. *Histochem. Cell Biol.* **2017**, *147*, 223–237. [CrossRef]
24. Catanzaro, E.; Calcabrini, C.; Bishayee, A.; Fimognari, C. Antitumor Potential of Marine and Freshwater Lectins. *Mar. Drugs* **2019**, *18*, 11. [CrossRef]
25. Kamei, R.; Devi, O.S.; Singh, S.J.; Singh, S.S. Roles and Biomedical Applications of Haemolymph Lectin. *Curr. Pharm. Biotechnol.* **2020**, *21*, 1444–1450. [CrossRef] [PubMed]
26. Gerdol, M.; Gomez-Chiarri, M.; Castillo, M.G.; Figueras, A.; Fiorito, G.; Moreira, R.; Novoa, B.; Pallavicini, A.; Ponte, G.; Roumbedakis, K.; et al. Immunity in Molluscs: Recognition and Effector Mechanisms, with a Focus on Bivalvia. In *Advances in Comparative Immunology*; Springer International Publishing: Cham, Switzerland, 2018; pp. 225–341.
27. Tran, N.H.; Rahman, M.Z.; He, L.; Xin, L.; Shan, B.; Li, M. Complete De Novo Assembly of Monoclonal Antibody Sequences. *Sci. Rep.* **2016**, *6*, 31730. [CrossRef] [PubMed]
28. Tunkijjanukij, S.; Mikkelsen, H.V.; Olafsen, J.A. A Heterogeneous Sialic Acid-Binding Lectin with Affinity for Bacterial LPS from Horse Mussel (Modiolus modiolus) Hemolymph. *Comp. Biochem. Physiol. Part B Biochem. Mol. Biol.* **1997**, *117*, 273–286. [CrossRef]
29. Tunkijjanukij, S.; Olafsen, J.A. Sialic acid-binding lectin with antibacterial activity from the horse mussel: Further characterization and immunolocalization. *Dev. Comp. Immunol.* **1998**, *22*, 139–150. [CrossRef]
30. Ghosh, S. Sialic acid binding lectins (SABL) from Mollusca, a review and insilico study of SABL from Solen grandis and Limax flavus. *J. Entomol. Zool. Stud.* **2017**, *5*, 1563–1572.
31. Allam, B.; Pales Espinosa, E.; Tanguy, A.; Jeffroy, F.; Le Bris, C.; Paillard, C. Transcriptional changes in Manila clam (Ruditapes philippinarum) in response to Brown Ring Disease. *Fish Shellfish Immunol.* **2014**, *41*, 2–11. [CrossRef] [PubMed]
32. Wang, K.; Espinosa, E.P.; Tanguy, A.; Allam, B. Alterations of the immune transcriptome in resistant and susceptible hard clams (Mercenaria mercenaria) in response to Quahog Parasite Unknown (QPX) and temperature. *Fish Shellfish Immunol.* **2016**, *49*, 163–176. [CrossRef]
33. Wang, L.; Wang, L.; Zhang, D.; Jiang, Q.; Sun, R.; Wang, H.; Zhang, H.; Song, L. A novel multi-domain C1qDC protein from Zhikong scallop Chlamys farreri provides new insights into the function of invertebrate C1qDC proteins. *Dev. Comp. Immunol.* **2015**, *52*, 202–214. [CrossRef]
34. Zhao, L.-L.; Jin, M.; Li, X.-C.; Ren, Q.; Lan, J.-F. Four C1q domain-containing proteins involved in the innate immune response in Hyriopsis cumingii. *Fish Shellfish Immunol.* **2016**, *55*, 323–331. [CrossRef] [PubMed]
35. Huang, Y.; Wang, W.; Ren, Q. Identification and function of a novel C1q domain-containing (C1qDC) protein in triangle-shell pearl mussel (Hyriopsis cumingii). *Fish Shellfish Immunol.* **2016**, *58*, 612–621. [CrossRef] [PubMed]
36. Huang, Y.; Wu, L.; Jin, M.; Hui, K.; Ren, Q. A C1qDC Protein (HcC1qDC6) with Three Tandem C1q Domains Is Involved in Immune Response of Triangle-Shell Pearl Mussel (Hyriopsis cumingii). *Front. Physiol.* **2017**, *8*, 521. [CrossRef]

37. Xu, T.; Xie, J.; Li, J.; Luo, M.; Ye, S.; Wu, X. Identification of expressed genes in cDNA library of hemocytes from the RLO-challenged oyster, Crassostrea ariakensis Gould with special functional implication of three complement-related fragments (CaC1q1, CaC1q2 and CaC3). *Fish Shellfish Immunol.* **2012**, *32*, 1106–1116. [CrossRef] [PubMed]
38. Leite, R.B.; Milan, M.; Coppe, A.; Bortoluzzi, S.; dos Anjos, A.; Reinhardt, R.; Saavedra, C.; Patarnello, T.; Cancela, M.; Bargelloni, L. mRNA-Seq and microarray development for the Grooved carpet shell clam, Ruditapes decussatus: A functional approach to unravel host -parasite interaction. *BMC Genomics* **2013**, *14*, 741. [CrossRef] [PubMed]
39. Grinchenko, A.V.; Sokolnikova, Y.N.; Ilyaskina, D.V.; Kumeiko, V.V. Seasonal Changes in Hemolymph Parameters of the Bivalve Modiolus kurilensis Bernard, 1983 from Vostok Bay, Sea of Japan. *Russ. J. Mar. Biol.* **2021**, *47*, 300–311. [CrossRef]
40. Liu, H.-H.; Xiang, L.-X.; Shao, J.-Z. A novel C1q-domain-containing (C1qDC) protein from Mytilus coruscus with the transcriptional analysis against marine pathogens and heavy metals. *Dev. Comp. Immunol.* **2014**, *44*, 70–75. [CrossRef]
41. Sokolnikova, Y.; Magarlamov, T.; Stenkova, A.; Kumeiko, V. Permanent culture and parasitic impact of the microalga Coccomyxa parasitica, isolated from horse mussel Modiolus kurilensis. *J. Invertebr. Pathol.* **2016**, *140*, 25–34. [CrossRef]
42. Baudelet, P.-H.; Ricochon, G.; Linder, M.; Muniglia, L. A new insight into cell walls of Chlorophyta. *Algal Res.* **2017**, *25*, 333–371. [CrossRef]
43. Chatterjee, B.P.; Adhya, M. Lectins with Varying Specificity and Biological Activity from Marine Bivalves. In *Marine Proteins and Peptides*; John Wiley & Sons, Ltd.: Chichester, UK, 2013; pp. 41–68.
44. Kobayashi, Y.; Tateno, H.; Ogawa, H.; Yamamoto, K.; Hirabayashi, J. Comprehensive List of Lectins: Origins, Natures, and Carbohydrate Specificities. In *Methods in Molecular Biology*; Humana Press Inc.: Totowa, NJ, USA, 2014; Volume 1200, pp. 555–577.
45. Iobst, S.T.; Wormald, M.R.; Weis, W.I.; Dwek, R.A.; Drickamer, K. Binding of sugar ligands to Ca(2+)-dependent animal lectins. I. Analysis of mannose binding by site-directed mutagenesis and NMR. *J. Biol. Chem.* **1994**, *269*, 15505–15511. [CrossRef]
46. Iobst, S.T.; Drickamer, K. Binding of sugar ligands to Ca(2+)-dependent animal lectins. II. Generation of high-affinity galactose binding by site-directed mutagenesis. *J. Biol. Chem.* **1994**, *269*, 15512–15519. [CrossRef]
47. Yabe, R.; Suzuki, R.; Kuno, A.; Fujimoto, Z.; Jigami, Y.; Hirabayashi, J. Tailoring a Novel Sialic Acid-Binding Lectin from a Ricin-B Chain-like Galactose-Binding Protein by Natural Evolution-Mimicry. *J. Biochem.* **2006**, *141*, 389–399. [CrossRef]
48. Dennis, J.W.; Laferte, S.; Fucuda, M.; Dell, A.; Carver, J.P. Asn-linked oligosaccharides in lectin-resistant tumor-cell mutants with varying metastatic potential. *Eur. J. Biochem.* **1986**, *161*, 359–373. [CrossRef] [PubMed]
49. Dennis, J.W.; Laferte, S. Tumor cell surface carbohydrate and the metastatic phenotype. *Cancer Metastasis Rev.* **1987**, *5*, 185–204. [CrossRef]
50. Yang, X.; Wu, L.; Duan, X.; Cui, L.; Luo, J.; Li, G. Adenovirus Carrying Gene Encoding Haliotis discus discus Sialic Acid Binding Lectin Induces Cancer Cell Apoptosis. *Mar. Drugs* **2014**, *12*, 3994–4004. [CrossRef]
51. Wu, B.; Mei, S.; Cui, L.; Zhao, Z.; Chen, J.; Wu, T.; Li, G. Marine Lectins DlFBL and HddSBL Fused with Soluble Coxsackie-Adenovirus Receptor Facilitate Adenovirus Infection in Cancer Cells BUT Have Different Effects on Cell Survival. *Mar. Drugs* **2017**, *15*, 73. [CrossRef]
52. Li, G.; Mei, S.; Cheng, J.; Wu, T.; Luo, J. Haliotis discus discus Sialic Acid-Binding Lectin Reduces the Oncolytic Vaccinia Virus Induced Toxicity in a Glioblastoma Mouse Model. *Mar. Drugs* **2018**, *16*, 141. [CrossRef]
53. Stavenhagen, K.; Laan, L.C.; Gao, C.; Mehta, A.Y.; Heimburg-Molinaro, J.; Glickman, J.N.; van Die, I.; Cummings, R.D. Tumor cells express pauci- and oligomannosidic N-glycans in glycoproteins recognized by the mannose receptor (CD206). *Cell. Mol. Life Sci.* **2021**, *78*, 5569–5585. [CrossRef]
54. Sugahara, T.; Ohama, Y.; Fukuda, A.; Hayashi, M.; Kawakubo, A.; Kato, K. The cytotoxic effect of Eucheuma serra agglutinin (ESA) on cancer cells and its application to molecular probe for drug delivery system using lipid vesicles. *Cytotechnology* **2001**, *36*, 93–99. [CrossRef] [PubMed]
55. Omokawa, Y.; Miyazaki, T.; Walde, P.; Akiyama, K.; Sugahara, T.; Masuda, S.; Inada, A.; Ohnishi, Y.; Saeki, T.; Kato, K. In vitro and in vivo anti-tumor effects of novel Span 80 vesicles containing immobilized Eucheuma serra agglutinin. *Int. J. Pharm.* **2010**, *389*, 157–167. [CrossRef]
56. Chen, S.; Zheng, T.; Shortreed, M.R.; Alexander, C.; Smith, L.M. Analysis of Cell Surface Carbohydrate Expression Patterns in Normal and Tumorigenic Human Breast Cell Lines Using Lectin Arrays. *Anal. Chem.* **2007**, *79*, 5698–5702. [CrossRef] [PubMed]
57. Qiu, Y.; Patwa, T.H.; Xu, L.; Shedden, K.; Misek, D.E.; Tuck, M.; Jin, G.; Ruffin, M.T.; Turgeon, D.K.; Synal, S.; et al. Plasma Glycoprotein Profiling for Colorectal Cancer Biomarker Identification by Lectin Glycoarray and Lectin Blot. *J. Proteome Res.* **2008**, *7*, 1693–1703. [CrossRef]
58. Wingfield, P. Protein Precipitation Using Ammonium Sulfate. In *Current Protocols in Protein Science*; John Wiley & Sons, Inc.: Hoboken, NJ, USA, 1998; pp. A.3F.1–A.3F.8.
59. Bradford, M.M. A rapid and sensitive method for the quantitation of microgram quantities of protein utilizing the principle of protein-dye binding. *Anal. Biochem.* **1976**, *72*, 248–254. [CrossRef]
60. Laemmli, U.K. Cleavage of Structural Proteins during the Assembly of the Head of Bacteriophage T4. *Nature* **1970**, *227*, 680–685. [CrossRef] [PubMed]
61. Edman, P.; Begg, G. A Protein Sequenator. *Eur. J. Biochem.* **1967**, *1*, 80–91. [CrossRef] [PubMed]
62. Turriziani, B.; Garcia-Munoz, A.; Pilkington, R.; Raso, C.; Kolch, W.; von Kriegsheim, A. On-Beads Digestion in Conjunction with Data-Dependent Mass Spectrometry: A Shortcut to Quantitative and Dynamic Interaction Proteomics. *Biology* **2014**, *3*, 320–332. [CrossRef]

63. Sievers, F.; Higgins, D.G. Clustal Omega for making accurate alignments of many protein sequences. *Protein Sci.* **2018**, *27*, 135–145. [CrossRef] [PubMed]
64. Grinchenko, A.; Sokolnikova, Y.; Korneiko, D.; Kumeiko, V. Dynamics of the Immune Response of the Horse Mussel *Modiolus kurilensis* (Bernard, 1983) Following Challenge with Heat-Inactivated Bacteria. *J. Shellfish. Res.* **2015**, *34*, 909–917. [CrossRef]
65. Yoshimizu, M.; Kimura, T. Study on the Intestinal Microflora of Salmonids. *Fish Pathol.* **1976**, *10*, 243–259. [CrossRef]
66. Yu, X.-Q.; Gan, H.R.; Kanost, M. Immulectin, an inducible C-type lectin from an insect, Manduca sexta, stimulates activation of plasma prophenol oxidase. *Insect Biochem. Mol. Biol.* **1999**, *29*, 585–597. [CrossRef]
67. Cooper, H.M.; Paterson, Y. Production of Polyclonal Antisera. *Curr. Protoc. Immunol.* **1995**, *13*. [CrossRef] [PubMed]
68. Grodzki, A.C.; Berenstein, E. Antibody Purification: Ammonium Sulfate Fractionation or Gel Filtration. In *Methods in Molecular Biology (Clifton, N.J.)*; Humana Press Inc.: Totowa, NJ, USA, 2010; Volume 588, pp. 15–26.
69. Lin, A.V. Indirect ELISA. In *Methods in Molecular Biology*; Humana Press Inc.: Totowa, NJ, USA, 2015; Volume 1318, pp. 51–59.

marine drugs

Review

Bioactive Properties of Peptides and Polysaccharides Derived from Peanut Worms: A Review

Yi Qi [1,2,†], Jingyi Zhou [1,†], Xiaoqin Shen [3], Meram Chalamaiah [4], Simin Lv [5], Hui Luo [1,2] and Liang Chen [1,*]

1 The Marine Biomedical Research Institute, Guangdong Medical University, Zhanjiang 524023, China; qiyi7272@gdmu.edu.cn (Y.Q.); nancy197@163.com (J.Z.); luohui@gdmu.edu.cn (H.L.)
2 Marine Chinese Medicine Branch, National Engineering Research Center for Modernization of Traditional Chinese Medicine, Zhanjiang 524023, China
3 College of Pharmacy, Guangdong Medical University, Zhanjiang 524023, China; sxq11192021@163.com
4 4-10 Ag/For Centre, Department of Agricultural, Food and Nutritional Science (AFNS), University of Alberta, Edmonton, AB T6G 2P5, Canada; chalamaiah.m@gmail.com
5 Guangdong Runyuan Zhongtian Biological Technology Co., Ltd., Dongguan 523808, China; lvsimin88@126.com
* Correspondence: cliang@gdmu.edu.cn
† These authors contributed equally to this work.

Abstract: Peanut worms (Sipunculids) are unsegmented marine worms that usually inhabit shallow waters. Peanut worms are good source of bioactive compounds including peptides and polysaccharides. Many recent studies have investigated the bioactive properties of peptides and polysaccharides derived from peanut worms in order to enhance their applications in food and pharmaceutical industries. The peptides and polysaccharides isolated from peanut worms have been reported to possess anti-hypertensive, anti-oxidant, immunomodulatory, anti-inflammatory, anti-cancer, anti-hypoxia and wound healing activities through the modulation of various molecular mechanisms. Most researchers used in vitro, cell culture and animal models for the determination of bioactivities of peanut worm derived compounds. However, studies in humans have not been performed considerably. Therefore, it is important to conduct more human studies for better utilization of marine bioactive compounds (peptides and polysaccharides) derived from peanut worms. This review mainly focuses on the bioactive properties of peptides and polysaccharides of peanut worms and their molecular mechanisms.

Keywords: marine worms; sipunculids; bioactive properties; peptides; polysaccharides

Citation: Qi, Y.; Zhou, J.; Shen, X.; Chalamaiah, M.; Lv, S.; Luo, H.; Chen, L. Bioactive Properties of Peptides and Polysaccharides Derived from Peanut Worms: A Review. *Mar. Drugs* **2022**, *20*, 10. https://doi.org/10.3390/md20010010

Academic Editors: Yuki Fujii, Marco Gerdol and Yasuhiro Ozeki

Received: 27 November 2021
Accepted: 15 December 2021
Published: 22 December 2021

1. Introduction

Peanut worms are a group of unsegmented marine worms that belong to the invertebrate phylum Sipuncula. There are about 162 species of peanut worms. They are mostly under 10 cm long and live in shallow waters. Peanut worms are edible and considered as a delicacy in several southeast Asian countries including China, Philippines and Vietnam [1]. Some peanut worms (e.g., *Sipunculus nudus*) have been used in traditional Chinese medicine for the treatment/management of various ailments including hypertension, neurosis, coughing with dyspnea, nocturia, carbuncles, sternalgia, physical weakness, tuberculosis and regulating the functions of stomach and spleen [1,2].

Nutritional composition of foods plays an important role in providing essential nutrients for maintaining good health. The nutritional/chemical composition of peanut worms has been investigated [1,3]. The major components of peanut worms are proteins, carbohydrates and ash. For example, the protein content of *P. esculenta* was found to be 74.5% with high quality amino acids, and carbohydrates were 6.2% [3]. Since peanut worms contain significant quantities of nutritional and bioactive compounds, it is important to exploit the peanut worms for identification and isolation of active bioactive compounds for various food and pharmaceutical applications.

Many recent investigations have demonstrated the isolation of diverse peptides and polysaccharides, with various biological properties such as anti-oxidant, anti-cancer, immunomodulatory, anti-inflammatory and anti-hypertensive activities, from a number of marine organisms with great potential for industrial and therapeutic applications [4–6]. Food derived bioactive peptides are usually short chains of amino acids and generally possess 2–20 amino acid residues. Bioactive peptides are inactive within the sequence of parent protein. However, the bioactive peptides can be generated by using in vitro enzymatic hydrolysis, fermentation and during food processing and gastrointestinal digestion. In vitro enzymatic hydrolysis is the most widely used method for the production of bioactive peptides from various food sources. Bioactive peptides are easily digestible and absorbable with less or no side effects [7]. Therefore, recently bioactive peptides have attracted great interest among scientists and consumers due to their health benefiting properties [8]. Numerous bioactive peptides with diverse health promoting properties have been reported from various marine species and other food sources including fish, molluscs, crustaceans, algae, milk, egg and plant sources [7,9,10].

Polysaccharides are complex natural macromolecular polymers composed of monosaccharide units linked by glycosidic bonds. Polysaccharides are widely present in plants, animals, microorganisms and marine organisms and play a vital role in the development of living organisms [11]. In recent years, polysaccharides derived from natural sources have attracted great attention due to their biocompatibility, non-toxicity, biodegradability and applications in food and pharmaceutical industries [12]. It has been documented that polysaccharides isolated from natural edible sources exhibit significant bioactive properties [13].

Several recent studies have demonstrated that the peptides and polysaccharides obtained from peanut worms possessed numerous health benefiting functions, namely, anti-inflammatory, anti-hypertensive, immunomodulatory, anti-cancer, wound healing, anti-oxidant and anti-hypoxia activities [14–20] (Figure 1). However, the biological properties of peptides and polysaccharides of peanut worms have not been reviewed in the literature so far. This prompted the authors to review the literature on various bioactivities of peptides and polysaccharides derived from peanut worms and their molecular mechanisms of action.

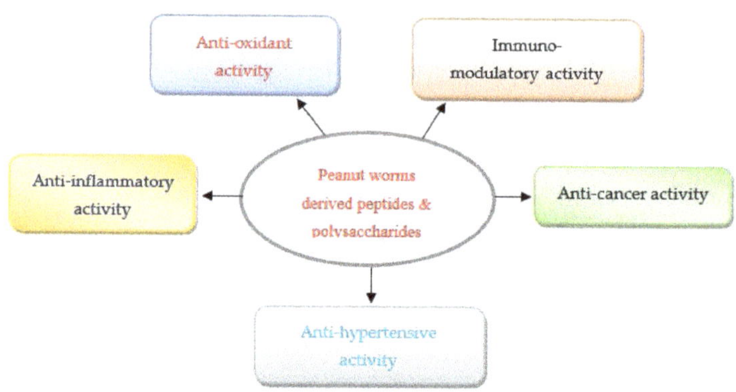

Figure 1. Various major bioactivities of peptides and polysaccharides derived from peanut worms.

2. Anti-Oxidative Properties of Peptides and Polysaccharides Derived from Peanut Worms

Cellular respiration in humans and other aerobic organisms generates reactive oxygen species (ROS) and free radicals. The anti-oxidant mechanisms (e.g., superoxide dismutase, catalase and glutathione peroxidase) in the body neutralize the excess production of free radicals and thereby maintain the balance between antioxidants and oxidative stress. The ROS and free radicals contain unpaired electrons and cause damage to important cellular components including lipids, proteins and DNA. The over production of ROS and free radicals has been linked to various diseases such as cancer, diabetes, atherosclerosis, Alzheimer disease and cardiovascular disorders [21–23]. Recently, there has been great interest in identification and isolation of anti-oxidant bioactive compounds from natural food sources (marine organisms, fish, milk, egg, algae, plant sources) due to their increased applications as safe alternatives to synthetic anti-oxidants in food and pharmaceutical industries.

Numerous peptides and polysaccharides with anti-oxidant activities have been identified and isolated from marine organisms for application as functional foods/health foods [24–26]. Peptides play an important role in the inhibition of free radicals and oxidative stress. It has been reported that the anti-oxidant mechanism of food originated peptides depends on composition, sequence and length [27,28]. Additionally, it was demonstrated that peptides with more hydrophobic (Ala, Val, Gly, Leu, Ile, Phe, Pro) amino acid residues can enhance anti-oxidant ability [15,29,30]. Most anti-oxidant peptides possess 3–15 amino acid residues [27,31–34]. Table 1 shows the molecular mechanisms of anti-oxidant peptides produced from various peanut worms.

Table 1. Molecular mechanisms of anti-oxidant peptides derived from peanut worms.

Peanut Worm Name	Enzyme Used to Produce Peptides	Peptide Sequence and Molecular Weight	In Vitro/Cell Culture/ Animals/Humans Used for the Study	Dose and Duration	Mechanism of Action/ Activities/Effects Showed	Ref.
Sipunculus nudus	Papain	Peptides 5868 Da	In vitro hydroxyl Radical scavenging activity	——	The polypeptide showed great hydroxyl radical scavenging activity with 95.42% inhibition.	[35]
Phascolosoma esculenta	Pancreatin	Peptides < 3 kDa	Mice	50, 100 and 150 mg/kg for 15 days	Peptides dose-dependently improved the oxidative stress status (GSH-Px, SOD, TAC andMDA) in mice	[15]
Phascolosoma esculenta	Papain	——	In vitro total anti-oxidant capacity	——	Collagen peptides from *Phascolosoma esculenta* showed total anti-oxidant capacity with 3.8 U/mg	[36]

There have been several investigations that reported the anti-oxidant capacity of peptides derived from peanut worms. The anti-oxidant activities exhibited by the peanut worms derived peptides were due to the inhibition of free radicals and enhancing the secretion of endogenous antioxidant enzymes such as SOD and glutathione peroxidase (GSH-Px). Zhu et al. [35] produced peptides from *Sipunculus nudus* by hydrolyzing with papain and found that peptides with molecular weight of 5868 Da showed excellent hydroxyl radical scavenging activity with 95% inhibition. Peptides generated from collagen of *Phascolosoma esculenta* have been reported to possess anti-oxidant capacity in vitro [36]. Liu et al. [15] prepared peptides (<3 kDa) from *Phascolosoma esculenta* by using pancreatin and investigated the anti-oxidant capacity of the peptides using mice model. It was found that peptide's administration at 50, 100 and 150 mg/kg for 15 days dose-dependently improved the oxidative stress parameters such as GSH-Px, SOD, TAC and MDA in mice.

Apart from peptides, many studies have demonstrated that polysaccharides derived from peanut worms exhibit antioxidant activity through several ways including inhibition of DPPH, hydroxyl, superoxide free radicals, having reducing power, enhancing the antioxidant enzymes such as superoxide dismutase, glutathione peroxidase and upregulation of Nrf2 signaling pathway [16,19,37,38]. Anti-oxidant properties of polysaccharides produced from various peanut worms are presented in Table 2.

Table 2. Molecular mechanisms of anti-oxidant activities of polysaccharides derived from various peanut worms.

Source	Composition of Polysaccharide Extract	Cell Culture/ Animal Models	Dose and Duration	Molecular Mechanisms/Effects	Ref.
Sipunculus nudus	Polysaccharide was com-posed of mannose, rhamnose, galacturonic acid, glucose, arabinose and fucose	In vitro hydroxyl radical activity.	0.25, 0.5, 1.0, 2.0, 5.0,10.0, 20.0 mg/mL for 30 min	Polysaccharide showed powerful scavenging activity on hydroxyl radical in a dose dependent manner.	[39]
Sipunculus nudus	————	In vitro reducing power, hydroxyl and superoxide radicals inhibition assay	200–1000 µg/mL for 30 min	*S. nudus* polysaccharides showed dose dependent inhibition of hydroxyl and superoxide radicals and exhibited great reducing power.	[37]
Phascolosoma esculenta	D-glucosyl, D-galactosyl, with small amount of D-mannosyl, D-arabinosyl and residues with a- and b-type linkage.	Mice	1, 10 and 5 mg/mL for 30 days	Oligosaccharides from *Phascolosoma esculenta* enhanced the enzyme activities of GSH-Px and SOD by upregulating Nrf2 mRNA expression in sepsis mice model.	[16]
Phascolosoma esculenta	Mannose, ribose, rhamnose, glucuronic acid, glucose, galactose, xylose, arabinose and fucose	In vitro DPPH, superoxide anion, hydroxyl radicals and ferrous ion chelating and mice model	0.2, 0.4 and 0.8 g/kgBW for 25 d	Polysaccharides from *Phascolosoma esculenta* scavenged free radicals dose-dependently and showed antioxidant activities in mice by enhancing superoxide dismutase (SOD) (10.2–22.2% and 18.8–26.9%), glutathione peroxidase (GSH-Px) (11.9–15.4% and 26.6–30.4%) activities in serum and liver.	[38]
Sipunculus nudus	———	In vitro DPPH and hydroxyl radical scavenging activities	0.2 mg/mL	Free radical scavenging rates increased significantly with the increase of concentration. The scavenging activities of hydroxyl radical and DPPH radical were found to be 12.58% at the concentration of 0.20 mg/mL.	[39]
Phascolosoma esculenta	polysaccharide contained glucose with acetylaminoand pyran rings and connected by α-glycosidic bonds	In vitro reducing power, DPPH and hydroxyl radical scavenging activities	1,5,10,15,20, and 25 mg/mL for 30 min	Polysaccharide showed DPPH and hydroxyl radical scavenging and reducing power with IC_{50} of 0.567 and 0.605, 2.976 mg/mL, respectively.	[19]

Researchers used various in vitro, and animal models to study the antioxidant capacity of polysaccharides derived from peanut worms. Studies conducted by Li et al. [39] and Qin et al. [37] documented that polysaccharides extracted from *Sipunculus nudus* showed a dose dependent inhibition of hydroxyl and superoxide radicals. Zhihao et al. [16] found that oligosaccharides from *Phascolosoma esculenta* exhibited anti-oxidative effects in sepsis mice model through the enhancement of enzyme activities of GSH-Px and SOD by activation of Nrf2 signaling pathway. Polysaccharide, composed of mannose, ribose, rhamnose, glucuronic acid, glucose, galactose, xylose, arabinose and fucose, derived from *Phascolosoma esculenta* have shown antioxidant activities in mice by enhancing superoxide dismutase (10–26%), glutathione peroxidase (11–30%) activities in serum and liver [38]. In a recent study, Yiqiao et al. [19] determined the radical scavenging activity and reducing power of polysaccharide isolated from *Phascolosoma esculenta* and reported the IC_{50} values of 0.567 and 0.605, 2.976 mg/mL respectively for DPPH and hydroxyl radicals and reducing power.

Most studies used in vitro and mice models for the determination of anti-oxidant activity of peptides and polysaccharides from peanut worms. However, cell culture and human studies are scanty. Therefore, further research is needed in humans and cells to verify the anti-oxidant benefits of these bioactive compounds of peanut worms.

3. Anti-Inflammatory Activities of Peptides and Polysaccharides Derived from Peanut Worms

Inflammation is a natural and complex defense mechanism of the immune system against a variety of harmful stimuli including pathogens, damaged cells, toxic chemicals, and irradiation. Generally, acute inflammation is beneficial to the body. However, chronic and uncontrolled inflammation can lead to various diseases such as cancer, diabetes, atherosclerosis, chronic kidney disease, arthritis and neurodegenerative disorders [40,41]. NSAIDs (non-steroidal anti-inflammatory drugs) are commonly used drugs for the treatment of inflammation. However, side effects associated with cardiovascular/renal/gastrointestinal systems have limited the application of NSAIDs. Therefore, recently there is a great interest in the use of food derived bioactive components (e.g., peptides and polysaccharides) for treatment or management of inflammation/inflammatory diseases due to their food origin and no/less side effects. Research has revealed that food derived bioactive compounds play an important role in the mitigation of inflammation/inflammatory diseases.

Numerous recent studies have shown that bioactive compounds (peptides, polyphenols, proteins, lipids and polysaccharides) derived from several food sources (fish, milk, egg, marine organisms and plant foods) possess anti-inflammatory properties [42–46]. Marine peanut worms have been investigated for anti-inflammatory compounds. Peptides and polysaccharides isolated from peanut worms have demonstrated to modulate the inflammatory response by inhibiting the production of pro-inflammatory mediators, TNF-α, IL-1β and TGF-β1, and reducing the activity of enzymes, cyclooxygenase-2 (COX-2) and inducible nitric oxide synthase (iNOS).

Recently, anti-inflammatory peptides have been identified and isolated from few peanut worms after hydrolysis with proteolytic enzymes. Peptides isolated from *Sipunculus nudus* have shown anti-inflammatory effects through the modulation of several mechanisms such as reduction of expression of pro-inflammatory mediators, TNF-α, IL-1β, IL-6, iNOS and COX-2 and inhibition of NO production [1,20]. The anti-inflammatory effects of peptides isolated from peanut worms are presented in Table 3.

The food originated anti-inflammatory peptides are usually 2–10 amino acids in length. In addition to the peptide's length, composition and sequence of peptides also play an important role in anti-inflammatory activity [47,48]. In recent years, researchers have documented the extraction of anti-inflammatory peptides from peanut worm, *Sipunculus nudus*. Sangtanoo et al. [1] isolated two peptides, LSPLLAAH (821.48 Da) and TVNLAYY (843.42 Da), from *Sipunculus nudus* after hydrolysis with neutrase, flavourzyme, and Alcalase and reported that both peptides showed strong anti-inflammatory activity in LPS-stimulated RAW264.7 macrophages by decreasing the expression of pro-inflammatory mediators, iNOS, IL-6, TNF-α and COX-2 after treatment with 30, 60, 120 mM for 12 h. In another recent study, Lin et al. [20] produced collagen peptides from *Sipunculus nudus* by using animal hydrolytic protease and flavor protease and demonstrated that peptides inhibited inflammation in the wound of mice skin through the reduction of mRNA levels of TGF-β1, TNF-α and IL-1β.

Table 3. Molecular mechanisms of anti-inflammatory peptides derived from peanut worms.

Peanut Worm Name	Enzyme Used to Produce Peptides	Peptide Sequence and Molecular Weight	In Vitro/Cell Culture/ Animals/Humans Used for the Study	Dose and Duration	Mechanism of Action/Activities/Effects Showed	Ref.
Sipunculus nudus	Neutrase, Flavourzyme, and Alcalase	LSPLLAAH (821.48 Da) and TVNLAYY (843.42 Da).	RAW 264.7 macrophages	30, 60, 120 mm for 12 h	Peptides (LSPLLAAH and TVNLAYY) inhibited NO production and decreased the expression of pro-inflammatory mediators, iNOS, IL-6, TNF-α, and COX-2, in LPS-stimulated RAW264.7 macrophages.	[1]
Sipunculus nudus	Animal hydrolytic protease (3000 U/g) and flavor protease	Collagen peptides < 5 kDa	Mice	2 g/mL for 7 days	Peptides showed anti-inflammatory activity through the reduction of mRNA levels of TGF-β1, TNF-α and IL-1β in the wound of mice skin.	[20]

Apart from the peptides, the anti-inflammatory effects of polysaccharides isolated from peanut worms have also been studied considerably in recent years (Table 4). Chen-Xiao et al. [2] investigated the anti-inflammatory activities of water soluble polysaccharides of *Sipunculus nudus* using different mouse and rat models of inflammation. It was found that polysaccharide treatment at 50, 100 and 200 mg/kgBW for 6 days dose-dependently reduced the inflammation of carrageenan-induced paw oedema, dextran-induced rat paw oedema, carrageenan-induced peritonitis, xylene-induced ear oedema, and acetic acid-induced vascular permeability in mice. Zhihao et al. [16] extracted oligosaccharides, composed of D-glucosyl, and D-galactosyl residues with α- and β-type linkages, from *Phascolosoma esculenta* and demonstrated that administration of oligosaccharides at 1, 10 and 5 mg/mL for 30 days noticeably decreased the production of IL-1β and TNF-α and enhanced IL-10 in mice with sepsis induced inflammation.

The anti-inflammatory properties shown by the peptides and polysaccharides of peanut worms could be due to the suppression of nuclear factor kappa B (NF-κB) activation. Inhibition of NF-κB activation decreases the expression of COX-2 and iNOS and production of TNF-α and IL-1β, and therefore reduces inflammation. Though the anti-inflammatory activities have been reported for the peptides and polysaccharides of peanut worms, the underlying molecular mechanisms of anti-inflammatory activity have not been explored extensively so far. Therefore, further research is needed to determine the exact molecular mechanisms of action of these anti-inflammatory bioactive components (peptides and polysaccharides) isolated from peanut worms. This would enhance the applications of these bioactive compounds in food and pharmaceutical industries.

Table 4. Molecular mechanisms of anti-inflammatory activity of polysaccharides derived from various peanut worms.

Source	Composition of Polysaccharide Extract	Cell Culture/ Animal Models	Dose and Duration	Molecular Mechanisms/Effects	Ref.
Sipunculus nudus	Water extract	Mouse and rat oedema paw models	50, 100 and 200 mg/kg for 6 days	Water extract from the body wall of *Sipunculus nudus* showed dose-dependent anti-inflammatory activity in the carrageenan-induced paw oedema, dextran-induced rat paw oedema, cotton pellet granuloma, carrageenan-induced peritonitis, xylene-induced ear oedema, and acetic acid-induced vascular permeability models.	[2]
Phascolosoma esculenta	D-glucosyl, D-galactosyl, with small amount of D-mannosyl, D-arabinosyl and residues with α- and β-type linkage.	Mice	1, 10 and 5 mg/mL for 30 days	Oligosaccharides from *Phascolosoma esculenta* considerably decreased the secretion of IL-1β and TNF-α and enhanced the IL-10 in sepsis mice.	[16]

4. Anti-Hypertensive Activity of Peptides from Peanut Worms

Hypertension (high blood pressure) is a serious medical condition that affects 1280 million people globally aged between 30–79 years. It has been reported that several diseases related to heart (myocardial infarction, coronary heart disease), brain (stroke) and kidney (kidney failure) are linked to hypertension. Angiotensin I-converting enzyme (ACE; EC 3.4.15.1) plays an important role in regulation of blood pressure. ACE cleaves the two C-terminal dipeptides of inactive angiotensin I to produce angiotensin II, which is a potent vasoconstrictor that prevents the catalytic function of bradykinin, a vasodilator. Recent research has shown that inhibition of ACE with food derived peptides could be an effective way in the prevention/management of hypertension. It has been demonstrated that numerous peptides isolated from various food sources (milk, egg, fish, meat, marine organisms and vegetables) have shown anti-hypertensive activity [49–57].

Recently, several researchers investigated the anti-hypertensive effects of peptides derived from peanut worms such as *Phascolosoma esculenta* and *Sipunculus nudus*. Anti-hypertensive activities of peptides originated from peanut worms have been widely studied due to their excellent bioactive properties and safety profiles. Pepsin and trypsin are the two proteolytic enzymes that are extensively used in the production of anti-hypertensive peptides from peanut worms [3,17,58–61]. These enzymes specifically cleave the large molecular weight proteins into smaller peptides. The large proteins are usually unable to bind the active site of ACE, however, the smaller peptides generated by enzymatic hydrolysis could easily bind to the active site of ACE as inhibitor and thereby prevent high blood pressure [3]. It has been demonstrated that peptides derived from peanut worms showed anti-hypertensive activity through several mechanisms including changing the secondary structure of ACE, competitive and non-competitive inhibition of ACE, reducing systolic blood pressure and binding to the active sites of ACE [17,58–60] (Table 5).

Table 5. Molecular mechanisms of action of ACE-inhibitory/anti-hypertensive peptides derived from various peanut worms.

Peanut Worm Name	Enzyme Employed to Produce Peptides	Peptide Sequence and Molecular Weight	In Vitro/Cell Culture/ Animals/Humans Used for the Study	IC_{50}/EC_{50} Values	Activity/Mechanisms of Action Showed	Ref.
Phascolosoma esculenta	Pepsin	AWLHPGAPKVF	In vitro ACE inhibition assay & spontaneously hypertensive rats	IC_{50} value of 135 M	Peptide inhibited ACE through competitive inhibition and exhibited anti-hypertensive effects in rats by significantly reducing the systolic blood pressure around 30 mmHg.	[60]
Phascolosoma esculenta	Pepsin and trypsin	————	In vitro ACE inhibition assay & spontaneously hypertensive rats	IC_{50} values of 0.67 and 0.24 mg/mL	Peptides significantly reduced both diastolic blood pressure (DBP) and systolic blood pressure (SBP) and inhibited ACE in vitro.	[3]
Phascolosoma esculenta	Pepsin, and trypsin	AYF, EL, GLR, HK, and ILK	In vitro ACE inhibition assay	IC_{50} values of 3.43–4.18 U/ml	Peptides exhibited ACE inhibitory activity with IC_{50} values in the range of 3.43–4.18 U/mL	[61]
Phascolosoma esculenta	Pepsin and trypsin	284 di- and tri-peptides	In vitro ACE inhibition assay	IC_{50} less than 50 µM	Peptides inhibited the ACE.	[58]
Sipunculus nudus	Protamex	IND, VEPG, LADEF	In vitro ACE inhibition assay	IC_{50} values for ACE inhibition were 34.72, 20.55 and 22.77 µmol/L	The peptides IND, VEPG, and LADEF showed ACE inhibition activity with IC_{50} values of 34.72, 20.55 and 22.77 µmol/L, respectively.	[62]
Phascolosoma esculenta	Pepsin and trypsin	RYDF, YASGR and GNGSGYVSR	In vitro ACE inhibition assay & spontaneously hypertensive rats	IC_{50} values of 235, 184 and 29 µM respectively for RYDF, YASGR and GNGSGYVSR	Three peptides inhibited ACE through non-competitive inhibition. GNGSGYVSR reduced systolic blood pressure 31 mmHg at 2 h after oral administration in spontaneously hypertensive rats.	[59]
Phascolosoma esculenta	Pepsin and trypsin	GNGSGYV and SR	In vitro ACE inhibition assay	IC_{50} value of 170 µM	GNGSGYV and SR showed ACE inhibition through synergistic effect. SR initially attacked the catalytic Zn of ACE and formed coordinate bond, and then GNGSGYV attached with the residues of ACE active site by hydrogen bonds.	[17]

The ACE inhibitory/antihypertensive activity of food derived peptides depends on the composition, sequence and length of the peptides. Peptides with more hydrophobic amino acids (aromatic and branched chain aliphatic) have been demonstrated to inhibit ACE efficiently [3]. Most of the ACE inhibitory peptides isolated from peanut worms contain 2–11 amino acids residues [17,58–60,62]. Du et al. [60] used pepsin to cleave the water soluble protein of *Phascolosoma esculenta* in order to obtain low molecular weight peptides with potent angiotensin I-converting enzyme (ACE) inhibitory activity. A novel ACE inhibitory peptide, Ala-Trp-Leu-His-Pro-Gly-Ala-Pro-Lys-Val-Phe, was isolated with IC_{50} value of 135 M. The authors investigated the inhibitory kinetics of the peptide and found that the identified peptide inhibited ACE through competitive inhibition. Furthermore, the peptide (Ala-Trp-Leu-His-Pro-Gly-Ala-Pro-Lys-Val-Phe) at 10 mg/kg dose showed anti-hypertensive effects in spontaneously hypertensive rats (SHR) by significantly decreasing the systolic blood pressure (SBP) around 30 mmHg. Water-soluble and insoluble proteins of *P. esculenta* were extracted, by Wu et al. [3], and the fractions hydrolyzed with pepsin and trypsin. The hydrolysates exhibited ACE inhibitory activity with IC_{50} values between 0.1 to 0.67 mg/mL. It was demonstrated that the peptides derived from water-soluble and insoluble proteins significantly reduced both diastolic blood pressure (DBP) (23–45 mmHg) and systolic blood pressure (SBP) (20–33 mmHg) in spontaneously hypertensive rats after single oral administration at a dose of 1 g/kgBW. Protein from *Sipunculus nudus* was hydrolysed with Protamex and three ACE inhibitory peptides IND, VEPG, and LADEF were isolated with IC_{50} values of 34.72, 20.55 and 22.77 μmol/L, respectively [62]. In another investigation, Guo et al. [59] identified and isolated three anti-hypertensive peptides, RYDF, YASGR and GNGSGYVSR, from *Phascolosoma esculenta*. It was found that the three peptides inhibited ACE through non-competitive inhibition with IC_{50} values between 29–235 μM. Furthermore, GNGSGYVSR reduced systolic blood pressure by 31 mmHg at 2 h after oral administration in spontaneously hypertensive rats. Recently, two novel ACE inhibitory peptides, GNGSGYV and SR, were reported from *Phascolosoma esculenta* after hydrolysis with pepsin and trypsin. The two peptides showed synergistic effect on ACE inhibition with IC_{50} value of 170 μM. The synergistic mechanism indicated that SR and GNGSGYV significantly changed the secondary structure of ACE. It was also found that the peptide SR initially formed a coordinate bond with the catalytic Zn of ACE and then GNGSGYV attached with arginine of dipeptide (SR) and the amino acids of ACE active site through hydrogen bonds and thereby prevent the substrate attachment with ACE [17]. In addition to the traditional hydrolysis, virtual hydrolysis of peanut worms has been performed by several researchers in order to identify the potential ACE inhibitory peptides. Hongxi et al. [61] virtually hydrolyzed *Phascolosoma esculenta* protein with pepsin and trypsin and identified five ACE inhibitory peptides, AYF, EL, GLR, HK and ILK, with IC_{50} values in the range of 3.43–4.18 U/mL. In another study, Liu et al. [58] conducted virtual hydrolysis of *Phascolosoma esculenta* with pepsin, trypsin and a mixture of pepsin and trypsin for identification of ACE inhibitory peptides and predicted that 284 di- and tri-peptides from *Phascolosoma esculenta* possess ACE inhibitory activity with IC_{50} less than 50 μM.

The research indicates that anti-hypertensive peptides obtained from peanut worms could be used as a natural component for application in health foods/nutraceuticals to prevent/treat hypertension. However, further research is needed on the exact molecular mechanisms of action of these anti-hypertensive peptides.

5. Immunomodulatory Activity of Polysaccharides Derived from Peanut Worms

Immune system is a complex defense network that protects the body against invading pathogens such as bacteria, viruses, fungi, protozoans, and prevents the growth of cancer cells. Immunomodulators are compounds that can potentially suppress or stimulate the immune system of host by modulating various immune cells and/or their signaling molecules (e.g., T cells, B cells, macrophages, NK cells, dendritic cells and cytokines) [63].

Recently, several polysaccharides with immunomodulatory activity have been identified from various food sources including mushrooms, fruits, cereals and algae [13].

Recent research has indicated that polysaccharides stimulate various immune cells (e.g., macrophages, NK cells, dendritic cells) by binding to the cell surface receptors such as Toll-like receptor 4 (TLR4), cluster of differentiation 14 (CD14), scavenger receptor (SR), complement receptor 3 (CR3), mannose receptor (MR) and Dectin-1 [13,63].

It has been demonstrated that polysaccharides derived from peanut worms exert immunomodulatory effects through various mechanisms such as activation of macrophages, increasing the indexes of immune organs (thymus and spleen), enhancing the secretion of cytokines, and improving the phagocytosis function of macrophages and NK cell activity [18,64–66]. Table 6 shows the immunomodulatory activities of polysaccharides derived from peanut worms.

A number of studies reported the immune stimulating activity of polysaccharides extracted from peanut worms, *Sipunculus nudus* and *Phascolosoma esculenta*. Liang et al. [64] investigated the immunomodulatory effects of polysaccharides from *P. esulenta* using a mice model and found that administration of polysaccharides at 3.0, 6.0, 9.0 mg/kgBW for 2 months considerably stimulated the Con-A activated mouse spleenocytes and increased the index of liver, spleen and thymus of mice. A water soluble polysaccharide, composed of rhamnose (28%), fucose (16%) and galactose (56%), extracted from *Sipunculus nudus* exhibited immunomodulatory effects by activating macrophages through the upregulation of expression of cytokines, IL-6 and TNF-α, and inducing the expression of iNOS and COX-2 [65]. Li et al. [66] reported that polysaccharides from *Sipunculus nudus* enhanced the cellular and humoral immunity in mice by increasing the phagocytosis function and NK cell activity. Su et al. [67] used hepatoma (HepG2) bearing mice model to study the immunomodulatory effects of polysaccharides derived from *Sipunculus nudus*. The mice were treated with 50, 100, and 200 mg/kg polysaccharide for 30 days and it was found that the polysaccharide composed of L-rhamnose, L-arabinose, D-ribose, D-glucose and D-galactose showed immune stimulating effects through increase of thymus and spleen indexes, and upregulated the IL-2, IFN-γ, and TNF-α cytokines in serum of mice. In another recent study, Su et al. [18] demonstrated that polysaccharide of *Sipunculus nudus* increased the index of immune organs and augmented the secretion of cytokines, IL-2, IFN-γ and TNF-α, in hepatoma mice after treatment with 50, 100 and 200 mg/kg polysaccharides for 16 days. These results suggest that immunomodulatory polysaccharides from peanut worms have great potential for application as nutraceutical/health foods.

Although immunomodulatory activities of polysaccharides derived from peanut worms have been documented, the molecular mechanisms of action and structure-function relationship of these polysaccharides have not been investigated. Thus, more research is needed on molecular mechanisms to decipher the immunomodulatory effects of these polysaccharides. Additionally, the immune stimulatory effects of peptides derived peanut worms need to be investigated considerably in order to enhance their application as health foods.

Table 6. Molecular mechanisms of immunomodulatory activity of polysaccharides derived from peanut worms.

Source	Composition of Polysaccharide Extract	Cell Culture/ Animal Models	Dose and Duration	Molecular Mechanisms/Effects	Ref.
Phascolosoma esculenta	———	Mice model	3.0, 6.0, 9.0 mg/kgBW for 2 months	Polysaccharides from *P. esulenta* significantly enhanced liver, spleen and thymus index of mice and increased Con A-stimulated mouse spleen cells.	[64]
Sipunculus nudus	Monosaccharide composition -rhamnose (28%), fucose (16%) and galactose (56%)	Murine macrophages from BALB/c mice and human macrophages	5–80 μg/mL for 24 h	The water soluble polysaccharide isolated from *S. nudus* showed immunostimulating activity by activating macrophages through the upregulation of expression of cytokines, IL-6 and TNF-α, and inducing the expression of iNOS and COX-2.	[65]
Sipunculus nudus	———	Mice	———	Polysaccharides from *Sipunculus nudus* promoted the cellular immunity and humoral immunity through the enhancing the phagocytosis function and NK cell activity in mice.	[66]
Sipunculus nudus	L-rhamnose, Larabinose, D-ribose, D-glucose and D-galactose	Hepatoma HepG2-bearing Mice	50,100, and 200 mg/kg, 1 month	Polysaccharide extract from *Sipunculus nudus* enhanced the immune response through increase of thymus and spleen indexes, and upregulating the IL-2, IFN-γ, and TNF-α cytokines in serum of mice.	[67]
Sipunculus nudus	Repeating units are →3,4-β-D-GlcpNAC (1→ and →4)-α-D-Glcp (1→ in the ratio of 15:1; →2)-α-D-Galp-(1→ as a side chain; and β-D-Galp-(1→ and α-D-Glcp-(1→ as end groups	hepatoma HepG2-bearingmice	50,100, and 200 mg/kg, 16 days	Polysaccharide increased the index of immune organs and augmented the secretion of cytokines IL-2, IFN-γ and TNF-α.	[18]

6. Anti-Cancer Activities of Polysaccharides Derived from Peanut Worms

Cancer is one of the major causes of death worldwide. Approximately 10 million deaths were occurred in 2020 due to various cancers [68]. Cancer is uncontrolled growth of cells in the body. The cancer cells attack the adjacent normal cells and spread to the other parts of the body. Currently, many different chemo drugs are used in the cancer treatment. However, side effects (neutropenia, nausea, vomiting, hair loss, blood clots etc.) associated with these drugs have limited their usage. Additionally, several chemotherapy drugs kill the normal cells along with the cancer cells. Therefore, recently there is great interest in finding new anti-cancer agents with less side effects from natural sources such as spices, vegetables, marine sources etc. [69,70]. Polysaccharides isolated from numerous natural resources including plant and marine sources have been reported to have inhibition activity on malignant cells primarily through the induction of apoptosis [71,72].

Polysaccharides extracted from peanut worms have been demonstrated to possess anti-cancer activity against cancer cells and in vivo animal models. Most researchers determined the anti-cancer effects of peanut worm derived polysaccharides in hepatocellular carcinoma. Scientific evidence from recent studies showed that polysaccharides derived from peanut worms exhibited anti-cancer effects by modulating numerous molecular mechanisms including preventing the DNA synthesis, increasing the expression of pro-apoptosis proteins, TNF-α, caspase-3, and Bax, decreasing the expression of the anti-apoptosis proteins, survivin, Bcl-2, and VEGF, up-regulation of caspase-3, caspase-8, and caspase-9, enhancing the expression of ATF4, DDIT3, and IkBα and reduction of CYR61, HSP90, and VEGF expression [18,67,73]. The mechanisms of action of anticancer activities of polysaccharides derived from peanut worms are presented in Table 7.

Polysaccharides isolated from peanut worm *Sipunculus nudus* have shown anti-cancer effects in cultured cancer cells and mice models with transplanted cancer cells. Jie et al. [73,74] studied the molecular mechanisms of anticancer effects of polysaccharides, extracted from *Sipunculus nudus*, after treating the Hepg2.2.15 cells with 0.13, 0.25, 0.5 and 1 mg/mL for 24 and 48 h and reported that *Sipunculus nudus* derived polysaccharides induced dose-dependent apoptosis on Hepg2.2.15 cells by increasing the expression of pro-apoptosis proteins, TNF-α, caspase-3, and Bax, and decreasing the expression of the anti-apoptosis proteins, survivin, Bcl-2, and VEGF. Su et al. [67] investigated the anti-cancer effects of polysaccharides of *Sipunculus nudus* using HepG2 cells-bearing mice model and demonstrated that administration of polysaccharides at 50, 100 and 200 mg/kg for 1 month significantly inhibited the growth of HepG2 cells through increasing the expression of ATF4, DDIT3 and IkBα and down-regulation of CYR61, HSP90 and VEGF expression. In another recent study, Su et al. [18] reported that polysaccharides extracted from *Sipunculus nudus* showed anti-cancer effects in HepG2-bearing mice by inducing the apoptosis of tumor cells through the up-regulation of caspase-3, caspase-8, caspase-9 and Bax, and down-regulation of B-cell lymphoma-2 and vascular endothelial growth factor protein expression.

The above scientific evidence confirmed that polysaccharides of peanut worms could inhibit the growth of the cancer cells. However, there are various limitations that could hamper the use of these polysaccharides for human applications as nutraceuticals/drugs. Firstly, more reliable and consistent scientific evidence from clinical studies about the beneficial effects of these polysaccharides is needed before its use for the treatment/management of cancer. Secondly, data about cytotoxic effects on normal cells and in-depth knowledge of exact molecular mechanisms of anti-cancer activity of peanut worms derived polysaccharides are required from cell culture, animal and clinical studies. Additionally, most studies investigated the anticancer activity of polysaccharides isolated from *Sipunculus nudus*, although there are many peanut worm species in phylum, Sipuncula. Therefore, anti-cancer activity for polysaccharides of other peanut worm species needs to be explored.

Table 7. Molecular mechanisms of anti-cancer activity of polysaccharides derived from various peanut worms.

Source	Composition of Polysaccharide Extract	Cell Culture/ AnimalModels	Dose and Duration	Molecular Mechanisms/Effects	Ref.
Sipunculus nudus	Extractcontains 35.3% neutral sugar, including Ara 10.7%, Rha 12.6%, Gal 16.4%, Glu 31.3%, Xyl 18.2%, and Man 10.8%.	Hepg2.2.15 cells	0.13, 0.25, 0.5, and 1 mg/mL for 24 and 48 h	Polysaccharides showed anti-cancer activities by preventing the DNA synthesis of Hepg2.2.15 cells and increasing the expression of pro-apoptosis proteins, TNF-α, caspase-3, and Bax, and decreasing the expression of the anti-apoptosis proteins survivin, Bcl-2, and VEGF.	[73]
Sipunculus nudus	L-rhamnose, Larabinose, D-ribose, D-glucose and D-galactose	Hepatoma HepG2-bearing Mice	50,100, and 200 mg/kg, 1 month	Polysaccharides showed anti-tumor activity by inhibiting the growth of HepG2 cells through increase of ATF4, DDIT3, and IkBα expression and decrease of CYR61, HSP90, and VEGF expression.	[67]
Sipunculus nudus	Repeating units of →3,4-β-D-GlcpNAC (1→ and →4)-α-D-Glcp(1→ in the ratio of 15:1; →2)-α-D-Galp-(1→ as a side chain; and β-D-Galp-(1→ and α-D-Glcp-(1→ as end groups	hepatoma HepG2-bearing mice	50,100, and 200 mg/kg, 16 days	Extracted polysaccharide enhanced the apoptosis of tumour cells through the mitochondrial apoptosis pathway by upregulating caspase-3, caspase-8, caspase-9 and BCL2-associated X, and downregulating B-cell lymphoma-2 and vascular endothelial growth factor protein expression.	[18]

7. Other Bioactivities of Peanut Worm Derived Peptides and Polysaccharides

Apart from above mentioned bioactive properties, many researchers documented various other bioactive properties including wound healing capacity, memory improvement and anti-hypoxia activity for peanut worm derived peptides and polysaccharides (Tables 8 and 9). Chen-Xiao and Zi-Ru [14] used various hypoxia mice models to determine the anti-hypoxia activity of polysaccharides extracted from *Sipunculus nudus* and revealed that polysaccharides showed significant anti-hypoxic activity on normobaric hypoxia, chemical intoxicant hypoxia and acute cerebral ischemia hypoxia models after treatment with 10, 30, 100 mg/kg for 6 days. Liu et al. [15] produced peptides < 3 kDa from *Phascolosoma esculenta* by using pancreatin and found that administration of peptides at 50, 100 and 150 mg/kg for 15 days improved the spatial learning and memory ability in mice through the up-regulation of NR2A, NR2B, BDNF and CREB mRNA expressions in hippocampus. Lin et al. [20] investigated the wound healing properties of collagen peptides derived from *Sipunculus nudus* and demonstrated that the peptides clearly improved the healing rate and inhibited scar formation in mice by enhancing collagen deposition and inhibition of TGF-β/Smads signaling pathway.

Table 8. Anti-hypoxia activity of polysaccharides derived from various peanut worms.

Source	Composition of Polysaccharide Extract	Cell Culture/ Animal Models	Dose and Duration	Molecular Mechanisms/Effects	Ref.
Sipunculus nudus	Rhamnose (28%), fucose (16%) and galactose (56%).	Mice model	10, 30, 100 mg/kg for 6 days	The extracted polysaccharide exhibited significant anti-hypoxic activity on normobarie hypoxia, chemical intoxicant hypoxia and acute cerebral ischemia hypoxia models in mice.	[14]

Table 9. Effects of peptides derived from peanut worms on wound healing and spatial learning and memory.

Peanut Worm Name	Enzyme Used to Produce Peptides	Peptide Sequence and Molecular Weight	In Vitro/Cell Culture/ Animals/Humans Used for the Study	Dose and Duration	Mechanism of Action/ Activities/Effects Showed	Ref.
Sipunculus nudus	Animal hydrolytic protease (3000 U/g) and flavor protease (3000 U/g)	Collagen peptides < 5 kDa	human umbilical vein endothelial cells (HUVEC), human immortalized keratinocytes (HaCaT) and human skin fibroblasts (HSF) and mice	2 g/mL for 28 days and 500 μg/mL for 12, 24, 30, 36 h	Collagen peptides derived from *Sipunculus nudus* exhibited great capacity to induce HUVEC, HaCaT and HSF cells proliferation and migration in vitro. Peptides noticeably improved the healing rate and inhibited scar formation in mice through the mechanisms of reducing inflammation, enhancing collagen deposition and recombination and blockade of the TGF-β/Smads signal pathway.	[20]
Phascolosoma esculenta	Pancreatin	Peptides < 3 kDa	Mice	50, 100 and 150 mg/kg for 15 days	Peptides improved the spatial learning and memory ability doses-dependently through the up-regulation of NR2A, NR2B, BDNF and CREB mRNA expressions in hippocampus of mice. 100 mg/kg group showed better performance in spatial learning and memory compared with 50, and 150 mg/kg.	[15]

8. Conclusions

In this review, the various bioactive properties of peptides and polysaccharides originated from peanut worms were summarized. Peanut worms derived bioactive compounds (peptides and polysaccharides) exhibited anti-oxidant, anti-inflammatory, immunomodulatory, anti-hypertensive, anti-cancer and wound healing activities through the modulation of various molecular mechanisms. Most researchers investigated these bioactivities using in vitro, cell culture and animal models. Clinical studies confirming these bioactivities of peptides and polysaccharides of peanut worms are scanty in literature. Therefore, more clinical investigations are needed to enhance the applications of these bioactive compounds of peanut worms. Additionally, in-depth molecular mechanisms of action, bioavailability and safety profiles of these bioactive compounds of peanut worms need to be investigated thoroughly. Moreover, peptides produced from peanut worms could be further explored for their potential anti-diabetic, anti-microbial, anti-obesity and anti-atherosclerosis activities. Although there are several species in phylum sipuncula, only two species (*Sipunculus nudus* and *Phascolosoma esculenta*) have been investigated extensively with regard to the bioactive compound's identification and isolation. Hence, other species of peanut worms could also be studied for possible identification of bioactive compounds for various human applications.

Author Contributions: Y.Q. and J.Z. prepared the original draft, X.S., S.L. and M.C. collected and analyzed the data, H.L. and L.C. revised manuscript and supervised the work. All authors have read and agreed to the published version of the manuscript.

Funding: This research was funded by the public service platform of South China Sea for R & D marine biomedicine resources (2017C8A), Zhanjiang Science and Technology Development Special Fund Project 2020A04005 and Guangdong province science and technology plan project (2HC19002).

Institutional Review Board Statement: Not applicable.

Data Availability Statement: Not applicable.

Conflicts of Interest: The authors declare no conflict of interest.

Abbreviations

ACE: Angiotensin I-converting enzyme; ATF4: Activating transcription factor 4; Bcl-2: B-cell lymphoma 2; BDNF: Brain Derived Neurotrophic Factor; CD14: cluster of differentiation 14; COX-2: cyclooxygenase-2; CR3: complement receptor 3; CREB: cAMP responsive element binding protein; CYR61: Cysteine-rich angiogenic inducer 61; DBP: diastolic blood pressure; DDIT3:DNA Damage Inducible Transcript 3; DNA: Deoxyribonucleic acid; DPPH: 2,2-diphenyl-1-picrylhydrazyl; GPx: Glutathione peroxidase; GSH-Px: glutathione peroxidase; HaCaT: human immortalized keratinocytes; HSF: human skin fibroblasts; HSP90: Heat shock protein 90; HUVEC: human umbilical vein endothelial cells; IFN-γ: Interferon gamma; IL: interleukin; iNOS: inducible nitric oxide synthase; MDA: malondialdehyde; mRNA: messenger ribonucleic acid; MR: mannose receptor; NF-κB: nuclear factor kappa B; NK cells: Natural killer cells; NO: Nitric oxide; Nrf2: nuclear factor erythroid 2-related factor; NSAIDs: non-steroidal anti-inflammatory drugs; ROS: reactive oxygen species; SBP: systolic blood pressure; SR: scavenger receptor; SOD: superoxide dismutase; TAC: Totalantioxidantcapacity; TGF-β1: Transforming growth factor beta1; TLR4: Toll-like receptor 4; TNF-α: Tumour Necrosis Factor alpha; VEGF: Vascular endothelial growth factor.

References

1. Sangtanoo, P.; Srimongkol, P.; Saisavoey, T.; Reamtong, O.; Karnchanatat, A. Anti-inflammatory action of two novel peptides derived from peanut worms (*Sipunculus nudus*) in lipopolysaccharide-induced RAW264.7 macrophages. *Food Funct.* **2020**, *11*, 552–560. [CrossRef] [PubMed]
2. Zhang, C.X.; Dai, Z.R.; Cai, Q.X. Anti-inflammatory and anti-nociceptive activities of *Sipunculus nudus* L. extract. *J. Ethnopharmacol.* **2011**, *137*, 1177–1182. [CrossRef] [PubMed]
3. Wu, Y.; Fang, M.; Du, L.; Wu, H.; Liu, Y.; Guo, M.; Xie, J.; Wei, D. The nutritional composition and anti-hypertensive activity on spontaneously hypertensive rats of sipuncula *Phascolosoma esculenta*. *Food Funct.* **2014**, *5*, 2317. [CrossRef] [PubMed]
4. Ruiz-Ruiz, F.; Mancera-Andrade, E.I.; Iqbal, H.M. Marine-Derived Bioactive Peptides for Biomedical Sectors: A Review. *Protein Pep. Lett.* **2017**, *24*, 109–117. [CrossRef]
5. Lee, Y.E.; Kim, H.; Seo, C.; Park, T.; Lee, K.B.; Yoo, S.Y.; Hong, S.C.; Kim, J.T.; Lee, J. Marine polysaccharides: Therapeutic efficacy and biomedical applications. *Arch. Pharm. Res.* **2017**, *40*, 1006–1020. [CrossRef] [PubMed]
6. Zhong, Q.; Wei, B.; Wang, S.; Ke, S.; Chen, J.; Zhang, H.; Wang, H. The Antioxidant Activity of Polysaccharides Derived from Marine Organisms: An Overview. *Mar. Drugs* **2019**, *17*, 674. [CrossRef]
7. Chalamaiah, M.; Yu, W.; Wu, J. Immunomodulatory and anticancer protein hydrolysates (peptides) from food proteins: A review. *Food Chem.* **2018**, *245*, 205–222. [CrossRef]
8. Chalamaiah, M.; Ulug, S.K.; Hong, H.; Wu, J. Regulatory requirements of bioactive peptides (protein hydrolysates) from food proteins. *J. Funct. Foods* **2019**, *58*, 123–129. [CrossRef]
9. Pádraigín, A.H.; Richard, J.G. Bioactive peptides from marine processing waste and shellfish: A review. *J. Funct. Foods* **2012**, *4*, 6–24.
10. Wang, X.; Yu, H.; Xing, R.; Li, P. Characterization, preparation, and purification of marine bioactive peptides. *BioMed Res. Int.* **2017**, *2017*, 9746720. [CrossRef]
11. Yu, Y.; Shen, M.; Song, Q.; Xie, J. Biological activities and pharmaceutical applications of polysaccharide from natural resources: A review. *Carbohydr. Polym.* **2018**, *183*, 91–101. [CrossRef] [PubMed]
12. Jun, L.; Stefan, W.; Chunlin, X. A review of bioactive plant polysaccharides: Biological activities, functionalization, and biomedical applications. *Bioact. Carbohydr. Diet. Fibre* **2015**, *5*, 31–61.
13. Barbosa, J.R.; de Carvalho Junior, R.N. Polysaccharides obtained from natural edible sources and their role in modulating the immune system: Biologically active potential that can be exploited against COVID-19. *Trends Food Sci. Technol.* **2021**, *108*, 223–235. [CrossRef] [PubMed]

14. Zhang, C.X.; Dai, Z.R. Anti-hypoxia activity of a polysaccharide extracted from the *Sipunculus nudus* L. *Int. J. Biol. Macromol.* **2011**, *49*, 523–526. [CrossRef] [PubMed]
15. Liu, L.; Cao, J.; Chen, J.; Zhang, X.; Wu, Z.; Xiang, H. Effects of peptides from *Phascolosoma esculenta* on spatial learning and memory via anti-oxidative character in mice. *Neurosci. Lett.* **2016**, *631*, 30–35. [CrossRef]
16. Yang, Z.; Pan, Y.; Chen, J.; Zhang, H.; Wei, H.; Wu, Z.; Liu, L. Anti-inflammatory, anti-oxidative stress effect of *Phascolosoma esculenta* oligosaccharides on Escherichia coli-induced sepsis mice. *Food Sci. Biotechnol.* **2019**, *28*, 1871–1879. [CrossRef]
17. Wu, J.; Xie, D.; Chen, X.; Tang, Y.J.; Wang, L.; Xie, J.; Wei, D. Inhibitory mechanism of a substrate-type angiotensin I-converting enzyme inhibitory peptide. *Proc. Biochem.* **2019**, *79*, 97–104. [CrossRef]
18. Su, J.; Liao, D.; Su, Y.; Liu, S.; Jiang, L.; Wu, J.; Liu, Z.; Wu, Y. Novel polysaccharide extracted from *Sipunculus nudus* inhibits HepG2 tumour growth in vivo by enhancing immune function and inducing tumour cell apoptosis. *J. Cell. Mol. Med.* **2021**, *25*, 8338–8351. [CrossRef] [PubMed]
19. Lu, Y.; Chi, H.; Li, G.; Chen, Q.; Qiu, J.; Yu, H.; Fang, X.; Chen, X. Study on alkaline extraction process optimization of polysaccharides from coelomic fluid of *Phasolosma esculenta* and its antioxidant activity in vitro. *Sci. Technol. Food Ind.* **2021**, *42*, 204–210.
20. Lin, H.; Zheng, Z.; Yuan, J.; Zhang, C.; Cao, W.; Qin, X. Collagen peptides derived from *Sipunculus nudus* accelerate wound healing. *Molecules* **2021**, *26*, 1385. [CrossRef] [PubMed]
21. Pizzino, G.; Irrera, N.; Cucinotta, M.; Pallio, G.; Mannino, F.; Arcoraci, V.; Squadrito, F.; Altavilla, D.; Bitto, A. Oxidative Stress: Harms and Benefits for Human Health. *Oxid. Med. Cell. Longev.* **2017**, *2017*, 8416763. [CrossRef]
22. Sharifi-Rad, M.; Anil Kumar, N.V.; Zucca, P.; Varoni, E.M.; Dini, L.; Panzarini, E.; Rajkovic, J.; Fokou, P.V.T.; Azzini, E.; Peluso, I.; et al. Lifestyle, Oxidative Stress, and Antioxidants: Back and Forth in the Pathophysiology of Chronic Diseases. *Front. Physiol.* **2020**, *11*, 694. [CrossRef] [PubMed]
23. Forman, H.J.; Zhang, H. Targeting oxidative stress in disease: Promise and limitations of antioxidant therapy. *Nat. Rev. Drug Discov.* **2021**, *20*, 689–709. [CrossRef]
24. Cheung, R.C.; Ng, T.B.; Wong, J.H. Marine peptides: Bioactivities and applications. *Mar. Drugs* **2015**, *13*, 4006–4043. [CrossRef]
25. Ruocco, N.; Costantini, S.; Guariniello, S.; Costantini, M. Polysaccharides from the marine environment with pharmacological, cosmeceutical and nutraceutical potential. *Molecules* **2016**, *21*, 551. [CrossRef]
26. Lafarga, T.; Acién-Fernández, F.G.; Garcia-Vaquero, M. Bioactive peptides and carbohydrates from seaweed for food applications: Natural occurrence, isolation, purification, and identification. *Algal Res.* **2020**, *48*, 101909. [CrossRef]
27. Nwachukwu, I.D.; Aluko, R.E. Structural and functional properties of food protein-derived antioxidant peptides. *J. Food Biochem.* **2019**, *43*, e12761. [CrossRef] [PubMed]
28. Kai, W.; Han, L.; Hong, H.; Pan, J.; Liu, H.; Luo, Y. Purification and identification of novel antioxidant peptides from silver carp muscle hydrolysate after simulated gastrointestinal digestion and transepithelial transport. *Food Chem.* **2021**, *342*, 128275.
29. Chalamaiah, M.; Dinesh Kumar, B.; Hemalatha, R.; Jyothirmayi, T. Fish protein hydrolysates: Proximate composition, amino acid composition, antioxidant activities and applications: A review. *Food Chem.* **2012**, *135*, 3020–3038. [CrossRef]
30. Liu, W.Y.; Zhang, J.T.; Miyakawa, T.; Li, G.M.; Gu, R.Z.; Tanokura, M. Antioxidant properties and inhibition of angiotensin-converting enzyme by highly active peptides from wheat gluten. *Sci. Rep.* **2021**, *11*, 5206. [CrossRef] [PubMed]
31. Tadesse, S.A.; Emire, S.A. Production and processing of antioxidant bioactive peptides: A driving force for the functional food market. *Heliyon* **2020**, *6*, e04765. [CrossRef]
32. Fernando, I.P.S.; Park, S.Y.; Han, E.J.; Kim, H.S.; Kang, D.S.; Je, J.Y.; Ahn, C.B.; Ahn, G. Isolation of an antioxidant peptide from krill protein hydrolysates as a novel agent with potential hepatoprotective effects. *J. Funct. Foods* **2020**, *67*, 103889. [CrossRef]
33. Tonolo, F.; Folda, A.; Cesaro, L.; Scalcon, V.; Marin, O.; Ferro, S.; Bindoli, A.; Rigobello, M.P. Milk-derived bioactive peptides exhibit antioxidant activity through the Keap1-Nrf2 signaling pathway. *J. Funct. Foods* **2020**, *64*, 103696. [CrossRef]
34. Zhang, J.; Li, M.; Zhang, G.; Tian, Y.; Kong, F.; Xiong, S.; Zhao, S.; Jia, D.; Manyande, A.; Du, H. Identification of novel antioxidant peptides from snakehead (*Channa argus*) soup generated during gastrointestinal digestion and insights into the anti-oxidation mechanisms. *Food Chem.* **2021**, *337*, 127921. [CrossRef]
35. Zhu, Y.L.; Li, S.D.; Liao, Y.; Li, J.M. Made polypeptide from *Sipunculus nudus* by papain enzymolysis and its effects on hydroxy radical cleaning. *Adv. Mater. Res.* **2013**, *699*, 360–366. [CrossRef]
36. Xiang, H.; Liu, L.; Lu, X.; Shi, Y.; Chen, J. Optimization of the preparation of collagen antioxidant peptide of *Phascolosoma esculenta* by response surface method. *Sci. Technol. Food Ind.* **2014**, *13*, 253–257.
37. Zhang, Q.; Dong, L.; Tong, T.; Wang, Q.; Xu, M. Polysaccharides in *Sipunculus nudus*: Extraction condition optimization and antioxidant activities. *J. Ocean Univ. China* **2017**, *16*, 74–80. [CrossRef]
38. Wu, Y.; Jiang, H.; Lin, J.S.; Liu, J.; Wu, C.J.; Xu, R. Antioxidant, hypolipidemic and hepatic protective activities of polysaccharides from *Phascolosoma esculenta*. *Mar. Drugs* **2020**, *18*, 158. [CrossRef]
39. Li, N.; Shen, X.; Liu, Y.; Zhang, J.; He, Y.; Liu, Q.; Jiang, D.; Zong, J.; Li, J.; Hou, D.; et al. Isolation, characterization, and radiation protection of *Sipunculus nudus* L. polysaccharide. *Int. J. Biol. Macromol.* **2016**, *83*, 288–296. [CrossRef] [PubMed]
40. Chen, L.; Deng, H.; Cui, H.; Fang, J.; Zuo, Z.; Deng, J.; Li, Y.; Wang, X.; Zhao, L. Inflammatory responses and inflammation-associated diseases in organs. *Oncotarget* **2018**, *9*, 7204–7218. [CrossRef]
41. Furman, D.; Campisi, J.; Verdin, E.; Carrera-Bastos, P.; Targ, S.; Franceschi, C.; Ferrucci, L.; Gilroy, D.W.; Fasano, A.; Miller, G.W.; et al. Chronic inflammation in the etiology of disease across the life span. *Nat. Med.* **2019**, *25*, 1822–1832. [CrossRef]

42. Lordan, R.; Zabetakis, I. Invited review: The anti-inflammatory properties of dairy lipids. *J. Dairy Sci.* **2017**, *100*, 4197–4212. [CrossRef] [PubMed]
43. Yahfoufi, N.; Alsadi, N.; Jambi, M.; Matar, C. The immunomodulatory and anti-inflammatory role of polyphenols. *Nutrients* **2018**, *10*, 1618. [CrossRef]
44. Sun, J.; Gou, Y.; Liu, J.; Chen, H.; Kan, J.; Qian, C.; Zhang, N.; Niu, F.; Jin, C. Anti-inflammatory activity of a water-soluble polysaccharide from the roots of purple sweet potato. *RSC Adv.* **2020**, *10*, 39673–39686. [CrossRef]
45. Liang, Q.; Chalamaiah, M.; Liao, W.; Ren, X.; Ma, H.; Wu, J. Zein hydrolysate and its peptides exert anti-inflammatory activity on endothelial cells by preventing TNF-α-induced NF-κB activation. *J. Funct. Foods* **2020**, *64*, 103598. [CrossRef]
46. Wang, Y.; Zhu, H.; Wang, X.; Yu, Y.; Xie, J. Natural food polysaccharides ameliorate inflammatory bowel disease and its mechanisms. *Foods* **2021**, *10*, 1288. [CrossRef]
47. Snigdha, G.; Kaustav, M. Structural-features of food-derived bioactive peptides with anti-inflammatory activity: A brief review. *J. Food Biochem.* **2018**, *43*, e12531.
48. Zhu, W.; Ren, L.; Zhang, L.; Qiao, Q.; Farooq, M.Z.; Xu, Q. The Potential of Food Protein-Derived Bioactive Peptides against Chronic Intestinal Inflammation. *Mediat. Inflamm.* **2020**, *2020*, 6817156. [CrossRef]
49. Seung, Y.L.; Sun, J.H. Antihypertensive peptides from animal products, marine organisms, and plants. *Food Chem.* **2017**, *228*, 506–517.
50. Mirzapour, M.; Rezaei, K.; Sentandreu, M.A. Identification of potent ACE inhibitory peptides from wild almond proteins. *J. Food Sci.* **2017**, *82*, 2421–2431. [CrossRef]
51. Ola, A.; Moncef, N. Basic and recent advances in marine antihypertensive peptides: Production, structure-activity relationship and bioavailability. *Trends Food Sci. Technol.* **2019**, *88*, 543–557.
52. Abachi, S.; Bazinet, L.; Beaulieu, L. Antihypertensive and angiotensin-I-converting enzyme (ACE)-inhibitory peptides from fish as potential cardioprotective compounds. *Mar. Drugs* **2019**, *17*, 613. [CrossRef] [PubMed]
53. Agustina, E.N.; Santiago, S.; Alejandra, V.Q.; Maria, C.A. Amaranth as a source of antihypertensive peptides. *Front. Plant Sci.* **2020**, *11*, 578631.
54. Yu, Z.; Guo, H.; Shiuan, D.; Xia, C.; Zhao, W.; Ding, L.; Zheng, F.; Liu, J. Interaction mechanism of egg white-derived ACE inhibitory peptide TNGIIR with ACE and its effect on the expression of ACE and AT1 receptor. *Food Sci. Hum. Wellness* **2020**, *9*, 52–57. [CrossRef]
55. Durak, M.Z.; Neslihan, A.T. Antihypertensive Peptides in Dairy Products. *Am. J. Biomed. Sci. Res.* **2020**, *7*, 191–195. [CrossRef]
56. Arshdeep, K.; Bababode, A.; Poorva, S.; Deepansh, S.; Sawinder, K. Recently isolated food-derived antihypertensive hydrolysates and peptides: A review. *Food Chem.* **2021**, *346*, 128719.
57. Ma, K.; Wang, Y.; Wang, M.; Wang, Z.; Wang, X.; Ju, X.; He, R. Antihypertensive activity of the ACE–renin inhibitory peptide derived from Moringa oleifera protein. *Food Funct.* **2021**, *12*, 8994–9006. [CrossRef]
58. Liu, Y.; Zhang, L.; Guo, M.; Wu, H.; Xie, J.; Wei, D. Virtual screening for angiotensin I-converting enzyme inhibitory peptides from *Phascolosoma esculenta*. *Bioresour. Bioprocess.* **2014**, *1*, 17. [CrossRef]
59. Guo, M.; Chen, X.; Wu, Y.; Zhang, L.; Huang, W.; Yuan, Y.; Fang, M.; Xie, J.; Wei, D. Angiotensin I-converting enzyme inhibitory peptides from Sipuncula (*Phascolosoma esculenta*): Purification, identification, molecular docking and antihypertensive effects on spontaneously hypertensive rats. *Proc. Biochem.* **2017**, *63*, 84–95. [CrossRef]
60. Du, L.; Fang, M.; Wu, H.; Xie, J.; Wu, Y.; Li, P.; Zhang, D.; Huang, Z.; Xia, Y.; Zhou, L.; et al. A novel angiotensin I-converting enzyme inhibitory peptide from *Phascolosoma esculenta* water-soluble protein hydrolysate. *J. Funct. Foods* **2013**, *5*, 475–483. [CrossRef]
61. Wu, H.; Liu, Y.; Guo, M.; Xie, J.; Jiang, X. A virtual screening method for inhibitory peptides of angiotensin I–converting enzyme. *J. Food Sci.* **2014**, *79*, C1635. [CrossRef]
62. Sun, X.; Wang, M.; Liu, B.; Sun, Z. Purification and characterization of angiotensin I converting enzyme inhibition peptides from Sandworm *Sipunculus nudus*. *J. Ocean Univ. China* **2017**, *16*, 911–915. [CrossRef]
63. Yin, M.; Zhang, Y.; Li, H. Advances in research on immunoregulation of macrophages by plant polysaccharides. *Front. Immunol.* **2019**, *10*, 145. [CrossRef]
64. Liang, R. Orthogonal test design for optimization of the extraction of polysaccharides from *Phascolosoma esulenta* and evaluation of its immunity activity. *Carbohydr. Polym.* **2008**, *73*, 558–563.
65. Zhang, C.X.; Dai, Z.R. Immunomodulatory activities on macrophage of a polysaccharide from *Sipunculus nudus* L. *Food Chem. Toxicol.* **2011**, *49*, 2961–2967. [CrossRef] [PubMed]
66. Li, K.X.; Shen, X.R.; He, Y.; Jiang, D.W.; Liu, Y.M.; Hou, D.Y.; Chen, W. Effect of *Sipunculus nudus* polysaccharides on immunological function in mice. *Chin. J. Mar. Drugs* **2012**, *31*, 46–49.
67. Su, J.; Jiang, L.; Wu, J.; Liu, Z.; Wu, Y. Immunologic effect of polysaccharides extracted from *Sipunculus nudus* (SNP) on Hepatoma HepG2-bearing Mice. *bioRxiv* **2017**, *10*, 175190. [CrossRef]
68. Ferlay, J.; Ervik, M.; Lam, F.; Colombet, M.; Mery, L.; Piñeros, M.; Znaor, A.; Soerjomataram, I.; Bray, F. *Global Cancer Observatory: Cancer Today*; International Agency for Research on Cancer: Lyon, France, 2020; Available online: https://gco.iarc.fr/today (accessed on 5 February 2021).
69. Pucci, C.; Martinelli, C.; Ciofani, G. Innovative approaches for cancer treatment: Current perspectives and new challenges. *eCancer Med. Sci.* **2019**, *13*, 961. [CrossRef]
70. Khalifa, S.; Elias, N.; Farag, M.A.; Chen, L.; Saeed, A.; Hegazy, M.F.; Moustafa, M.S.; Abd El-Wahed, A.; Al-Mousawi, S.M.; Musharraf, S.G.; et al. Marine natural products: A source of novel anticancer drugs. *Mar. Drugs* **2019**, *17*, 491. [CrossRef] [PubMed]

71. Alves, C.; Silva, J.; Pinteus, S.; Gaspar, H.; Alpoim, M.C.; Botana, L.M.; Pedrosa, R. From marine origin to therapeutics: The antitumor potential of marine algae-derived compounds. *Front. Pharmacol.* **2018**, *9*, 777. [CrossRef]
72. Khan, T.; Date, A.; Chawda, H.; Patel, K. Polysaccharides as potential anticancer agents-A review of their progress. *Carbohydr. Polym.* **2019**, *210*, 412–428. [CrossRef] [PubMed]
73. Su, J.; Jiang, L.L.; Wu, J.N.; Liu, Z.Y.; Wu, Y.P. Anti-tumor and anti-virus activity of polysaccharides extracted from *Sipunculus nudus* (SNP) on Hepg2.2.15. *Int. J. Biol. Macromol.* **2016**, *87*, 597–602. [CrossRef] [PubMed]
74. Chen, W.; Wang, X.; Shao, D.; Li, L.; Huang, Y.; Sun, H.; Li, L.; Lin, Y. Study on extraction process of *Sipunculus nudus* polysaccharide and its antioxidant activity. *Agric. Biotechnol.* **2018**, *7*, 112–118.

marine drugs

Article

First Insights into the Repertoire of Secretory Lectins in Rotifers

Marco Gerdol

Department of Life Sciences, University of Trieste, Via Giorgieri 5, 34128 Trieste, Italy; mgerdol@units.it

Abstract: Due to their high biodiversity and adaptation to a mutable and challenging environment, aquatic lophotrochozoan animals are regarded as a virtually unlimited source of bioactive molecules. Among these, lectins, i.e., proteins with remarkable carbohydrate-recognition properties involved in immunity, reproduction, self/nonself recognition and several other biological processes, are particularly attractive targets for biotechnological research. To date, lectin research in the Lophotrochozoa has been restricted to the most widespread phyla, which are the usual targets of comparative immunology studies, such as Mollusca and Annelida. Here we provide the first overview of the repertoire of the secretory lectin-like molecules encoded by the genomes of six target rotifer species: *Brachionus calyciflorus*, *Brachionus plicatilis*, *Proales similis* (class Monogononta), *Adineta ricciae*, *Didymodactylos carnosus* and *Rotaria sordida* (class Bdelloidea). Overall, while rotifer secretory lectins display a high molecular diversity and belong to nine different structural classes, their total number is significantly lower than for other groups of lophotrochozoans, with no evidence of lineage-specific expansion events. Considering the high evolutionary divergence between rotifers and the other major sister phyla, their widespread distribution in aquatic environments and the ease of their collection and rearing in laboratory conditions, these organisms may represent interesting targets for glycobiological studies, which may allow the identification of novel carbohydrate-binding proteins with peculiar biological properties.

Keywords: rotifera; pattern recognition receptors; microbe-associated molecular patterns; innate immunity; C-type lectins; C1q domain-containing proteins; galectins

Citation: Gerdol, M. First Insights into the Repertoire of Secretory Lectins in Rotifers. *Mar. Drugs* **2022**, *20*, 130. https://doi.org/10.3390/md20020130

Academic Editor: Hitoshi Sashiwa

Received: 20 January 2022
Accepted: 7 February 2022
Published: 9 February 2022

1. Introduction

Lectin-like molecules play a fundamental role in several physiological processes shared by all animals, including, critically, the discrimination between "self" and "nonself" through the specific recognition of carbohydrate moieties exposed on cellular surfaces. These glycans, when associated with microorganisms, are generally referred to as microbe-associated molecular patterns (MAMPs) or, in the case of potentially pathogenic microbes, pathogen-associated molecular patterns (PAMPs) [1].

The proteins expressed by the host that are involved in carbohydrate recognition are collectively known as pattern recognition receptors (PRRs), which may exert their function at different levels, i.e., in the extracellular environment, at the plasma membrane or within the cell. In the context of immune response, the activity of a heterogeneous group of small secretory PRRs usually leads to the coating of invading microbes. This process may in turn trigger a complex response involving several additional molecular and cellular players which vary widely along the metazoan tree of life. These include, among others, the melanization cascade (typically observed in arthropods and other invertebrates) [2], the production of a large arsenal of antimicrobial peptides [3], the activation of the complement system (well-described in vertebrates and present in a primitive form also in many invertebrates) [4] and the recruitment of specialized phagocytic cells [5]. Furthermore, the ability to recognize MAMPs and to modulate immune responses has been linked with the maintenance of gut microbiome homeostasis [6], as well as the establishment of beneficial bacterial symbioses [7], which are particularly relevant in aquatic environments [8].

Besides their key role in immune recognition, lectins are involved in a number of other physiological processes, to which they contribute thanks to their remarkable ability to recognize glycan moieties with high specificity. For example, some lectins play an important role in reproduction and gamete recognition [9,10], in the clearance of apoptotic cells thanks to the recognition of damage-associated molecular patterns (DAMPs) [11], in larval settlement and metamorphosis [12] and in the recognition of food particles in filter-feeding bivalves [6,13].

Aquatic invertebrates have been a preferred target for lectin identification and purification during the past three decades, as revealed by the fact that many of the best functionally characterized lectins from non-vertebrate metazoans derive from corals, echinoderms and mollusks [14–16]. Among the Lophotrochozoa, one of the two clades of spiralian animals together with Ecdysozoa, most glycobiological and immunological studies have been so far focused on species belonging to the phyla Mollusca or Annelida, amenable for research due to their relatively large body size and the ease of sampling and laboratory handling [17]. Other lophotrochozoan phyla have been nearly completely neglected up to now, leaving a remarkable gap of knowledge concerning the main molecular players involved in carbohydrate recognition.

Among these, the phylum Rotifera, which comprises over 2000 described species with a widespread distribution in freshwater environments, but occasionally found also in brackish and saltwater habitats, represents a particularly intriguing unexplored resource for lectin research. Rotifera are classically subdivided between two classes, namely, Monogononta (the most species-rich class) and Bdelloidea, even though phylogenetic evidence suggests that Seisonidea and Acanthocephala also belong to the very same monophyletic group. Bdelloids display a few peculiar features compared with all other lophotrochozoans, such as a remarkable ability to withstand extreme temperatures [18] and ionizing radiations, which is thought to derive from efficient DNA double-strand break repair [19], and obligatory parthenogenetic reproduction, which results from a long-term asexual evolutionary history [20]. Another interesting feature of rotifers lies in their remarkable genetic divergence from the other major lophotrochozoan phyla. Indeed, monogonont and bdelloid rotifer genomes differ greatly, both in terms of size and architecture, which in bdelloids is significantly impacted by the presence of transposable elements [21], massive horizontal gene transfer [22] and signatures of long-term asexual reproduction [23].

Rotifers often belong to cryptic species complexes, which can only be correctly identified through DNA barcoding, and have in most cases a cosmopolitan distribution [24,25]. These organisms, which usually have a very small size (100–1000 μm), constitute a significant fraction of microzooplankton and their biomass can be particularly relevant in certain environments, such as coastal lagoons or shallow, acidified, metal-contaminated lakes [26–28]. During the 1970s and 1980s, some rotifer species, such as the eurhyaline *Brachionus plicatilis*, were successfully established as live feeds in marine fish aquaculture, thanks to their fast population growth and ease of intensive culture [29–31] (i.e., up to two billion individuals can be obtained in one day per cubic meter of culture [32]). This would undoubtedly represent an interesting opportunity for glycobiology studies, as a sufficient biomass could be readily available for lectin isolation and purification.

The successful adaptation of rotifers to a challenging environment, where they are potentially exposed to a broad range of microorganisms, suggests that these small animals might have developed carbohydrate-binding strategies similar to those described in other aquatic invertebrates in which multiple biomolecules with high biotechnological potential have been previously identified. Moreover, due to their peculiar features and their high tendency to acquire novel genes by horizontal gene transfer, these small metazoans might be considered as a potential source of isolation for a number of novel lectins with unusual and interesting biological properties.

This work preliminarily explores the repertoire of secretory lectins from six rotifer species belonging to the classes Monogononta and Bdelloidea. The publicly available genomes of these species were screened to look for annotated genes encoding proteins bearing known carbohydrate-binding domains (CRDs). Unlike other lophotrochozoan phyla, in which lectin-like proteins are often encoded by tandemly duplicated paralogous genes displaying high pairwise sequence homology, rotifers do not show evidence of massive gene family expansion events. However, they display a highly diversified arsenal of carbohydrate-binding proteins whose biological properties could be explored and biotechnologically exploited in the near future.

2. Results

The screening of six rotifer genomes allowed the identification of a relatively small number of secretory lectin-like molecules compared with other lophotrochozoans, which are often characterized by massive gene family expansion events that involve carbohydrate-binding proteins, as exemplified by the case of C1qDC proteins in bivalves [33–35]. Based on available data in the literature, only lectin-like proteins displaying a canonical signal peptide for secretion and which display no significant primary sequence conservation among the different lectin families will be here described; the only exception is represented by galectins, which rely on unconventional secretion.

In the class Monogononta, *Brachionus calyciflorus* was the species in which the highest number of lectins was identified (38), followed by the congeneric species *Brachionus plicatilis* (25) and *Proales similis* (14). In the class Bdelloidea, *Rotaria sordida* and *Adineta ricciae* displayed a similar number of lectins (27 and 22, respectively), whereas the third rotifer species, *Didymodactylos carnosus*, had the lowest number of associated lectin sequences in this study (eight) (Table 1; the full list of gene accession IDs is provided in Table S1). Based on these observations, it can be estimated that just a very tiny fraction of all protein-coding genes in rotifers (i.e., 0.02–0.15%) encode secretory lectins characterized by the presence of previously described conserved domains. Nevertheless, despite the lack of evident lectin family expansions, the lectin-like proteins identified in all rotifer species displayed a remarkable molecular diversity, as revealed by their classification within nine different families (Table 1): (i) fibrinogen-related domain-containing proteins (FReDs) (Section 2.1); (ii) C-type lectins (Section 2.2); (iii) C1q-domain containing (C1qDC) proteins (Section 2.3); (iv) galectins (Section 2.4); (v) R-type lectins (Section 2.5); (vi) F-type lectins (Section 2.6); (vii) SUEL-type lectins; (viii) H-type lectins; (ix) jacalin-like lectins (Section 2.7).

Table 1. Number of secretory lectins identified in the six rotifer species analyzed in this study. The full list of gene accession IDs is provided in Table S1.

	Bdelloidea			Monogononta		
	Adineta ricciae	*Rotaria sordida*	*Didymodactylos carnosus*	*Proales similis*	*Brachionus calyciflorus*	*Brachionus plicatilis*
FReDs	6	6	1	3	2	6
C-type lectins	3	2	1	3	25	17
C1qDC proteins	4	6	2	1	1	1
Galectins	4	8	2	1	1	1
R-type lectins	0	1	0	0	3	0 [a]
F-type lectins	3	3	1	2	3	0 [b]
SUEL-type lectins	0	0	0	4	3	0 [c]
H-type lectins	2	1	0	0	0	0
Jacalin-like lectins	0	0	1	0	0	0
Apextrins	0	0	0	0	0	0
DUF3011 lectins	0	0	0	0	0	0

[a] Two partial BPBT lectins (see Section 2.5), lacking a signal peptide, likely due to incorrect annotation, were detected. [b] A single FTL with three CRDs, lacking a signal peptide, likely due to incorrect annotation, was detected. [c] Two short single-domain SUEL-type lectins, encoded by two paralogous genes, were likely incorrectly fused in a single gene model.

2.1. FReD-Containing Proteins

Fibrinogen-related domain-containing proteins (FReDs) share structural similarity with the C-terminal domain of vertebrate ficolins, i.e., *N*-acetylglucosamine (GlcNAc)-specific carbohydrate-binding proteins, which play a key role in the lectin pathway of the complement system [36,37]. The fibrinogen C-terminal domain is associated with a number of metazoan lectins with widespread taxonomic distribution, from cnidarians to vertebrates, which hold remarkable glycan-binding properties and often play an important role in the context of immune recognition, as revealed by several studies carried out in Mollusca [38–40].

A subgroup of FReDs named fibrinogen-related proteins (FREPs), which combine one or two N-terminal immunoglobulin domains with a single C-terminal fibrinogen domain, have been implicated in the resistance of snails to trematode infections [41,42]. Comparative immunogenomics studies have previously revealed that bona fide FREPs [43], as well as GREPs and CREPs (i.e., FReDs associated with galectin and CTL domains, respectively [44]), are restricted to the gastropod subclass Heterobranchia. Nevertheless, other mollusks display a high number of proteins with a simpler architecture, comprising a signal peptide and the fibrinogen-like domain, often paired with a coiled-coil region of variable length, which may allow their oligomerization, in a similar fashion to collagen in vertebrate ficolins [45]. Single-domain FReDs, which retain significant glycan-binding properties in the Lophotrochozoa [46], underwent a significant expansion in bivalves, where they are often found with hundreds of paralogous gene copies [47] encoding inducible proteins with marked bacteria-agglutinating properties [48,49]. Similar expansions have certainly occurred in other lophotrochozoan phyla, such as brachiopods, even though the functional implications of these events are presently unclear [50].

All the rotifer species analyzed in this study had FReD genes in varying numbers, ranging from one (in *D. carnosus*) to six (in *R. sordida*, *B. plicatilis* and *A. ricciae*) (Table 1). The encoded proteins from bdelloid and monogonont rotifers displayed different architectures: while all FReDs shared a single peptide and displayed a fibrinogen-like domain in a C-terminal position, they were characterized by the presence of an N-terminal region of variable length (Figure 1). This region was markedly shorter in bdelloid FReDs, which usually displayed a relatively high (55–50%) primary sequence identity with horseshoe crab tachylectins [39], and much longer in monogonont FReDs, which, on the other hand, had a lower homology (i.e., 25–35%) with tachylectins. In all rotifer FReDs, this region lacked detectable conserved domains and structural homologies but displayed a low level of complexity and the occasional presence of threonine-rich amino acid stretches.

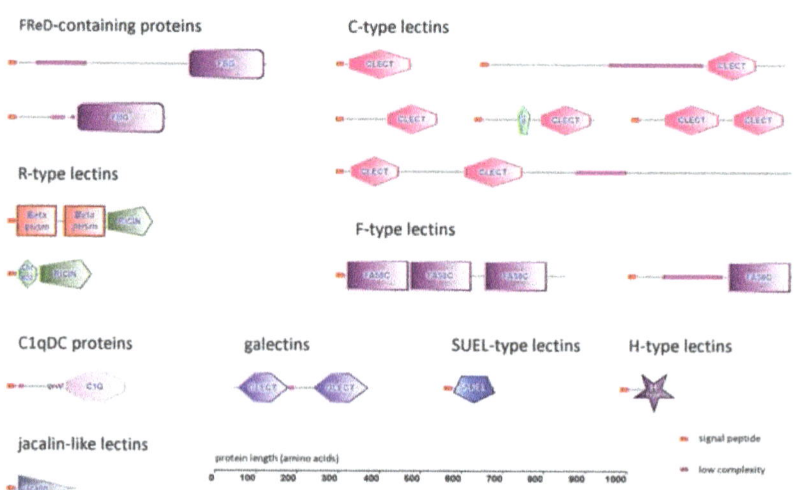

Figure 1. Schematic representation of the main type of secretory lectin-like molecules identified in Rotifera. FBG: Fibrinogen C-terminal domain; GLECT: galectin domain; SUEL: D-galactoside/L-rhamnose-binding SUEL lectin domain; CLECT: C-type lectin domain; FA58C: coagulation factor 5/8 C-terminal domain; EGF: epidermal growth factor domain; Cht BD2: chitin binding domain.

2.2. C-Type Lectins

C-type lectins (CTLs) are one of the largest and most studied families of lectins in lophotrochozoans, with several dozen proteins having been functionally characterized in mollusks and segmented worms [51–53]. Their characterizing CRD, which displays a broad calcium-dependent binding specificity, is often found in large multidomain membrane-bound proteins which may or may not have a lectin function [54,55]. Their remarkable structural diversity has led to the development of a complex classification system, which has been subjected to multiple updates over the years [54,56,57]. Since such a classification still appears to be strongly biased towards vertebrates, it is not fully adequate to describe the variegate domain combinations found in animal CTLs.

Compared with their membrane-bound counterparts, secretory CTLs usually display a simpler structure, which comprises a signal peptide, followed by either one or two tandemly repeated CRDs. In addition, the N-terminal region may also include coiled-coil or collagen repeats with effector functions [58,59]. Besides having a role in MAMP recognition, the CTLs of invertebrate metazoans can regulate different aspects of the innate immune response, including microbial opsonization, the activation of the prophenoloxidase-mediated melanization cascade and possibly also the activation of the complement system, mirroring the role of the mannan-binding lectin in the lectin pathway of the vertebrate complement system [52,60,61].

As far as the Lophotrochozoa are concerned, multiple studies have previously revealed that CTLs belong to highly expanded gene families in Mollusca [33,62], Annelida and Brachiopoda [50]. The investigations carried out here in Rotifera revealed a highly variable number of CTLs among species. While CTLs represented the largest group of secretory lectins in the genus *Brachionus* (i.e., 25 in *B. calyciflorus* and 17 in *B. plicatilis*), only a few proteins of this type (i.e., one to four) could be identified in the four other species (Table 1). Most of the proteins identified in *Brachionus* spp. had a single CRD (Figure 1), which often followed a relatively long (i.e., ~100 amino acids) N-terminal region with no recognizable conserved domains. In addition, both *Brachionus* species displayed a few CTLs with two consecutive CRDs, whose architecture resembled those of insect immulectins [63]. Another type of domain combination included the presence of an epidermal growth factor (EGF)-like domain, placed immediately before the CRD. EGF domains are often found in association with certain large vertebrate CTLs found in the extracellular matrix or bound to the cell membrane, such as selectins and lecticans. However, the combination of a single EGF domain and the CTL CRD has never been described before in the Lophotrochozoa. The third analyzed monogonont rotifer species, *P. similis*, only displayed three CTL genes: two encoded short, single-domain lectins, whereas the third one had an additional EGF-like domain, as previously described in *Brachionus* spp. (Figure 1).

The three bdelloid species had a smaller number of genes encoding secretory CTLs: three were identified in *A. ricciae*, two in *R. sordida* and a single one in *D. carnosus*. Two CTLs from *R. sordida* and one from *A. ricciae* were short single-domain CTLs. *D. carnosus* and *A. ricciae* shared the presence of an orthologous sequence with two recognizable CRDs located at the N-terminal end, followed by a long region with no detectable primary sequence or structural homologies. The third CTL identified in *A. ricciae* showed an unusually long N-terminal low-complexity region, highly enriched in threonine and serine residues, followed by a C-terminal CRD (Figure 1).

In general, rotifer CTLs only showed a poor primary sequence homology (i.e., 20–30%) with functionally characterized molluscan CTLs, which prevented the ascertainment of clear orthology relationships. It is worth mentioning that a single protein belonging to the CTL family had been previously described and functionally characterized in Rotifera. Nevertheless, the sequences orthologous with this protein, which serves as the mate recognition pheromone in the male individuals of *Brachionus manjavacas* [64], are not reported in the present study due to the presence of a transmembrane domain.

2.3. *C1q Domain-Containing Proteins*

C1q domain-containing (C1qDC) proteins belong to a widespread family of highly versatile globular proteins with remarkable binding properties [65,66]. Besides their well-characterized involvement in the vertebrate complement system, C1qDC proteins carry out important functions in other biological processes which have only recently started to be unveiled [67]. For example, thanks to the carbohydrate-binding properties demonstrated in several metazoan phyla [68,69], C1qDC proteins should be regarded as PRRs involved in immune recognition. This role has been investigated in detail in Mollusca [70,71], where C1qDC proteins are associated with massive gene family expansions [34,35,72]. In bivalves, such expansions involve C1qDC proteins that either have a very simple architecture (signal peptide + C1q domain) or contain an additional N-terminal coiled-coil region. Moreover, in some gastropod species, such as *Littorina littorea*, the C1q domain is combined with one or two immunoglobulin-like domains, originating a small class of proteins known as QREPs, which are upregulated in response to *Himasthla elongata* infections [73].

Unlike bivalves but similar to other lophotrochozoan phyla, such as annelids and brachiopods [50,73], rotifers only display a very few secretory C1qDC proteins (Table 1). In detail, a single orthologous C1qDC gene could be identified in the three monogonont rotifer species, whereas the three bdelloid rotifers had a variable number of C1qDC genes, ranging from two (in *D. carnosus*) to six (in *R. sordida*), with evidence of a few nearly identical paralogs (further supported by phylogenetic evidence; see below). In all cases, rotifer C1qDC proteins were relatively short (<350 aa) and displayed a single C-terminal C1q domain (Figure 1). All proteins had a short (~30 aa long) collagen-like region placed immediately before the start of the C1q domain, which was characterized by the presence of nine highly conserved glycine residues (Figure 2A). This domain organization denotes the typical structure of C1q-like proteins, which represent the most common type of C1qDC proteins in vertebrates [50]. C1q-like proteins are present (but rare) in the lophotrochozoan species characterized by C1qDC gene family expansions, in which collagen repeats are usually replaced by coiled-coil regions [34].

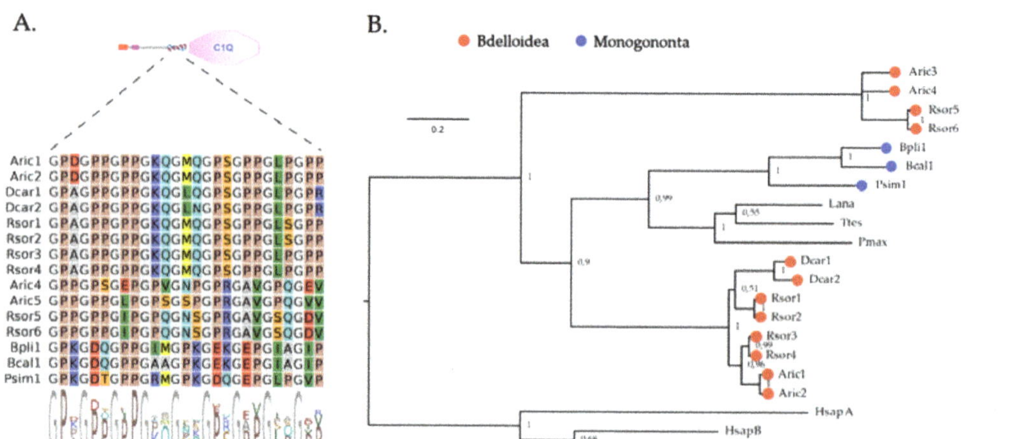

Figure 2. (**A**) Schematic structure of the C1qDC proteins identified in rotifers, with a zoom on the collagen region. (**B**) Bayesian phylogeny of C1qDC proteins from rotifers, obtained with 500,000 generations of an MCMC analysis, run under an LG+I+G model of molecular evolution. The numbers shown close to each node represent posterior probability support values. Aric: *A. ricciae;* Rsor: *R. sordida;* Psim: *P. similis;* Dcar: *D. carnosus;* Bcal: *B. calyciflorus;* Bpli: *B. plicatilis;* Lana: *Lingula anatina;* Pmax: *Pecten maximus;* Ttes: *Testudinalia testudinalis;* Hsap: *Homo sapiens*. Human sequences were used as an outgroup for tree-rooting purposes; A, B and C indicate the human C1qA, C1qB and C1qC chains.

From a phylogenetic point of view, the C1qDC proteins of rotifers were subdivided into three distinct groups (Figure 2B): the first included the C1qDC proteins from Monogononta, which were clustered with high support (posterior probability = 0.99) with a few C1q-like proteins previously identified in other lophotrochozoans and hypothesized to play a key role in the proto-complement system [73]. The C1qDC proteins from bdelloid rotifers were clustered in two groups: the first one, which included a few highly similar paralogous genes in each species (two in *D. carnosus* and *A. ricciae*, four in *R. sordida*), comprised proteins with high sequence homology relative to the group of C1qDC proteins from Monogononta and other lophotrochozoans described above. These proteins displayed, as a peculiar feature, an N-terminal low complexity Ser- and Gln-rich region. The second group of C1qDC proteins from bdelloids only comprised sequences from *A. ricciae* and *R. sordida*, which displayed a high divergence with all the other sequences mentioned above and may therefore represent bdelloid innovations.

2.4. Galectins

Galectins are taxonomically widespread and structurally well-conserved β-galactosyl-binding lectins which carry out a multitude of different functions, including cell adhesion, cellular homeostasis and self/non-self and microbial recognition [74]. Based on their structural organization, lophotrochozoan galectins can generally be considered as belonging to the "tandem-repeat" subtype and contain either two or four CRDs [75–78], with rare occurrences of galectins with three CRDs [50]. Although phylogenetic analyses have previously revealed a monophyletic origin for all molluscan galectins [79], it is presently unclear whether this consideration also applies to the galectins from other lophotrochozoan phyla.

This investigation allowed the identification of galectin genes in all the six analyzed rotifer genomes, even though their number significantly varied among species. While all Monogononta only had a single galectin, bdelloid genomes encoded multiple galectin genes, ranging from two (*D. carnosus*) to eight (*R. sordida*) (Table 1). All rotifer galectins displayed two tandemly repeated CRDs, separated by a connecting region of variable length (Figure 1). No galectins with four CRDs could be identified, confirming the previous observation that, within the Lophotrochozoa, this subtype is restricted to brachiopods, phoronids and annelids [50]. Primary sequence homology with other members of the galectin family from non-rotifer lophotrochozoans was generally in the range of 30–35%. Consistently with previous observations in other metazoans, the encoded proteins lacked a canonical signal peptide and might therefore use an alternative secretion route [80].

2.5. Ricin β-Trefoil Lectins

The R-type lectin (RTL) domain, originally described in the plant toxin ricin, is found in a number of metazoan multidomain proteins with different functions, including hydrolases, glycosyltransferases and membrane-bound receptors [81]. Nevertheless, smaller proteins with no additional domains can serve as lectins in the extracellular environment, playing a role in PAMP recognition. A number of secretory R-type lectins with different glycan-binding properties, containing either one or two consecutive CRDs, have been previously isolated in annelids [82–84] and mollusks [85]. A second family of lectins, named mytilectins, which share the same β-trefoil three-dimensional structure but do not conform with the canonical R-type lectin primary sequence signature, show a discontinuous distribution among the Lophotrochozoa and have only been described so far in a few bivalve mollusks and brachiopods [50,86,87].

While rotifer genomes encoded several proteins with R-type lectin domains, in most cases these were associated with other domains known to exert catalytic activities (e.g., glycosylases, hydrolases, etc.) or with transmembrane domains. Strong evidence in support of the existence of secretory RTLs could be collected only for two out of the six rotifer species analyzed in this study, i.e., *R. sordida*, among bdelloids, and *B. calyciflorus*, among Monogononta (Table 1).

In detail, the three secretory RTLs identified in *B. calyciflorus* displayed an unusual domain architecture, never before reported in other metazoans. Indeed, these proteins showed the presence of two consecutive VOMI (vitelline membrane outer layer protein I) domains, followed by a C-terminal ricin-like CRD (Figure 1). Although the VOMI domain is typically found in proteins found in the outer layer of the egg vitelline membrane [88], it shares a β-prism fold that has been previously identified in other carbohydrate-binding proteins, including jacalins, a class of plant-specific lectins [89–91], as well as in the *B. thuringiensis* delta endotoxin [92]. Due to the simultaneous presence of these two structurally different CRDs, which clearly presents an interesting path for exploration in glycobiological studies, we defined these unusual proteins as BPBT (β-prism, β-trefoil) lectins. Two BTBP lectins, orthologous to those found in *B. calyciflorus* but lacking a signal peptide (possibly due to an incorrect annotation), were also found in the congeneric species *B. plicatilis*, but not in the other species, suggesting that this domain combination may be exclusively present in *Brachionus* spp.

On the other hand, the single secretory RTL found in *R. sordida* was unrelated to BPBT lectins, since this protein was relatively short (i.e., 200 amino acids) and included a chitin-binding domain in an N-terminal position [93] (Figure 1). This domain is shared by several chitinases and other smaller chitin-binding proteins, which include some with demonstrated effector activity in the context of invertebrate innate immunity, such as horseshoe crab tachycytin [94] and mussel mytichitin [95], and others with presumed lectin-like functions [96].

No sequence orthologous to brachiopod and molluscan mytilectins could be found in rotifers, confirming the discontinuous taxonomic distribution of these β-trefoil lectins in the Lophotrochozoa.

2.6. F-Type Lectins

F-type lectins are characterized by the presence of a β-barrel jellyroll fold which allows fucose recognition [97] and which is also found in the C-terminal domain of coagulation factors 5/8. Despite being associated with relatively short secretory proteins with a lectin function, the typical CRD of FTLs is often found in large multidomain proteins with different catalytic activities [98]. The frequent combination of this domain with several other non-lectin domains mirrors the previously mentioned functional plasticity of the CRDs of CTLs and RTLs. Previous studies have reported that FTLs underwent expansion in some gastropod species [77], and some functional evidence collected in bivalves has linked these proteins to bacterial recognition [99], in addition to the well-established role of the FTL domain-containing proteins bindins in gamete recognition [100]. This observation is consistent with the detection of the FTL domain in a relatively high number of rotifer proteins, only a few of which were characterized by the presence of a signal peptide or displayed a domain organization consistent with a lectin function (Table 1).

Two different types of secretory F-type lectin sequences were detected in rotifers. The first type, present as a single-copy gene in the three monogonont species but missing in the three bdelloids, was a protein displaying a low-complexity threonine- and serine-rich N-terminal region, followed by a single CRD lacking any significant primary sequence homology with known FTLs but showing high predicted structural similarity with human coagulation factors [101] and discoidins [102] (Figure 1). The second type, shared by all rotifer species (even though *B. plicatilis* only displayed a protein lacking the signal peptide, likely due to incorrect annotation), displayed three consecutive FTL CRDs (Figure 1). While the first and the second ones were well recognizable, the third one did not conform with the canonical F-type lectin signature. These triple-CRD FTLs displayed a relatively high (i.e., 40%) sequence identity with several proteins encoded by the genomes of other lophotrochozoans, including mollusks and annelids, suggesting a high degree of evolutionary conservation.

2.7. Other Types of Lectins

In sea urchins, a group of lectins, characterized by the presence of a D-galactoside/L-rhamnose-binding SUEL (acronym for sea urchin egg lectin) domain, carry out egg-protecting functions [103,104]. While this type of lectins is also found in lophotrochozoan genomes, they have been better functionally characterized in deuterostome invertebrates [105–107]. To date, their role in lophotrochozoans remains elusive, even though a bivalve SUEL-type lectin was shown to promote the agglutination of Gram-negative bacteria through LPS binding [108]. The SUEL domain was present in some large multidomain membrane-associated proteins of bdelloids but not in secretory proteins. On the other hand, the three monogonont rotifer genomes encoded a few short SUEL-type lectins (Table 1), which lacked accessory conserved domains (Figure 1) and did no bear any detectable primary sequence homology with other metazoan sequences with known functions. With the exception of two sequences detected in *P. similis*, these proteins displayed high pairwise primary sequence homology and clearly belonged to a monophyletic family.

H-type lectins (HTLs) represent a poorly functionally characterized family of *N*-acetylgalactosamine-binding lectins, which are believed to carry out a defensive role against bacterial infections in fertilized snail eggs [109,110]. Although very little information is available about the involvement of HTLs in lophotrochozoan immunity, comparative genomics analyses indicate that they do not belong to expanded gene families, neither in Mollusca [111,112] nor in Brachiopoda [50]. Nevertheless, transcriptome scans carried out in gastropod mollusks revealed the presence of a novel domain combination between immunoglobulin-like domains and HTL domains in the so-called HREPs [73]. The analysis of rotifer genomes revealed the presence of secretory H-type lectins in just two out of the three bdelloid species (i.e., *A ricciae* and *R. sordida*). On the other hand, no HTL was identified in Monogononta (Table 1). These proteins had a similar simple architecture, with a single CRD, placed immediately after a well-recognizable signal peptide (Figure 1). Rotifer HTLs were encoded by open reading frames with a relatively small size (i.e., 120 codons) and displayed poor sequence homology with other known sequences (i.e., less than 40% primary sequence homology vs. *L. anatina*). This may suggest that the apparent lack of secretory HTLs in four out of six target species derives from missing gene models that could not be included in the annotation tracks of the respective genomes due to poor supporting evidence.

Section 2.5 reports the presence of BPBT lectins, which bear a jacalin-like β-prism structural domain in combination with the RTL CRD, in *Brachionus* spp. The screening for additional proteins bearing a canonical jacalin domain led to the identification of a single protein in *D. carnosus* with no orthologs in other rotifer species. This lectin, which displayed a well recognizable signal peptide for secretion, lacked significant primary sequence homology with other previously characterized proteins, but displayed a highly significant structural match with a number of jacalin-like lectins from plants and with a few metazoan proteins. These included, as the only lophotrochozoan representative, the PPL3 lectin from the bivalve mollusk *Pteria penguin*, which is involved in shell mineralization [113]. Other relevant metazoan proteins which display the same structural fold are the human pancreatic secretory protein ZG16b, important for the condensation of pancreatic enzymes [114], the WGA16 protein from boar sperm [115] and the zebrafish pore-forming protein Dln1 [116].

No gene encoding proteins homologous to the egg-protecting lectins from *Aplysia dactylomela*, characterized by the presence of the orifera lack lectins homologous to her metazoan sequences with known function.ked significant similarity withology DUF3011 domain [117], could be detected in Rotifera. Likewise, no apextrin-like proteins were detected in any of the studied species. Proteins carrying an apextrin C-terminal domain (ApeC) have been previously shown to mediate bacterial recognition in amphioxus [118]. While they are also present in bivalve mollusks and brachiopods [50,62], this study rules out their possible involvement in immune recognition in rotifers.

3. Discussion

Genome- or transcriptome-wide screening approaches have previously been successfully used to identify lectin-like proteins in several different eukaryote species, both in the plant and animal kingdoms [77,96,112,119–122], and can be effectively used as a preliminary step to investigate the repertoire of lectin-like molecules present in non-model species. This approach might be particularly intriguing for understudied animal phyla, which, despite the lack of previous glycobiological investigations, might be endowed with a particularly rich repertoire of lectin-like molecules as the result of their adaptation to a challenging environment. Rotifers, like other aquatic organisms, are potentially highly exposed to ubiquitous waterborne microorganisms, which may in some cases be pathogenic. Considering the well-described complex innate immune systems of other lophotrochozoans, such as mollusks and annelids, rotifers appear to be good candidates for the bioinformatics-assisted discovery of carbohydrate-binding proteins involved in MAMP recognition. No information is presently available concerning the glycans expressed by rotifer tissues and only a single membrane-bound C-type lectin has been described in *B. manjavacas* [66] prior to this work. Hence, this represents the first investigation of this type carried out in this relatively large and widespread but poorly studied phylum of small aquatic animals.

Although this approach obviously suffers from some limitations, which will be described in detail below, it has allowed: (i) the identification of the presence of secretory lectin-like molecules belonging to at least nine different families, characterized by distinct structural folds (Figure 3), in rotifers; (ii) the highlighting of significant differences in terms of distribution and domain organization between the two major classes of rotifers, as well as among species; and (iii) the ruling out of the possibility that known lectin-encoding gene families underwent significant expansion in Rotifera, marking a clear difference with other lophotrochozoan phyla.

Even though this in silico screening approach allowed the identification of several proteins that are likely to be secreted to the extracellular environment and have significant carbohydrate-binding properties, the list of the putative rotifer lectin-like proteins here provided (detailed in Table S1) should be considered as strictly preliminary. The glycan-binding properties of the identified proteins, as well as their possible involvement in MAMP recognition, should be validated through functional data collected with classical biochemical, glycobiological and immunological approaches.

Some limitations of this genome-wide bioinformatics screening approach reside in the fact that the correct identification of lectin-like proteins depends on the accuracy of gene annotations. While all the genomes analyzed in this work had a high quality, both in terms of assembly metrics and in terms of BUSCO completeness [123], a few chimeric gene models, as well as models of ORFs which were clearly subjected to 5′ or 3′ truncation compared with other full-length orthologs and paralogs, were occasionally observed. For the sake of consistency, these gene models were disregarded, even though the presence of incomplete gene models in a given species was reported, whenever relevant, in Table 1. In addition, some lectin-like proteins identified in this work were rather short, with an ORF barely exceeding 100 codons, and lacked at the same time significant primary sequence homology with other known sequences deposited in public databases. These factors might have negatively impacted the annotation of other orthologous and paralogous genes, which may therefore be apparently missing in some of the target genomes, as discussed in Section 2.7. Hence, the completeness of our report may suffer from these uncertainties, and the number of lectins reported in Table 1 should be considered as inherently subjected to slight underestimates.

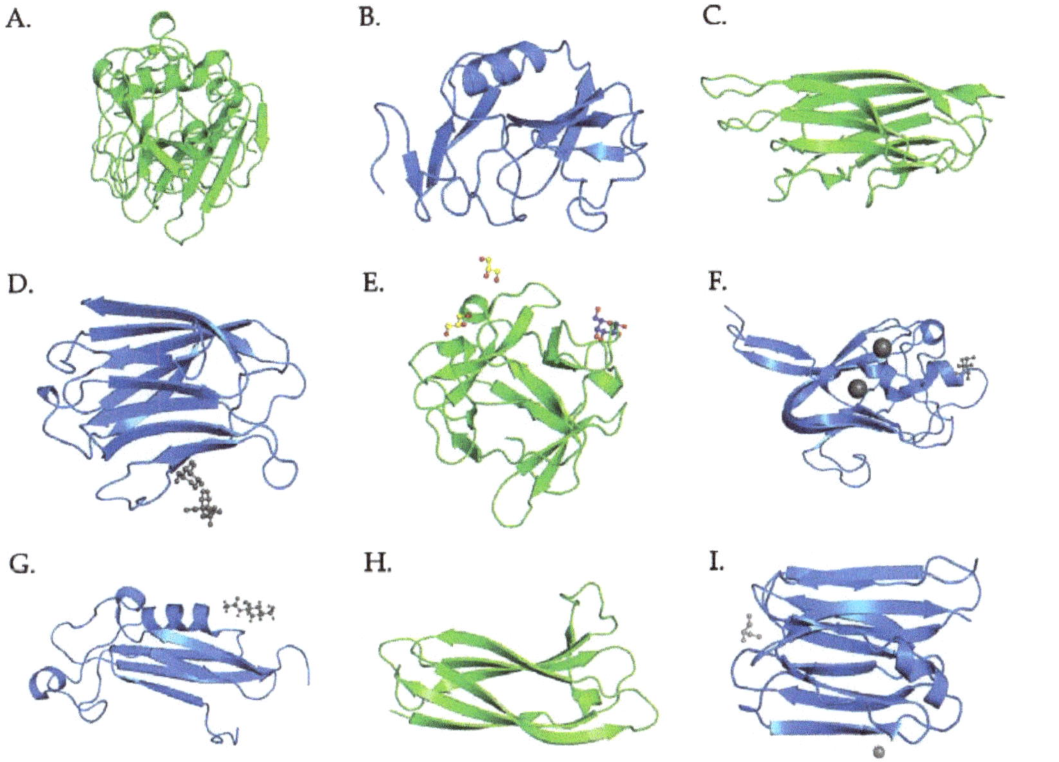

Figure 3. Three-dimensional folding of the carbohydrate-recognition domains of representative members of the nine lectin families identified in Rotifera. (**A**) C-type lectin domain, pdb entry: 1b6e; (**B**) fibrinogen C-terminal domain, pdb entry: 1fib; (**C**) C1q domain, pdb entry: 4ous; (**D**) galectin domain, pdb entry: 1a3k; (**E**) R-type lectin β-trefoil domain, pdb entry: 4iyb; (**F**) F-type lectin domain, pdb entry: 1k12; (**G**) SUEL-type lectin domain, pdb entry: 2jx9; (**H**) H-type lectin domain, pdb entry: 2ces; (**I**) jacalin β-prism domain, pdb entry: 3apa. The figures are reproduced courtesy of PDBe (https://www.ebi.ac.uk/pdbe/, accessed on 20 January 2022).

Another possible limitation of this work was the impossibility of proceeding with a reliable in silico screening of candidate lectin molecules characterized by the presence of domains which are not primarily or exclusively linked with a carbohydrate-binding function. This was the case, for example, for I-type lectins (also known as siglecs), which share an immunoglobulin-like fold with a very high number of other proteins involved in a very broad range of functions [124] and which have been previously identified in some lophotrochozoans [125]. Similarly, some chitin-binding lectins, mostly from plants [126], are characterized by the presence of a chitinase-like domain that includes a few key mutations that ablate its catalytic function. However, since these two lectin families are either membrane-bound (siglecs) or uncommon in metazoans (chitinase-like lectins), their exclusion from the set of domains included in in silico searches was unlikely to have an impact on the identification of secretory lectins in rotifers.

Finally, it needs to be stressed that rotifer genomes are extremely gene-rich and encode several thousand proteins which lack any significant primary sequence homology with other sequences deposited in public databases and which have no detectable conserved domain. We cannot exclude that some as yet uncharacterized protein families may carry out important carbohydrate-binding functions in these animals. Nevertheless, the combination of classical biochemical and glycobiological approaches with bioinformatics approaches should enable the identification of the full-length sequence of the lectins isolated from rotifers starting from small peptide fragments, as previously carried out on several occasions with other aquatic invertebrates [68,85,127].

In summary, this work allowed the confirmation of the potential interest of rotifers as future targets for glycobiological studies focused on the identification of novel lectins. These, based on the significant diversity of the associated structural folding, might be endowed with different carbohydrate-binding properties, which may support the development of new biotechnological tools, such as lectin-based biosensors with potential applications in cancer research. Besides the interest that such molecules might have in terms of biotechnological applications, another aspect that remains to be clarified is whether these rotifer secretory lectins carry out biological functions similar to those previously described in other lophotrochozoan phyla.

4. Materials and Methods

4.1. Identification of Lectin-Like Molecules

Six rotifer species with a publicly available fully sequenced genome and an associated gene annotation track were selected (Table 2). *Didymodactylos carnosus* Milne 1916 [21], *Rotaria sordida* Western, 1893 [21] and *Adineta ricciae* Segers & Shiel, 2005 [21] were selected for the class Bdelloidea; *Brachionus plicatilis* Müller, 1786 [128], *Brachionus calyciflorus* Pallas, 1776 [129] and *Proales similis* de Beauchamp, 1907 [130] were selected for the class Monogononta.

Table 2. List of the six rotifer species analyzed in this study, with genome size and number of annotated protein-coding genes.

Species Name	Class	Genome Size (Mb)	Protein-Coding Genes
Adineta ricciae	Bdelloidea	173	49,015
Rotaria sordida	Bdelloidea	361	61,901
Didymodactylos carnosus	Bdelloidea	356	46,863
Proales similis	Monogononta	33	10,785
Brachionus calyciflorus	Monogononta	30	24,328
Brachionus plicatilis	Monogononta	107	52,502

The proteome of each of the six target species was screened and a search was made for secretory proteins, i.e., those including either a highly supported canonical signal peptide, detected with SignalP v.5.0 [131], or, in the case of galectins, which are known to use a non-canonical secretion signal, with SecretomeP v.2.0 [132]. At the same time, candidate proteins needed to lack transmembrane regions, which were detected with TMHMM v.2.0 [133]. Signal peptide and transmembrane region predictions were further checked with Phobius v.1.01 [134]. Putative lectins were identified based on the presence of the following PFAM conserved domains, detected with HMMER [135] based on default threshold e-values: fibrinogen beta and gamma chains, C-terminal globular domain (PF00147), C-type lectin domain (PF00059), C1q domain (PF00386), galactose-binding lectin domain (PF02140), ricin-type beta-trefoil domain (PF00652 and PF14200), F-type lectin/discoidin domain (PF00754), galactoside-binding lectin domain (PF00337), H-type lectin domain (PF09458), jacalin-like lectin domain (PF01419), DUF3011 (PF11218) and the C-terminal domain of apextrin (PF16977).

The presence of other conserved domains was checked with InterProScan v.5 [136] and the possible presence of conserved structural folds in regions lacking conserved domains was investigated with HHpred [137]. To avoid the inclusion of proteins carrying lectin-like domains but likely involved in non-lectin functions, sequences displaying conserved domains with known catalytic action (e.g., glucanases, hydrolases, kinases, peptidases, etc.) were disregarded. Protein sequences deriving from gene models annotated as "partial", as well as those that displayed obviously truncated characterizing domains and which might therefore derive either from pseudogenes or from mis-annotations, were disregarded.

4.2. Bayesian Phylogenetic Inference

All rotifer C1qDC proteins were included in a multiple sequence alignment (MSA), prepared with MUSCLE [138], together with three selected lophotrochozoan C1q-like sequences (i.e., XP_013399541.1 from *Lingula anatina*, QBA18422.1 from *Testudinalia testudinalis* and XP_033760315.1 from *Pecten maximus*). The human C1qA (NP_057075.1), C1qB (NP_000482.3) and C1qc (NP_758957.2) chains were also added to the alignment for tree-rooting purposes. The MSA was refined with Gblocks v.0.91b [139] to remove unalignable, poorly informative regions. Bayesian phylogenetic analysis was carried out with MrBayes v.3.2.7a [140], running two parallel MCMC analyses for 500,000 generations each, sampling one tree every 1,000 generations. The analysis was run under an LG+G+I model of molecular evolution, which was estimated to be the best fitting for this dataset, with ModelTest-NG [141]. Run convergence was checked with Tracer v.1.7.1 [142] by determining that all estimated parameters reached an effective sample size value higher than 100.

Supplementary Materials: The following supporting information can be downloaded at: https://www.mdpi.com/article/10.3390/md20020130/s1, Table S1: List of the gene accession IDs of secretory lectin-like sequences identified in Rotifera.

Funding: This research received no external funding.

Institutional Review Board Statement: Not applicable.

Data Availability Statement: All the genomic data analyzed in this study are publicly available as supplementary materials of the original publications, referenced in the materials and methods section.

Acknowledgments: The author would like to thank Yasuhiro Ozeki and Yuki Fujii for their helpful suggestions regarding the contents of this manuscript.

Conflicts of Interest: The author declares no conflict of interest.

References

1. Stuart, L.M.; Paquette, N.; Boyer, L. Effector-Triggered versus Pattern-Triggered Immunity: How Animals Sense Pathogens. *Nat. Rev. Immunol.* **2013**, *13*, 199–206. [CrossRef] [PubMed]
2. Wang, J.-L.; Zhang, Q.; Tang, L.; Chen, L.; Liu, X.-S.; Wang, Y.-F. Involvement of a Pattern Recognition Receptor C-Type Lectin 7 in Enhancing Cellular Encapsulation and Melanization Due to Its Carboxyl-Terminal CRD Domain in the Cotton Bollworm, *Helicoverpa Armigera*. *Dev. Comp. Immunol.* **2014**, *44*, 21–29. [CrossRef] [PubMed]
3. Wang, X.-W.; Xu, J.-D.; Zhao, X.-F.; Vasta, G.R.; Wang, J.-X. A Shrimp C-Type Lectin Inhibits Proliferation of the *Hemolymph Microbiota* by Maintaining the Expression of Antimicrobial Peptides. *J. Biol. Chem.* **2014**, *289*, 11779–11790. [CrossRef] [PubMed]
4. Brouwer, N.; Dolman, K.M.; van Zwieten, R.; Nieuwenhuys, E.; Hart, M.; Aarden, L.A.; Roos, D.; Kuijpers, T.W. Mannan-Binding Lectin (MBL)–Mediated Opsonization Is Enhanced by the Alternative Pathway Amplification Loop. *Mol. Immunol.* **2006**, *43*, 2051–2060. [CrossRef] [PubMed]
5. Kerrigan, A.M.; Brown, G.D. C-Type Lectins and Phagocytosis. *Immunobiology* **2009**, *214*, 562. [CrossRef]
6. Pales Espinosa, E.; Allam, B. Reverse Genetics Demonstrate the Role of Mucosal C-Type Lectins in Food Particle Selection in the Oyster *Crassostrea Virginica*. *J. Exp. Biol.* **2018**, *221*, jeb174094. [CrossRef]
7. Nyholm, S.V.; Graf, J. Knowing Your Friends: Invertebrate Innate Immunity Fosters Beneficial Bacterial Symbioses. *Nature Rev. Microbiol.* **2012**, *10*, 815–827. [CrossRef]
8. Grossart, H.-P.; Riemann, L.; Tang, K. Molecular and Functional Ecology of Aquatic Microbial Symbionts. *Front. Microbiol.* **2013**, *4*, 59. [CrossRef]
9. Li, H.-H.; Cai, Y.; Li, J.-C.; Su, M.P.; Liu, W.-L.; Cheng, L.; Chou, S.-J.; Yu, G.-Y.; Wang, H.-D.; Chen, C.-H. C-Type Lectins Link Immunological and Reproductive Processes in *Aedes Aegypti*. *iScience* **2020**, *23*, 101486. [CrossRef]
10. Springer, S.A.; Moy, G.W.; Friend, D.S.; Swanson, W.J.; Vacquier, V.D. Oyster Sperm Bindin Is a Combinatorial *Fucose Lectin* with Remarkable Intra-Species Diversity. *Int. J. Dev. Biol.* **2008**, *52*, 759–768. [CrossRef]
11. Nauta, A.J.; Castellano, G.; Xu, W.; Woltman, A.M.; Borrias, M.C.; Daha, M.R.; van Kooten, C.; Roos, A. Opsonization with C1q and Mannose-Binding Lectin Targets Apoptotic Cells to Dendritic Cells. *J. Immunol.* **2004**, *173*, 3044–3050. [CrossRef] [PubMed]
12. Maki, J.S.; Mitchell, R. Involvement of Lectins in the Settlement and Metamorphosis of Marine *Invertebrate Larvae*. *Bull. Mar. Sci.* **1985**, *37*, 675–683.
13. Emmanuelle, P.E.; Mickael, P.; Evan, W.; Shumway, S.E.; Bassem, A. Lectins Associated with the Feeding Organs of the Oyster *Crassostrea Virginica* Can Mediate Particle Selection. *Biol. Bull.* **2009**, *217*, 130–141.

14. Kvennefors, E.C.E.; Leggat, W.; Hoegh-Guldberg, O.; Degnan, B.M.; Barnes, A.C. An Ancient and Variable Mannose-Binding Lectin from the Coral *Acropora Millepora* Binds Both Pathogens and Symbionts. *Dev. Comp. Immunol.* **2008**, *32*, 1582–1592. [CrossRef] [PubMed]
15. Giga, Y.; Ikai, A.; Takahashi, K. The Complete Amino Acid Sequence of Echinoidin, a Lectin from the Coelomic Fluid of the Sea Urchin *Anthocidaris Crassispina.* Homologies with Mammalian and Insect Lectins. *J. Biol. Chem.* **1987**, *262*, 6197–6203. [CrossRef]
16. Yamaura, K.; Takahashi, K.G.; Suzuki, T. Identification and Tissue Expression Analysis of C-Type Lectin and Galectin in the Pacific Oyster, *Crassostrea Gigas*. *Comp. Biochem. Physiol. B Biochem. Mol. Biol.* **2008**, *149*, 168–175. [CrossRef]
17. Gerdol, M.; Gomez-Chiarri, M.; Castillo, M.G.; Figueras, A.; Fiorito, G.; Moreira, R.; Novoa, B.; Pallavicini, A.; Ponte, G.; Roumbedakis, K.; et al. Immunity in Molluscs: Recognition and Effector Mechanisms, with a Focus on Bivalvia. In *Advances in Comparative Immunology*; Cooper, E.L., Ed.; Springer International Publishing: Cham, Switzerland, 2018; pp. 225–234. ISBN 978-3-319-76768-0.
18. Shain, D.H.; Halldórsdóttir, K.; Pálsson, F.; Aðalgeirsdóttir, G.; Gunnarsson, A.; Jónsson, Þ.; Lang, S.A.; Pálsson, H.S.; Steinþórssson, S.; Arnason, E. Colonization of Maritime Glacier Ice by *Bdelloid Rotifera*. *Mol. Phylogenet. Evol.* **2016**, *98*, 280–287. [CrossRef] [PubMed]
19. Hespeels, B.; Knapen, M.; Hanot-Mambres, D.; Heuskin, A.-C.; Pineux, F.; Lucas, S.; Koszul, R.; van Doninck, K. Gateway to Genetic Exchange? DNA Double-Strand Breaks in the Bdelloid Rotifer *Adineta Vaga* Submitted to Desiccation. *J. Evol. Biol.* **2014**, *27*, 1334–1345. [CrossRef]
20. Welch, J.L.M.; Welch, D.B.M.; Meselson, M. Cytogenetic Evidence for Asexual Evolution of *Bdelloid Rotifers*. *Proc. Natl. Acad. Sci. USA* **2004**, *101*, 1618–1621. [CrossRef]
21. Nowell, R.W.; Wilson, C.G.; Almeida, P.; Schiffer, P.H.; Fontaneto, D.; Becks, L.; Rodriguez, F.; Arkhipova, I.R.; Barraclough, T.G. Evolutionary Dynamics of Transposable Elements in *Bdelloid Rotifers*. *eLife* **2021**, *10*, e63194. [CrossRef]
22. Eyres, I.; Boschetti, C.; Crisp, A.; Smith, T.P.; Fontaneto, D.; Tunnacliffe, A.; Barraclough, T.G. Horizontal Gene Transfer in Bdelloid Rotifers Is Ancient, Ongoing and More Frequent in Species from Desiccating Habitats. *BMC Biol.* **2015**, *13*, 90. [CrossRef] [PubMed]
23. Gladyshev, E.A.; Arkhipova, I.R. Genome Structure of Bdelloid Rotifers: Shaped by Asexuality or Desiccation? *J. Hered.* **2010**, *101* (Suppl. S1), S85–S93. [CrossRef] [PubMed]
24. Gabaldón, C.; Fontaneto, D.; Carmona, M.J.; Montero-Pau, J.; Serra, M. Ecological Differentiation in Cryptic Rotifer Species: What We Can Learn from the *Brachionus Plicatilis* Complex. *Hydrobiologia* **2017**, *796*, 7–18. [CrossRef]
25. Malekzadeh-Viayeh, R.; Pak-Tarmani, R.; Rostamkhani, N.; Fontaneto, D. Diversity of the Rotifer Brachionus Plicatilis Species Complex (*Rotifera: Monogononta*) in Iran through Integrative Taxonomy. *Zool. J. Linn. Soc.* **2014**, *170*, 233–244. [CrossRef]
26. Lam-Höai, T.; Rougier, C.; Lasserre, G. Tintinnids and Rotifers in a Northern Mediterranean Coastal Lagoon. Structural Diversity and Function through Biomass Estimations. *Mar. Ecol. Prog. Ser.* **1997**, *152*, 13–25. [CrossRef]
27. Assefa, E.; Mengidtou, S. Seasonal Variation of Biomass and Secondary Production of Thermocyclops (*Cyclopoida*) and Brachionus (*Rotifera*) Spp. in a Shallow Tropical Lake Kuriftu, Ethiopia. *SINET Ethiop. J. Sci.* **2011**, *34*, 73–88. [CrossRef]
28. Yan, N.D.; Geiling, W. Elevated Planktonic Rotifer Biomass in Acidified Metal-Contaminated Lakes near Sudbury, Ontario. *Hydrobiologia* **1985**, *120*, 199–205. [CrossRef]
29. Lubzens, E.; Tandler, A.; Minkoff, G. Rotifers as Food in Aquaculture. *Hydrobiologia* **1989**, *186*, 387–400. [CrossRef]
30. Dhont, J.; Dierckens, K.; Støttrup, J.; Van Stappen, G.; Wille, M.; Sorgeloos, P. 5–Rotifers, Artemia and Copepods as Live Feeds for Fish Larvae in Aquaculture. In *Advances in Aquaculture Hatchery Technology*; Allan, G., Burnell, G., Eds.; Woodhead Publishing Series in Food Science, Technology and Nutrition; Woodhead Publishing: Sawston, UK, 2013; pp. 157–202. ISBN 978-0-85709-119-2.
31. Hagiwara, A.; Marcial, H.S. The Use of Non-Brachionus Plicatilis Species Complex Rotifer in Larviculture. *Hydrobiologia* **2019**, *844*, 163–172. [CrossRef]
32. Hagiwara, A.; Kim, H.-J.; Marcial, H. Mass Culture and Preservation of *Brachionus Plicatilis* Sp. Complex. In *Rotifers: Aquaculture, Ecology, Gerontology, and Ecotoxicology*; Hagiwara, A., Yoshinaga, T., Eds.; Fisheries Science Series; Springer: Singapore, 2017; pp. 35–45. ISBN 978-981-10-5635-2.
33. Zhang, L.; Li, L.; Guo, X.; Litman, G.W.; Dishaw, L.J.; Zhang, G. Massive Expansion and Functional Divergence of Innate Immune Genes in a Protostome. *Sci. Rep.* **2015**, *5*, 8693. [CrossRef]
34. Gerdol, M.; Venier, P.; Pallavicini, A. The Genome of the Pacific Oyster Crassostrea Gigas Brings New Insights on the Massive Expansion of the C1q Gene Family in Bivalvia. *Dev. Comp. Immunol.* **2015**, *49*, 59–71. [CrossRef] [PubMed]
35. Gerdol, M.; Greco, S.; Pallavicini, A. Extensive Tandem Duplication Events Drive the Expansion of the C1q-Domain-Containing Gene Family in Bivalves. *Mar. Drugs* **2019**, *17*, 583. [CrossRef] [PubMed]
36. Matsushita, M. Ficolins: Complement-Activating Lectins Involved in Innate Immunity. *J. Innate Immun.* **2010**, *2*, 24–32. [CrossRef] [PubMed]
37. Matsushita, M.; Fujita, T. The Role of Ficolins in Innate Immunity. *Immunobiology* **2002**, *205*, 490–497. [CrossRef]
38. Imamichi, Y.; Yokoyama, Y. Purification, Characterization and CDNA Cloning of a Novel Lectin from the Jellyfish *Nemopilema Nomurai*. *Comp. Biochem. Physiol. B Biochem. Mol. Biol.* **2010**, *156*, 12–18. [CrossRef]

39. Gokudan, S.; Muta, T.; Tsuda, R.; Koori, K.; Kawahara, T.; Seki, N.; Mizunoe, Y.; Wai, S.N.; Iwanaga, S.; Kawabata, S. Horseshoe Crab Acetyl Group-Recognizing Lectins Involved in Innate Immunity Are Structurally Related to Fibrinogen. *Proc. Natl. Acad. Sci. USA* **1999**, *96*, 10086–10091. [CrossRef]
40. Gout, E.; Garlatti, V.; Smith, D.F.; Lacroix, M.; Dumestre-Pérard, C.; Lunardi, T.; Martin, L.; Cesbron, J.-Y.; Arlaud, G.J.; Gaboriaud, C.; et al. Carbohydrate Recognition Properties of Human Ficolins: Glycan Array Screening Reveals the Sialic Acid Binding Specificity of M-Ficolin. *J. Biol. Chem.* **2010**, *285*, 6612–6622. [CrossRef]
41. Adema, C.M. Fibrinogen-Related Proteins (FREPs) in Mollusks. *Results Probl. Cell Differ.* **2015**, *57*, 111–129. [CrossRef]
42. Gordy, M.A.; Pila, E.A.; Hanington, P.C. The Role of Fibrinogen-Related Proteins in the Gastropod Immune Response. *Fish Shellfish Immunol.* **2015**, *46*, 39–49. [CrossRef]
43. Gorbushin, A.M.; Panchin, Y.V.; Iakovleva, N.V. In Search of the Origin of FREPs: Characterization of *Aplysia Californica* Fibrinogen-Related Proteins. *Dev. Comp. Immunol.* **2010**, *34*, 465–473. [CrossRef]
44. Dheilly, N.M.; Duval, D.; Mouahid, G.; Emans, R.; Allienne, J.-F.; Galinier, R.; Genthon, C.; Dubois, E.; Du Pasquier, L.; Adema, C.M.; et al. A Family of Variable Immunoglobulin and Lectin Domain Containing Molecules in the Snail *Biomphalaria Glabrata*. *Dev. Comp. Immunol.* **2015**, *48*, 234–243. [CrossRef] [PubMed]
45. Skazina, M.A.; Gorbushin, A.M. Characterization of the Gene Encoding a Fibrinogen-Related Protein Expressed in *Crassostrea Gigas* Hemocytes. *Fish Shellfish Immunol.* **2016**, *54*, 586–588. [CrossRef] [PubMed]
46. Kurachi, S.; Song, Z.; Takagaki, M.; Yang, Q.; Winter, H.C.; Kurachi, K.; Goldstein, I.J. Sialic-Acid-Binding Lectin from the Slug Limax Flavus–Cloning, Expression of the Polypeptide, and Tissue Localization. *Eur. J. Biochem.* **1998**, *254*, 217–222. [CrossRef] [PubMed]
47. Huang, B.; Zhang, L.; Li, L.; Tang, X.; Zhang, G. Highly Diverse Fibrinogen-Related Proteins in the Pacific Oyster *Crassostrea Gigas*. *Fish Shellfish Immunol.* **2015**, *43*, 485–490. [CrossRef]
48. Xiang, Z.; Qu, F.; Wang, F.; Li, J.; Zhang, Y.; Yu, Z. Characteristic and Functional Analysis of a Ficolin-like Protein from the Oyster *Crassostrea Hongkongensis*. *Fish Shellfish Immunol.* **2014**, *40*, 514–523. [CrossRef]
49. Yang, C.; Wang, L.; Zhang, H.; Wang, L.; Huang, M.; Sun, Z.; Sun, Y.; Song, L. A New Fibrinogen-Related Protein from Argopecten Irradians (AiFREP-2) with Broad Recognition Spectrum and Bacteria Agglutination Activity. *Fish Shellfish Immunol.* **2014**, *38*, 221–229. [CrossRef]
50. Gerdol, M.; Luo, Y.-J.; Satoh, N.; Pallavicini, A. Genetic and Molecular Basis of the Immune System in the Brachiopod *Lingula Anatina*. *Dev. Comp. Immunol.* **2018**, *82*, 7–30. [CrossRef]
51. Jin, Q.; Sun, Q.; Zhang, J.; Sun, L. First Characterization of Two C-Type Lectins of the Tubeworm *Alaysia Sp.* from a Deep-Sea Hydrothermal Vent. *Dev. Comp. Immunol.* **2018**, *86*, 17–25. [CrossRef]
52. Yang, J.; Wang, L.; Zhang, H.; Qiu, L.; Wang, H.; Song, L. C-Type Lectin in Chlamys Farreri (CfLec-1) Mediating Immune Recognition and Opsonization. *PLoS ONE* **2011**, *6*, e17089. [CrossRef]
53. Huang, M.; Song, X.; Zhao, J.; Mu, C.; Wang, L.; Zhang, H.; Zhou, Z.; Liu, X.; Song, L. A C-Type Lectin (AiCTL-3) from Bay Scallop Argopecten Irradians with Mannose/Galactose Binding Ability to Bind Various Bacteria. *Gene* **2013**, *531*, 31–38. [CrossRef]
54. Zelensky, A.N.; Gready, J.E. The C-Type Lectin-like Domain Superfamily. *FEBS J.* **2005**, *272*, 6179–6217. [CrossRef] [PubMed]
55. Pees, B.; Yang, W.; Zárate-Potes, A.; Schulenburg, H.; Dierking, K. High Innate Immune Specificity through Diversified C-Type Lectin-Like Domain Proteins in Invertebrates. *J. Innate Immun.* **2016**, *8*, 129–142. [CrossRef] [PubMed]
56. Drickamer, K. Evolution of Ca^{2+}-Dependent Animal Lectins11Abbreviations: CRD, Carbohydrate-Recognition Domain; EGF, Epidermal Growth Factor. In *Progress in Nucleic Acid Research and Molecular Biology*; Cohn, W.E., Moldave, K., Eds.; Academic Press: San Diego, CA, USA, 1993; Volume 45, pp. 207–232.
57. Drickamer, K.; Fadden, A.J. Genomic Analysis of C-Type Lectins. *Biochem. Soc. Symp.* **2002**, *69*, 59–72. [CrossRef] [PubMed]
58. Holmskov, U.L. Collectins and Collectin Receptors in Innate Immunity. *APMIS Suppl.* **2000**, *100*, 1–59. [CrossRef]
59. Sun, J.-J.; Lan, J.-F.; Zhao, X.-F.; Vasta, G.R.; Wang, J.-X. Binding of a C-Type Lectin's Coiled-Coil Domain to the Domeless Receptor Directly Activates the JAK/STAT Pathway in the Shrimp Immune Response to Bacterial Infection. *PLoS Pathog.* **2017**, *13*, e1006626. [CrossRef] [PubMed]
60. Vasta, G.R.; Ahmed, H.; Tasumi, S.; Odom, E.W.; Saito, K. Biological Roles of Lectins in Innate Immunity: Molecular and Structural Basis for Diversity in Self/Non-Self Recognition. In *Current Topics in Innate Immunity*; Lambris, J.D., Ed.; Springer: New York, NY, USA, 2007; pp. 389–406.
61. Li, H.; Zhang, H.; Jiang, S.; Wang, W.; Xin, L.; Wang, H.; Wang, L.; Song, L. A Single-CRD C-Type Lectin from Oyster Crassostrea Gigas Mediates Immune Recognition and Pathogen Elimination with a Potential Role in the Activation of Complement System. *Fish Shellfish Immunol.* **2015**, *44*, 566–575. [CrossRef]
62. Gerdol, M.; Venier, P. An Updated Molecular Basis for Mussel Immunity. *Fish Shellfish Immunol.* **2015**, *46*, 17–38. [CrossRef]
63. Yu, X.-Q.; Gan, H.; Kanost, R.M. Immulectin, an Inducible C-Type Lectin from an Insect, *Manduca Sexta*, Stimulates Activation of Plasma Prophenol Oxidase. *Insect Biochem. Mol. Biol.* **1999**, *29*, 585–597. [CrossRef]
64. Couser, L.M. A Mannose Receptor-like Molecule Likely Serves as the Mate Recognition Pheromone Receptor in the Male Rotifer *Brachionus Manjavacas*. Bachelor's Thesis, Georgia Institute of Technology, Atlanta, GA, USA, 2010.
65. Kishore, U.; Gaboriaud, C.; Waters, P.; Shrive, A.K.; Greenhough, T.J.; Reid, K.B.M.; Sim, R.B.; Arlaud, G.J. C1q and Tumor Necrosis Factor Superfamily: Modularity and Versatility. *Trends Immunol.* **2004**, *25*, 551–561. [CrossRef]

66. Carland, T.M.; Gerwick, L. The C1q Domain Containing Proteins: Where Do They Come from and What Do They Do? *Dev. Comp. Immunol.* **2010**, *34*, 785–790. [CrossRef]
67. Ghai, R.; Waters, P.; Roumenina, L.T.; Gadjeva, M.; Kojouharova, M.S.; Reid, K.B.M.; Sim, R.B.; Kishore, U. C1q and Its Growing Family. *Immunobiology* **2007**, *212*, 253–266. [CrossRef]
68. Hasan, I.; Gerdol, M.; Fujii, Y.; Ozeki, Y. Functional Characterization of OXYL, A SghC1qDC LacNAc-Specific Lectin from The Crinoid Feather Star Anneissia Japonica. *Mar. Drugs* **2019**, *17*, 136. [CrossRef] [PubMed]
69. Gerlach, D.; Schlott, B.; Schmidt, K.-H. Cloning and Expression of a Sialic Acid-Binding Lectin from the *Snail Cepaea* Hortensis. *FEMS Immunol. Med. Microbiol.* **2004**, *40*, 215–221. [CrossRef]
70. Xiong, X.; Li, C.; Zheng, Z.; Du, X. Novel Globular C1q Domain-Containing Protein (PmC1qDC-1) Participates in Shell Formation and Responses to Pathogen-Associated Molecular Patterns Stimulation in Pinctada *Fucata Martensii*. *Sci. Rep.* **2021**, *11*, 1105. [CrossRef] [PubMed]
71. Huang, Y.; Wu, L.; Jin, M.; Hui, K.; Ren, Q. A C1qDC Protein (HcC1qDC6) with Three Tandem C1q Domains Is Involved in Immune Response of Triangle-Shell Pearl Mussel (*Hyriopsis Cumingii*). *Front. Physiol.* **2017**, *8*, 521. [CrossRef]
72. Xu, K.; Wang, Y.; Lian, S.; Hu, N.; Chen, X.; Dai, X.; Zhang, L.; Wang, S.; Hu, J.; Hu, X.; et al. Expansion of C1Q Genes in Zhikong Scallop and Their Expression Profiling After Exposure to the Toxic Dinoflagellates. *Front. Mar. Sci.* **2021**, *8*, 453. [CrossRef]
73. Gorbushin, A.M. Derivatives of the Lectin Complement Pathway in Lophotrochozoa. *Dev. Comp. Immunol.* **2019**, *94*, 35–58. [CrossRef]
74. Vasta, G.R.; Ahmed, H.; Nita-Lazar, M.; Banerjee, A.; Pasek, M.; Shridhar, S.; Guha, P.; Fernández-Robledo, J.A. Galectins as Self/Non-Self Recognition Receptors in Innate and Adaptive Immunity: An Unresolved Paradox. *Front. Immun.* **2012**, *3*, 199. [CrossRef]
75. Song, X.; Zhang, H.; Wang, L.; Zhao, J.; Mu, C.; Song, L.; Qiu, L.; Liu, X. A Galectin with Quadruple-Domain from Bay Scallop *Argopecten Irradians* Is Involved in Innate Immune Response. *Dev. Comp. Immunol.* **2011**, *35*, 592–602. [CrossRef]
76. Zhang, D.; Jiang, S.; Hu, Y.; Cui, S.; Guo, H.; Wu, K.; Li, Y.; Su, T. A Multidomain Galectin Involved in Innate Immune Response of Pearl Oyster *Pinctada Fucata*. *Dev. Comp. Immunol.* **2011**, *35*, 1–6. [CrossRef]
77. Gorbushin, A.M.; Borisova, E.A. Lectin-like Molecules in Transcriptome of *Littorina Littorea* Hemocytes. *Dev. Comp. Immunol.* **2015**, *48*, 210–220. [CrossRef] [PubMed]
78. Ruan, L.; Xu, H.; Lin, W.; Shi, H.; Cui, Z.; Xu, X. A Novel Beta-Galactose-Specific Lectin of the Tubeworm, *Ridgeia Piscesae*, from the Hydrothermal Vent. *Acta Oceanologica Sinica* **2017**, *36*, 61–67. [CrossRef]
79. Vasta, G.R.; Feng, C.; Bianchet, M.A.; Bachvaroff, T.R.; Tasumi, S. Structural, Functional, and Evolutionary Aspects of Galectins in Aquatic Mollusks: From a Sweet Tooth to the Trojan Horse. *Fish Shellfish Immunol.* **2015**, *46*, 94–106. [CrossRef] [PubMed]
80. Popa, S.J.; Stewart, S.E.; Moreau, K. Unconventional Secretion of Annexins and Galectins. *Semin. Cell Dev. Biol.* **2018**, *83*, 42. [CrossRef] [PubMed]
81. Cummings, R.D.; Schnaar, R. R-Type Lectins. In *Essentials of Glycobiology*, 3rd ed.; Cold Spring Harbor Laboratory Press: New York, NY, USA, 2017; Chapter 31.
82. Kawsar, S.M.A.; Takeuchi, T.; Kasai, K.; Fujii, Y.; Matsumoto, R.; Yasumitsu, H.; Ozeki, Y. Glycan-Binding Profile of a D-Galactose Binding Lectin Purified from the Annelid, *Perinereis Nuntia Ver. Vallata*. *Comp. Biochem. Physiol. B Biochem. Mol. Biol.* **2009**, *152*, 382–389. [CrossRef]
83. Hirabayashi, J.; Dutta, S.K.; Kasai, K. Novel Galactose-Binding Proteins in Annelida. Characterization of 29-KDa Tandem Repeat-Type Lectins from the Earthworm *Lumbricus Terrestris*. *J. Biol. Chem.* **1998**, *273*, 14450–14460. [CrossRef]
84. Kawsar, S.M.A.; Hasan, I.; Rajia, S.; Koide, Y.; Fujii, Y.; Hayashi, R.; Yamada, M.; Ozeki, Y. Diverse Localization Patterns of an R-Type Lectin in Marine Annelids. *Molecules* **2021**, *26*, 4799. [CrossRef]
85. Fujii, Y.; Gerdol, M.; Kawsar, S.M.A.; Hasan, I.; Spazzali, F.; Yoshida, T.; Ogawa, Y.; Rajia, S.; Kamata, K.; Koide, Y.; et al. A GM1b/Asialo-GM1 Oligosaccharide-Binding R-Type Lectin from Purplish Bifurcate Mussels *Mytilisepta Virgata* and Its Effect on MAP Kinases. *FEBS J.* **2020**, *287*, 2612–2630. [CrossRef]
86. Hasan, I.; Sugawara, S.; Fujii, Y.; Koide, Y.; Terada, D.; Iimura, N.; Fujiwara, T.; Takahashi, K.G.; Kojima, N.; Rajia, S.; et al. MytiLec, a Mussel R-Type Lectin, Interacts with Surface Glycan Gb3 on Burkitt's Lymphoma Cells to Trigger Apoptosis through Multiple Pathways. *Mar. Drugs* **2015**, *13*, 7377–7389. [CrossRef]
87. Fujii, Y.; Dohmae, N.; Takio, K.; Kawsar, S.M.A.; Matsumoto, R.; Hasan, I.; Koide, Y.; Kanaly, R.A.; Yasumitsu, H.; Ogawa, Y.; et al. A Lectin from the Mussel Mytilus Galloprovincialis Has a Highly Novel Primary Structure and Induces Glycan-Mediated Cytotoxicity of Globotriaosylceramide-Expressing Lymphoma Cells. *J. Biol. Chem.* **2012**, *287*, 44772–44783. [CrossRef]
88. Shimizu, T.; Vassylyev, D.; Kido, S.; Doi, Y.; Morikawa, K. Crystal Structure of Vitelline Membrane Outer Layer Protein I (VMO-I): A Folding Motif with Homologous Greek Key Structures Related by an Internal Three-Fold Symmetry. *EMBO J.* **1994**, *13*, 1003–1010. [CrossRef] [PubMed]
89. Sankaranarayanan, R.; Sekar, K.; Banerjee, R.; Sharma, V.; Surolia, A.; Vijayan, M. A Novel Mode of Carbohydrate Recognition in Jacalin, a Moraceae Plant Lectin with a Beta-Prism Fold. *Nat. Struct. Biol.* **1996**, *3*, 596–603. [CrossRef] [PubMed]
90. Lee, X.; Thompson, A.; Zhang, Z.; Ton-that, H.; Biesterfeldt, J.; Ogata, C.; Xu, L.; Johnston, R.A.Z.; Young, N.M. Structure of the Complex of Maclura PomiferaAgglutinin and the T-Antigen Disaccharide, Galβ1,3GalNAc*. *J. Biol. Chem.* **1998**, *273*, 6312–6318. [CrossRef] [PubMed]

91. Rosa, J.C.; De Oliveira, P.S.; Garratt, R.; Beltramini, L.; Resing, K.; Roque-Barreira, M.C.; Greene, L.J. KM+, a Mannose-Binding Lectin from *Artocarpus Integrifolia*: Amino Acid Sequence, Predicted Tertiary Structure, Carbohydrate Recognition, and Analysis of the Beta-Prism Fold. *Protein Sci.* **1999**, *8*, 13–24. [CrossRef] [PubMed]
92. Rao, K.N.; Suresh, C.G.; Katre, U.V.; Gaikwad, S.M.; Khan, M.I. Two Orthorhombic Crystal Structures of a Galactose-Specific Lectin from *Artocarpus Hirsuta* in Complex with Methyl-Alpha-D-Galactose. *Acta Crystallogr. D Biol. Crystallogr.* **2004**, *60*, 1404–1412. [CrossRef]
93. Suetake, T.; Tsuda, S.; Kawabata, S.; Miura, K.; Iwanaga, S.; Hikichi, K.; Nitta, K.; Kawano, K. Chitin-Binding Proteins in Invertebrates and Plants Comprise a Common Chitin-Binding Structural Motif. *J. Biol. Chem.* **2000**, *275*, 17929–17932. [CrossRef]
94. Kawabata, S.; Nagayama, R.; Hirata, M.; Shigenaga, T.; Agarwala, K.L.; Saito, T.; Cho, J.; Nakajima, H.; Takagi, T.; Iwanaga, S. Tachycitin, a Small Granular Component in Horseshoe Crab Hemocytes, Is an Antimicrobial Protein with Chitin-Binding Activity. *J. Biochem.* **1996**, *120*, 1253–1260. [CrossRef]
95. Meng, D.-M.; Dai, H.-X.; Gao, X.-F.; Zhao, J.-F.; Guo, Y.-J.; Ling, X.; Dong, B.; Zhang, Z.-Q.; Fan, Z.-C. Expression, Purification and Initial Characterization of a Novel Recombinant Antimicrobial Peptide Mytichitin-A in *Pichia Pastoris*. *Protein Expr. Purif.* **2016**, *127*, 35–43. [CrossRef]
96. Bauters, L.; Naalden, D.; Gheysen, G. The Distribution of Lectins across the *Phylum Nematoda*: A Genome-Wide Search. *Int. J. Mol. Sci.* **2017**, *18*, 91. [CrossRef]
97. Odom, E.W.; Vasta, G.R. Characterization of a Binary Tandem Domain F-Type Lectin from Striped Bass (*Morone Saxatilis*). *J. Biol. Chem.* **2006**, *281*, 1698–1713. [CrossRef]
98. Bianchet, M.A.; Odom, E.W.; Vasta, G.R.; Amzel, L.M. A Novel Fucose Recognition Fold Involved in Innate Immunity. *Nat. Struct. Biol.* **2002**, *9*, 628–634. [CrossRef] [PubMed]
99. Arivalagan, J.; Marie, B.; Sleight, V.A.; Clark, M.S.; Berland, S.; Marie, A. Shell Matrix Proteins of the Clam, *Mya Truncata*: Roles beyond Shell Formation through Proteomic Study. *Mar. Genom.* **2016**, *27*, 69–74. [CrossRef] [PubMed]
100. Moy, G.W.; Springer, S.A.; Adams, S.L.; Swanson, W.J.; Vacquier, V.D. Extraordinary Intraspecific Diversity in Oyster Sperm Bindin. *Proc. Natl. Acad. Sci. USA* **2008**, *105*, 1993–1998. [CrossRef] [PubMed]
101. Ruben, E.A.; Rau, M.J.; Fitzpatrick, J.A.J.; Di Cera, E. Cryo-EM Structures of Human Coagulation Factors V and Va. *Blood* **2021**, *137*, 3137–3144. [CrossRef]
102. Mathieu, S.V.; Aragão, K.S.; Imberty, A.; Varrot, A. Discoidin I from Dictyostelium Discoideum and Interactions with Oligosaccharides: Specificity, Affinity, Crystal Structures, and Comparison with Discoidin II. *J. Mol. Biol.* **2010**, *400*, 540–554. [CrossRef] [PubMed]
103. Tateno, H. SUEL-Related Lectins, a Lectin Family Widely Distributed throughout Organisms. *Biosci. Biotechnol. Biochem.* **2010**, *74*, 1141–1144. [CrossRef]
104. Ozeki, Y.; Matsui, T.; Suzuki, M.; Titani, K. Amino Acid Sequence and Molecular Characterization of a D-Galactoside-Specific Lectin Purified from Sea Urchin (*Anthocidaris Crassispina*) Eggs. *Biochemistry* **1991**, *30*, 2391–2394. [CrossRef]
105. Carneiro, R.F.; Teixeira, C.S.; de Melo, A.A.; de Almeida, A.S.; Cavada, B.S.; de Sousa, O.V.; da Rocha, B.A.M.; Nagano, C.S.; Sampaio, A.H. L-Rhamnose-Binding Lectin from Eggs of the *Echinometra Lucunter*: Amino Acid Sequence and Molecular Modeling. *Int. J. Biol. Macromol.* **2015**, *78*, 180–188. [CrossRef]
106. Gasparini, F.; Franchi, N.; Spolaore, B.; Ballarin, L. Novel Rhamnose-Binding Lectins from the Colonial Ascidian Botryllus Schlosseri. *Dev. Comp. Immunol.* **2008**, *32*, 1177–1191. [CrossRef]
107. Franchi, N.; Schiavon, F.; Carletto, M.; Gasparini, F.; Bertoloni, G.; Tosatto, S.C.E.; Ballarin, L. Immune Roles of a Rhamnose-Binding Lectin in the Colonial Ascidian Botryllus Schlosseri. *Immunobiology* **2011**, *216*, 725–736. [CrossRef]
108. Naganuma, T.; Ogawa, T.; Hirabayashi, J.; Kasai, K.; Kamiya, H.; Muramoto, K. Isolation, Characterization and Molecular Evolution of a Novel Pearl Shell Lectin from a Marine Bivalve, Pteria Penguin. *Mol. Divers.* **2006**, *10*, 607–618. [CrossRef] [PubMed]
109. Uhlenbruck, G.; Prokop, O. An Agglutinin from Helix Pomatia, Which Reacts with Terminal N-Acetyl-D-Galactosamine. *Vox Sang.* **1966**, *11*, 519–520. [CrossRef] [PubMed]
110. Sanchez, J.-F.; Lescar, J.; Chazalet, V.; Audfray, A.; Gagnon, J.; Alvarez, R.; Breton, C.; Imberty, A.; Mitchell, E.P. Biochemical and Structural Analysis of Helix Pomatia Agglutinin: A hexameric lectin with a novel fold. *J. Biol. Chem.* **2006**, *281*, 20171–20180. [CrossRef] [PubMed]
111. Gerdol, M. Immune-Related Genes in Gastropods and Bivalves: A Comparative Overview. *Invertebr. Surviv. J.* **2017**, *14*, 95–111.
112. Gerdol, M.; Fujii, Y.; Hasan, I.; Koike, T.; Shimojo, S.; Spazzali, F.; Yamamoto, K.; Ozeki, Y.; Pallavicini, A.; Fujita, H. The Purplish Bifurcate Mussel Mytilisepta Virgata Gene Expression Atlas Reveals a Remarkable Tissue Functional Specialization. *BMC Genom.* **2017**, *18*, 590. [CrossRef]
113. Ogawa, T.; Sato, R.; Naganuma, T.; Liu, K.; Lakudzala, A.E.; Muramoto, K.; Osada, M.; Yoshimi, K.; Hiemori, K.; Hirabayashi, J.; et al. Glycan Binding Profiling of Jacalin-Related Lectins from the Pteria Penguin Pearl Shell. *Int. J. Mol. Sci.* **2019**, *20*, 4629. [CrossRef]
114. Kanagawa, M.; Satoh, T.; Ikeda, A.; Nakano, Y.; Yagi, H.; Kato, K.; Kojima-Aikawa, K.; Yamaguchi, Y. Crystal Structures of Human Secretory Proteins ZG16p and ZG16b Reveal a Jacalin-Related β-Prism Fold. *Biochem. Biophys. Res. Commun.* **2011**, *404*, 201–205. [CrossRef]

115. Garénaux, E.; Kanagawa, M.; Tsuchiyama, T.; Hori, K.; Kanazawa, T.; Goshima, A.; Chiba, M.; Yasue, H.; Ikeda, A.; Yamaguchi, Y.; et al. Discovery, Primary, and Crystal Structures and Capacitation-Related Properties of a Prostate-Derived Heparin-Binding Protein WGA16 from Boar Sperm. *J. Biol. Chem.* **2015**, *290*, 5484–5501. [CrossRef]
116. Jia, N.; Liu, N.; Cheng, W.; Jiang, Y.; Sun, H.; Chen, L.; Peng, J.; Zhang, Y.; Ding, Y.; Zhang, Z.; et al. Structural Basis for Receptor Recognition and Pore Formation of a Zebrafish Aerolysin-like Protein. *EMBO Rep.* **2016**, *17*, 235–248. [CrossRef]
117. Carneiro, R.F.; Torres, R.C.F.; Chaves, R.P.; de Vasconcelos, M.A.; de Sousa, B.L.; Goveia, A.C.R.; Arruda, F.V.; Matos, M.N.C.; Matthews-Cascon, H.; Freire, V.N.; et al. Purification, Biochemical Characterization, and Amino Acid Sequence of a Novel Type of Lectin from Aplysia Dactylomela Eggs with Antibacterial/Antibiofilm Potential. *Mar. Biotechnol.* **2017**, *19*, 49–64. [CrossRef]
118. Huang, G.; Huang, S.; Yan, X.; Yang, P.; Li, J.; Xu, W.; Zhang, L.; Wang, R.; Yu, Y.; Yuan, S.; et al. Two Apextrin-like Proteins Mediate Extracellular and Intracellular Bacterial Recognition in Amphioxus. *Proc. Natl. Acad. Sci. USA* **2014**, *111*, 13469–13474. [CrossRef] [PubMed]
119. Wang, W.; Gong, C.; Han, Z.; Lv, X.; Liu, S.; Wang, L.; Song, L. The Lectin Domain Containing Proteins with Mucosal Immunity and Digestive Functions in Oyster Crassostrea Gigas. *Fish Shellfish Immunol.* **2019**, *89*, 237–247. [CrossRef] [PubMed]
120. Dang, L.; Van Damme, E.J.M. Genome-Wide Identification and Domain Organization of Lectin Domains in Cucumber. *Plant Physiol. Biochem.* **2016**, *108*, 165–176. [CrossRef] [PubMed]
121. Lubyaga, Y.A.; Dolgikh, A.V.; Drozdova, P.B.; Nazarova, A.A.; Timofeyev, M.A. Transcriptome-Based Analysis of the Diversity of Membrane-Bound Lectins in Baikal Amphipods *Eulimnogammarus* Sp. and the Holarctic Amphipod *Gammarus Lacustris*. *Limnol. Freshw. Biol.* **2020**, 797–798. [CrossRef]
122. Zárate-Potes, A.; Ocampo, I.D.; Cadavid, L.F. The Putative Immune Recognition Repertoire of the Model Cnidarian Hydractinia Symbiolongicarpus Is Large and Diverse. *Gene* **2019**, *684*, 104–117. [CrossRef]
123. Simão, F.A.; Waterhouse, R.M.; Ioannidis, P.; Kriventseva, E.V.; Zdobnov, E.M. BUSCO: Assessing Genome Assembly and Annotation Completeness with Single-Copy Orthologs. *Bioinformatics* **2015**, *31*, 3210–3212. [CrossRef]
124. Angata, T.; Brinkman-Van der Linden, E. I-Type Lectins. *Biochim. Biophys. Acta* **2002**, *1572*, 294–316. [CrossRef]
125. Liu, C.; Jiang, S.; Wang, M.; Wang, L.; Chen, H.; Xu, J.; Lv, Z.; Song, L. A Novel Siglec (CgSiglec-1) from the Pacific Oyster (*Crassostrea Gigas*) with Broad Recognition Spectrum and Inhibitory Activity to Apoptosis, Phagocytosis and Cytokine Release. *Dev. Comp. Immunol.* **2016**, *61*, 136–144. [CrossRef]
126. Sulzenbacher, G.; Roig-Zamboni, V.; Peumans, W.J.; Henrissat, B.; van Damme, E.J.M.; Bourne, Y. Structural Basis for Carbohydrate Binding Properties of a Plant Chitinase-like Agglutinin with Conserved Catalytic Machinery. *J. Struct. Biol.* **2015**, *190*, 115–121. [CrossRef]
127. Hasan, I.; Gerdol, M.; Fujii, Y.; Rajia, S.; Koide, Y.; Yamamoto, D.; Kawsar, S.M.A.; Ozeki, Y. CDNA and Gene Structure of MytiLec-1, A Bacteriostatic R-Type Lectin from the Mediterranean Mussel (*Mytilus Galloprovincialis*). *Mar. Drugs* **2016**, *14*, 92. [CrossRef]
128. Han, J.; Park, J.C.; Choi, B.-S.; Kim, M.-S.; Kim, H.-S.; Hagiwara, A.; Park, H.G.; Lee, B.-Y.; Lee, J.-S. The Genome of the Marine Monogonont Rotifer Brachionus Plicatilis: Genome-Wide Expression Profiles of 28 Cytochrome P450 Genes in Response to Chlorpyrifos and 2-Ethyl-Phenanthrene. *Aquat. Toxicol.* **2019**, *214*, 105230. [CrossRef] [PubMed]
129. Kim, H.-S.; Lee, B.-Y.; Han, J.; Jeong, C.-B.; Hwang, D.-S.; Lee, M.-C.; Kang, H.-M.; Kim, D.-H.; Kim, H.-J.; Papakostas, S.; et al. The Genome of the Freshwater Monogonont Rotifer Brachionus Calyciflorus. *Mol. Ecol. Resour.* **2018**, *18*, 646–655. [CrossRef] [PubMed]
130. Kim, D.-H.; Kim, M.-S.; Hagiwara, A.; Lee, J.-S. The Genome of the Minute Marine Rotifer Proales Similis: Genome-Wide Identification of 401 G Protein-Coupled Receptor (GPCR) Genes. *Comp. Biochem. Physiol. D Genom. Proteom.* **2021**, *39*, 100861. [CrossRef] [PubMed]
131. Petersen, T.N.; Brunak, S.; von Heijne, G.; Nielsen, H. SignalP 4.0: Discriminating Signal Peptides from Transmembrane Regions. *Nat. Meth.* **2011**, *8*, 785–786. [CrossRef]
132. Bendtsen, J.D.; Jensen, L.J.; Blom, N.; Von Heijne, G.; Brunak, S. Feature-Based Prediction of Non-Classical and Leaderless Protein Secretion. *Protein Eng. Des. Sel.* **2004**, *17*, 349–356. [CrossRef]
133. Krogh, A.; Larsson, B.; von Heijne, G.; Sonnhammer, E.L. Predicting Transmembrane Protein Topology with a Hidden Markov Model: Application to Complete Genomes. *J. Mol. Biol.* **2001**, *305*, 567–580. [CrossRef]
134. Käll, L.; Krogh, A.; Sonnhammer, E.L.L. A Combined Transmembrane Topology and Signal Peptide Prediction Method. *J. Mol. Biol.* **2004**, *338*, 1027–1036. [CrossRef]
135. Finn, R.D.; Clements, J.; Eddy, S.R. HMMER Web Server: Interactive Sequence Similarity Searching. *Nucl. Acids Res.* **2011**, *39*, W29–W37. [CrossRef]
136. Jones, P.; Binns, D.; Chang, H.-Y.; Fraser, M.; Li, W.; McAnulla, C.; McWilliam, H.; Maslen, J.; Mitchell, A.; Nuka, G.; et al. InterProScan 5: Genome-Scale Protein Function Classification. *Bioinformatics* **2014**, *30*, 1236–1240. [CrossRef]
137. Söding, J.; Biegert, A.; Lupas, A.N. The HHpred Interactive Server for Protein Homology Detection and Structure Prediction. *Nucleic Acids Res.* **2005**, *33*, W244–W248. [CrossRef]
138. Edgar, R.C. MUSCLE: Multiple Sequence Alignment with High Accuracy and High Throughput. *Nucleic Acids Res.* **2004**, *32*, 1792–1797. [CrossRef] [PubMed]
139. Talavera, G.; Castresana, J. Improvement of Phylogenies after Removing Divergent and Ambiguously Aligned Blocks from Protein Sequence Alignments. *Syst. Biol.* **2007**, *56*, 564–577. [CrossRef] [PubMed]

140. Huelsenbeck, J.P.; Ronquist, F. MRBAYES: Bayesian Inference of Phylogenetic Trees. *Bioinformatics* **2001**, *17*, 754–755. [CrossRef] [PubMed]
141. Darriba, D.; Posada, D.; Kozlov, A.M.; Stamatakis, A.; Morel, B.; Flouri, T. ModelTest-NG: A New and Scalable Tool for the Selection of DNA and Protein Evolutionary Models. *Mol. Biol. Evol.* **2020**, *37*, 291–294. [CrossRef]
142. Rambaut, A.; Drummond, A.J.; Xie, D.; Baele, G.; Suchard, M.A. Posterior Summarization in Bayesian Phylogenetics Using Tracer 1.7. *Syst. Biol.* **2018**, *67*, 901–904. [CrossRef]

marine drugs

MDPI

Article

Taxonomic and Bioactivity Characterizations of *Mameliella alba* Strain LZ-28 Isolated from Highly Toxic Marine Dinoflagellate *Alexandrium catenella* LZT09

Cheng-Zhe Ren [1], Hui-Min Gao [1], Jun Dai [2], Wen-Zhuo Zhu [1], Fei-Fei Xu [3], Yun Ye [4], Xiao-Ling Zhang [1,5,*] and Qiao Yang [1,5,*]

1 Department of Marine Chemistry, College of Marine Science and Technology, Zhejiang Ocean University, Zhoushan 316022, China; renchengzhe@zjou.edu.cn (C.-Z.R.); gaohuimin@zjou.edu.cn (H.-M.G.); zhuwenzhuo@zjou.edu.cn (W.-Z.Z.)

2 Natural "111" Center for Cellular Regulation and Molecular Pharmaceutics, Key Laboratory of Fermentation Engineering, Ministry of Education, College of Bioengineering, Hubei University of Technology, Wuhan 430068, China; jundai@hbut.edu.cn

3 Zhejiang Yimeiyuan Testing Technology Co., Ltd., Zhoushan 316022, China; smexu713@163.com

4 Zhoushan Natural Resources Surveying and Mapping Design Centre, Zhoushan 316021, China; smeye918@gmail.com

5 ABI Group, Zhejiang Ocean University, Zhoushan 316022, China

* Correspondence: zhangxiaoling@zjou.edu.cn (X.-L.Z.); qiaoyang1979@whu.edu.cn (Q.Y.)

Citation: Ren, C.-Z.; Gao, H.-M.; Dai, J.; Zhu, W.-Z.; Xu, F.-F.; Ye, Y.; Zhang, X.-L.; Yang, Q. Taxonomic and Bioactivity Characterizations of *Mameliella alba* Strain LZ-28 Isolated from Highly Toxic Marine Dinoflagellate *Alexandrium catenella* LZT09. *Mar. Drugs* **2022**, *20*, 321. https://doi.org/10.3390/md20050321

Academic Editors: Yuki Fujii, Marco Gerdol, Yasuhiro Ozeki and Zeinab Khalil

Received: 5 April 2022
Accepted: 10 May 2022
Published: 12 May 2022

Abstract: Microalgae host varied microbial consortium harboring cross-kingdom interactions with fundamental ecological significance in aquatic ecosystems. Revealing the complex biofunctions of the cultivable bacteria of phycosphere microbiota is one vital basis for deeply understanding the mechanisms governing these dynamic associations. In this study, a new light-yellow pigmented bacterial strain LZ-28 was isolated from the highly-toxic and harmful algal bloom-forming dinoflagellate *Alexandrium catenella* LZT09. Collective phenotypic and genotypic profiles were obtained to confidently identify this strain as a new *Mameliella alba* member. Comparative genomic analysis showed that strain LZ-28 shared highly similar functional features with other four marine algae-derived *M. alba* strains in spite of their distinctive isolation sources. Based on the bioactivity assaying, the mutual growth-promoting effects between bacterial strain LZ-28 and algal strain LZT09 were observed. After the culture conditions were optimized, strain LZ-28 demonstrated an extraordinary production ability for its bioflocculanting exopolysaccharides (EPS). Moreover, the portions of two monosaccharides glucose and fucose of the EPS were found to positively contribute to the bioflocculanting capacity. Therefore, the present study sheds light on the similar genomic features among the selected *M. alba* strains, and it also reveals the potential pharmaceutical, environmental and biotechnological implications of active EPS produced by this new *Mameliella alba* strain LZ-28 recovered from toxic bloom-forming marine dinoflagellate.

Keywords: bacterial exopolysaccharides; bioflocculanting activity; microalgae growth-promoting bacterium; harmful algal bloom-forming dinoflagellate; *Alexandrium catenella*; *Mameliella alba*

1. Introduction

The term "phycosphere" was initially coined in 1972 to describe the region immediately surrounding an individual algal cell that is enriched in organic matters, which are exuded by the cell into its surroundings [1]. It is known as the aquatic analogue of the rhizosphere in the soil ecosystem [2]. The aquatic phycosphere niche is regarded as the boundary of phytoplankton holobionts and the ecological interface for algae–bacteria interactions (ABI) [3]. The relationships that usually span mutualism, commensalism, antagonism, parasitism, and competition, usually involve cross-kingdom exchanges of nutrients, secondary metabolites, infochemicals, and gene-transfer agents (GTA) [4]. It has subsequently become apparent that these microscale interactions are far more sophisticated than previously thought. It usually requires the close spatial proximity of two sides involved in the interactions, which are usually facilitated by bacterial colonization in the phycosphere niches [5]. Within this microscopic interface, the exopolysaccharides or extracellular polysaccharides (EPS) contributed by both partners are the main components of the matrix, which embeds the proliferating cells and promotes colony formation [5]. It is also the essential chemical intermedia connecting the microscale interactions [1,2,6]. Presently, it is becoming increasingly clear that these complex inter-species or intra-species associations usually exert further ecosystem-scale influences on fundamental processes including primary production, phycotoxin biosynthesis, biogeochemical cycles and the microbial loop in the oceans [7].

Currently, the rapidly increasing observations show that phytoplankton-associated bacterial communities are often limited to a small handful of taxa groups, including specific members of the *Roseobacter* group (*Rhodobacteraceae*), *Flavobacteraceae*, and *Alteromonadaceae* [8–10]. These observations also provide emerging evidence for the species-specific associations between the phytoplanktons and their closely-associated bacterial consortium [2–6]. *Rhodobacteraceae* usually accounts for an overwhelming proportion of the bacterioplankton communities in marine environments [8,9]. Within this family, the genus *Mameliella* has become one key and representative member in the *Roseobacter* group, although it was initially established one decade ago with *Mameliella alba* JLT354-W^{T} as the type species [11]. Moreover, *Mameliella* members widely contributed to the marine environments in aromatic compound degradation, biogeochemical cycles of carbon and sulfur, dimethylsulfoniopropionate demethylation, and the production of diverse secondary metabolites [12]. In recent decades, four type species, *Mameliella phaeodactyli*, *Mameliella atlantica*, *Ponticoccus lacteus*, and *Alkalimicrobium pacificum*, were reclassified later as heterotypic synonyms of *M. alba* based on their phylogenomic characterizations [13]. Nowadays, the genus *Mameliella* contains only two type species with correctly published names, including *M. alba* and *M. sediminis* (https://lpsn.dsmz.de/genus/mameliella, accessed on 10 May 2022). Currently, over 1500 species of free-living dinoflagellates have been described, and the harmful algal blooms (HABs)-causing group occupies over 300 species [14]. Among them, over one quarter are toxic species that produce diverse types of phycotoxins, such as paralytic shellfish-poisoning (PSP) toxins [15,16]. Within the toxic dinoflagellate species, the globally distributed genus *Alexandrium*, in which the members were widely found in sub-polar, temperate and tropical coastal water environments, is a particularly well-known bloom-forming group due to its large and widespread threat to seafood production and human health [17,18].

The importance of the phycosphere has been postulated for four decades, yet only recent new technological progress in high-throughput pyrosequencing made it possible to start teasing apart the complex nature of the microbial composition within this unique microbial habitat [3–6,19–24]. The culture-independent isolation of the cultivable strains is still the first critical step for exploring the dynamics of algae–bacteria interactions [3–6]. Previously, we conducted the Phycosphere Microbiome Project (PMP) to convey the compositions of varied microbial consortiums that were closely-associated with diverse HAB-forming dinoflagellates [25–30]. During our investigation, the genus *Mameliella* was found to be one predominant group and was widely distributed in the bacterial communities associated with various marine dinoflagellates [10,19–23]. Consequently, a new red-pigmented bacterial strain, LZ-28, was isolated from the highly toxic *Alexandrium catenella* LZT09. Strain LZ-28 produced active bioflocculanting exopolysaccharides, which were also discovered in other cultivable bacterial strains isolated from marine dinoflagellates [27–31]. Additionally, it demonstrated the obvious growth-promoting ability of the algal strain LZT09, as previously verified by other marine bacteria strains derived from the same algal host [21,30]. Thus, the present work aims to characterize the taxonomic status of this new bacterial strain using combined taxonomic and phylogenomic approaches. Moreover, comparative genomic analysis is then emphatically performed for the five marine algae-derived strains among the ten selected *M. alba* strains to reveal their shared functional nature. Based on bioactivity assaying after culture condition optimization, the contribution of the individual monosaccharide portion of the EPS produced by strain LZ-28 to the bioflocculanting bioactivity is also characterized.

2. Results and Discussion

2.1. Characterization of the Composition of the Bacterial Community of Algal Strain LZT09

To reveal the microbial composition of the bacterial community associated with LZT09, the high-throughput sequencing of the V3–V4 variable region of bacterial 16S rRNA was performed. Based on the obtained result, the phylum Pseudomonadota represented the overwhelming dominant bacterial fraction (53.9%) of the total bacterial community of LZT09, followed by Bacteroidota (26.7%) and Cyanobacteria (15.2%). Other lineages, including Actinomycetota, Bacillota, and Spirochaetota, were also present, but each constituted less than 1.0% of the total (Figure 1). Similarly, previous studies showed that two phyla, Pseudomonadota and Bacteroidota, were the most major bacterial groups affiliated with the investigated bacterial community associated with marine dinoflagellates [20,21]. Moreover, members of *Rhodobacteraceae* were reported to be frequently present during various stages of HABs [32,33], and also regarded as a major taxa associated with some diatom cultures [34,35].

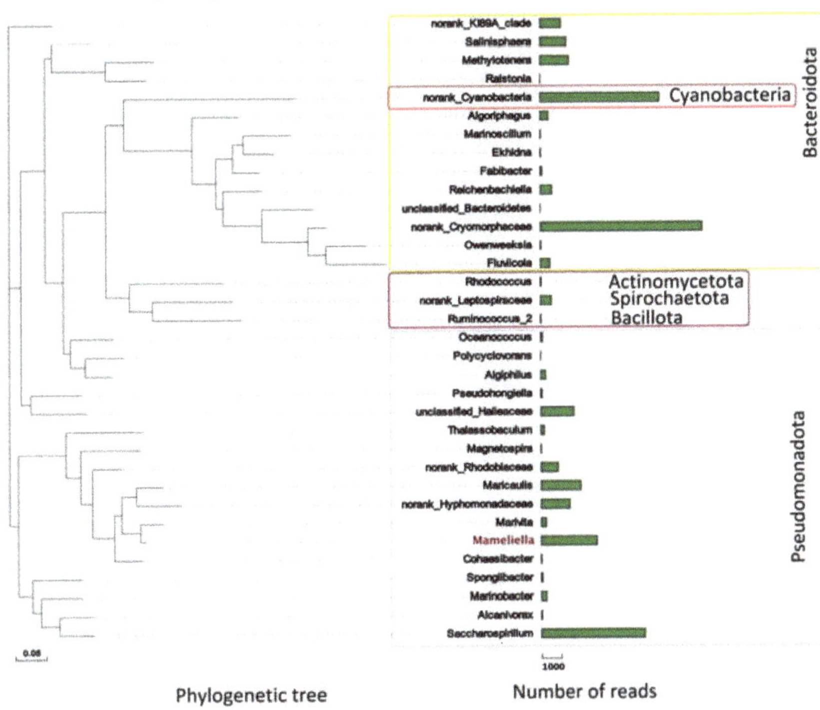

Figure 1. Phylogenetic tree and the number of reads of bacterial abundance at the genus level of the bacterial community associated with algal strain LZT09 by high-throughput sequencing of the V3-V4 variable region of bacterial 16S rRNA gene. Bar: 0.05 substitutions per nucleotide position in the phylogenetic tree.

In this study, *Rhodobacteraceae* was found to occupy only 8.7% of the total bacterial community of LZT09, which held the fifth dominant group following the family *Cryomorphaceae* (22.9%), one unidentified Cyanobacteria (16.8%), *Saccharospirillaceae* (14.7%), and the *Hyphomonadaceae* (9.7%). *Cryomorphaceae* is a member of the order Flavobacteriales within the phylum Bacteroidota, and has been found to distribute in a wide range of marine and terrestrial systems from tropical to polar regions [36]. Moreover, molecular phylogenetic studies show that the phylotypes related to the family *Cryomorphaceae* belong to the abundant coastal phytoplankton bloom-responding flavobacterial group [37]. In addition, at the genus level, five dominant bacterial groups, including one unidentified *Cryomorphaceae* (22.9%), one unidentified Cyanobacteria (16.8%), *Saccharospirillum* (14.6%), *Maricaulis* (5.6%), and *Mameliella* (5.1%), were identified, and totally accounted for 65.0% of the total bacterial community of LZT09 (Figure 1). It is worth noting that only one genus, *Microcystis*, was found in the Cyanobacteria group, which indicated that some cryptic cyanobacterial lineages were potentially hidden by the dinoflagellate host [37].

2.2. Phenotypic and Biochemical Characteristics of Bacterial Strain LZ-28

Previously, nine cultivable bacterial strains, including strain LZ-28, were isolated from the cultivable PM of LZT09 [10]. However, no strain was found to belong to the dominant group *Cryomorphaceae*. Previous studies have shown that the most cultured *Cryomorphaceae* species were mainly isolated from the low-temperature ecosystems, although the family *Cryomorphaceae* was widespread in a wide range of non-extreme ecosystems. These findings indicate the resistance of the *Cryomorphaceae* members towards be cultivation [36]. Thus, the constant optimization of the isolation procedure for the cultivable bacteria strains isolated from marine dinoflagellates still needs to be performed, especially by means of the emergence and prevalence of modern multi-omics data [38,39]. Based on the phenotypic characterization, the cellular colonies of strain LZ-28 grown on marine agar (MA) were circular, smooth, and convex with light-yellow colors. Cells of strain LZ-28 were Gram-stain-negative, rod shaped, and non-motile with the sizes of 0.7–1.0 μm for the width and 2.0–2.9 μm for the length (Figure 2).

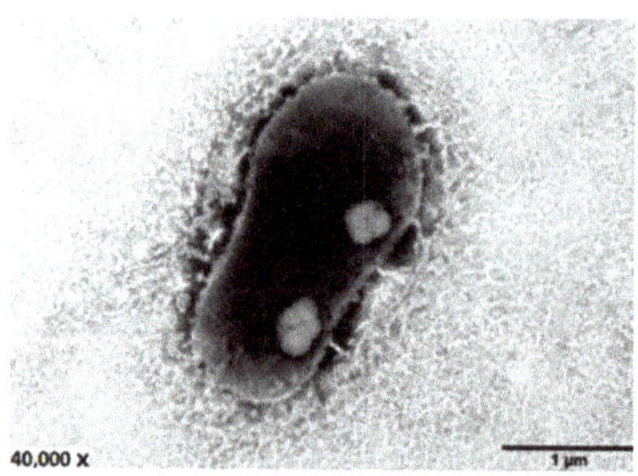

Figure 2. Transmission electron micrograph of the cell of the bacterial strain LZ-28 after being grown on MA at 28 °C for 2 days. Bar: 1 μm.

The physiological characterization revealed that strain LZ-28 grew at the pH ranging from 5.0 to 10.0 with the optimum growth at pH 7.0. The growth temperature ranged from 15 °C to 40 °C with the optimum value of 25–28 °C, at the presence of 1–10% (*w/v*) NaCl with the optimum value of 2.5%. No bacterial growth was observed under anaerobic conditions when grown on MA, even after three week-long incubations. The hydrolysis of starch, L-tyrosine, and gelatin were observed. Nitrate was reduced to nitrite, but the reduction of nitrite to nitrogen was not observed. Chemotaxonomic analysis showed that the major cellular fatty acids of strain LZ-28 consisted of $C_{18:0}$, $C_{18:1}\omega7c$ 11-methyl, and $C_{19:0}$ cyclo. Detailed comparisons of the morphological, biochemical, and physiological characteristics of strain LZ-28 and the other five *M. alba* strains are summarized in Table 1. The differential characteristics between strain LZ-28 and other *M. alba* members can be easily found, even though they share very high 16S rRNA gene-sequence similarities, for example, the distinguished colony color, the ability for citrate utilization, the presence/absence of a minor fatty acid of $C_{17:1}\,\omega8c$, as well as the apparently varied polar lipid profiles among the six compared *M. alba* strains.

Table 1. Phenotypic and genotypic characteristics of the bacterial strain LZ-28 and the other five *M. alba* strains.

Characteristic	LZ-28	JLT354-W^T	KD53	L6M1-5	JL351	F15
Isolation source	Marine dinoflagellate	Seawater	Marine diatom	Deep-sea sediment	Surface seawater	Deep-sea sediment
Colony color	Light-Yellow	White	Yellow	Yellow–white	Yellow	Cream–yellow
Cell size (μm)	(0.7–1.0) × (2.0–2.9)	(0.7–0.9) × (2.0–3.2)	(0.5–1.0) × (2.2–2.9)	(0.7–0.8) × (1.0–2.0)	(0.5–0.8) × (0.7–1.5)	(0.7–1.2) × (1.6–3.4)
NaCl range (optimum, *w/v*, %)	1.0–10.0 (2.5)	1.0–10.0 (1.0–3.0)	1.0–6.0 (3.0)	0.5–15.0 (3.0–5.0)	0.5–9.0 (1.0–4.0)	0–10.0 (3.5)
pH range (optimum)	5.0–10.0 (7.0)	6.0–9.0 (8.0)	6.0–9.5 (7.5–8.0)	5.0–10.5 (7.0)	7.0–10.0 (8.0)	6.0–11.0 (7.0–8.0)
Temperature range (optimum, °C)	15–40 (25–28)	10–30 (25)	16–37 (28)	10–41 (28–30)	15–40 (30)	4–50 (35–37)
Oxidase activity	+	+	+	−	+	+
Catalase activity	+	+	−	+	+	+
Poly-β-hydroxybutyrate	+	+	−	−	+	−
API 20E test						
Citrate utilization	−	−	−	w	w	+
Gelatinase	+	+	−	+	+	+
API 20NE test						
Reduction of nitrite	+	+	+	+	−	+
Gelatin hydrolysis	+	+	−	+	+	+
Phenylacetic acid utilization	−	+	+	+	+	−
Polar lipid profile [a]	PC, DPG, PE	AL	PC, PE, AL	PE, AL	PC, DPG, PE, AL, GL	PC, PE, GL
Fatty acid profile [b]						
$C_{16:0}$	3.2	5.7	5.3	4.8	5.3	4.5
$C_{17:0}$	1.1	*tr*	1.0	*tr*	*tr*	ND
$C_{18:0}$	7	9.2	11.1	10	8.6	8.5
$C_{12:1}$ 3OH	4.7	3.3	2.8	3.1	3.2	3.1
$C_{17:1}\,\omega8c$	−	*tr*	*tr*	*tr*	*tr*	1
$C_{18:1}\,\omega7c$ 11-methyl	10.8	6.6	7.9	7.5	6.1	5.4
$C_{19:0}$ cyclo $\omega8c$	3.2	5.7	5.3	4.8	5.3	4.5
Summed feature 8	65.5	70.8	69.5	70.7	72.5	74.3
16S rRNA gene similarity (%)	−	99.70	99.77	99.92	99.62	99.40
DNA G+C content (mol%) [c]	64.9	65.2	65.0	65.0	65.1	65.0
ANI value of LZ-28 (%)	−	98.0	98.2	98.0	98.0	97.9
AAI value of LZ-28 (%)	−	98.4	98.1	98.1	98.5	98.2
dDDH value of LZ-28 (%)	−	84.3	83.5	83.9	84.3	83.1

All data were obtained from this study unless otherwise indicated. +, positive; W, weakly positive; −, negative. [a] PC, phosphatidylcholine; DPG, diphosphatidylglycerol; PE, phosphatidylethanolamine; AL, aminolipids; GL, glycolipid. [b] Values represent percentage (%) of total fatty acid contents. *tr*, trace amount (less than 1.0%). ND, not detected. Summed feature 8 comprises $C_{18:1}\,\omega7c$ and/or $C_{18:1}\,\omega6c$. [c] Data were calculated from the respective genome sequences.

2.3. Phylogenetic Analysis Based on 16S rRNA Gene Sequences

In the phylogenetic tree constructed by the maximum-likelihood (ML) algorithm using the 16S rRNA gene sequences, strain LZ-28 formed a monophyletic branch together with the type and other non-type strains of *M. alba*, as well as the selected representative *Rhodobacteraceae* members (Figure 3). Previously, *M. phaeodactyli* (type-strain KD53) [12], *M. atlantica* (type-strain L6M1-5) [40], *Ponticoccus lacteus* (type-strain JL351) [41], and *Alkalimicrobium pacificum* (type-strain F15) [42] were later reclassified as heterotypic synonyms for *M. alba* based on the phylogenomic characterizations [13]. In this study, the phylogenetic analysis showed that strain LZ-28 shared high 16S rRNA gene similarity values of 99.70, 99.77, 99.92, 99.60, and 99.40% with five other *M. alba* strains, JLT354-W^T; KD53; L6M1-5; JL351; and F15, respectively, although they were obviously isolated from different sources (Table 1). These values were all exceeding the threshold values (98.65%) generally accepted for species delineation [43]. Thus, it collectively indicated that strain LZ-28 was affiliated with the genus *Mameliella*, and was probably a new member of *M. alba*.

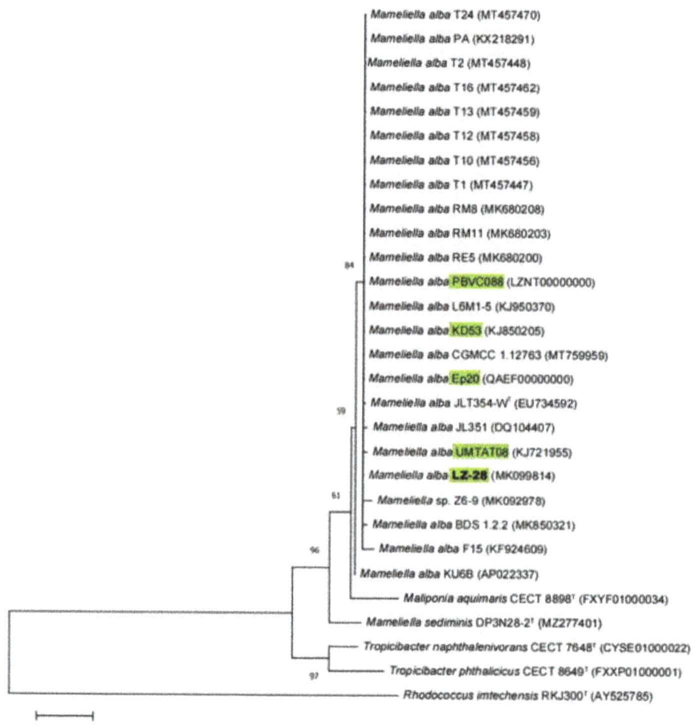

Figure 3. Phylogenetic tree using the maximum-likelihood (ML) algorithm constructed based on 16S rRNA gene sequences showing the relationship between *Mameliella* strains and other representative type species in the family *Rhodobacteraceae*. Bootstrap values (≥50%) based on 1000 replications are shown at the branch points. The five selected *M. alba* strains isolated from marine algae are marked in green. *Rhodococcus imtechensis* RKJ300^{T} (AY525785) was used as an outgroup. *Bar*, 0.02 substitutions per site.

2.4. Genomic Features of the Selected M. alba Strains

Due to the high-similarity values of the 16S rRNA gene sequence between strain LZ-28 and its phylogenetic neighbors in the family *Rhodobacteraceae*, the extra phylogenomic characterization was performed to ensure its taxonomic status. The genome sequence of strain LZ-28 was obtained by our previous study [10], and the other nine *M. alba* strains with available genomes were obtained from the NCBI database (https://www.ncbi.nlm.nih.gov, accessed on 2 April 2022). The isolation sources and the general genomic characteristics of the ten selected *M. alba* strains are summarized in Table 2. Among the ten selected *M. alba* strains, only strain KU6B has a complete genome containing a 5.386 Mbp circular chromosome and three circular plasmids of 256, 112, and 76 Kbp, respectively [44]. The genomic size of strain LZ-28 was 5.66 Mb with a G+C content of 64.94%. It contained 5497 protein-coding DNA sequences (CDSs) and 61 RNA genes, including 8 rRNA, 50 tRNA, and 3 other RNA genes, respectively, as well as 86 pseudogenes. The genome of strain LZ-28 was the fifth largest, with a genome range of 5.26–5.90 Mb. However, the genomic DNA G+C content of strain LZ-28 was the lowest (64.94 mol%) among the ten selected *M. alba* strains.

Table 2. The isolation sources and general genomic features of the ten selected *Mameliella alba* strains in this study.

Strain	Isolation Source and Year	Genome Size (Mb)	G+C Content (mol%) [a]	Protein	CDS [b]	GenBank Accession No.
LZ-28	Toxic marine dinoflagellate *Alexandrium catenella* LZT09, East China Sea, 2018	5.66	64.94	5502	5497	JAANYX000000000
KD53	Marine diatom *Phaeodactylum tricornutum*, Xiamen, China, 2013	5.42	65.10	5278	5223	NIWC00000000
Ep20	Marine dinoflagellate, *Symbiodinium* sp., 2015	5.51	65.04	5345	5343	QAEF00000000
PBVC088	Toxic marine dinoflagellate, *Pyrodinium bahamense* var. *Compressum*, 2012	5.51	64.96	5689	5632	LZNT00000000
UMTAT08	Marine dinoflagellate *Alexandrium tamiyavanichii* AcMS01, 1997	5.84	65.01	5761	5719	JSUQ00000000
JLT354-W^T	Seawater, South China Sea, 2006	5.26	65.21	5126	5132	FMZI00000000
JL351	Surface seawater (111°00′ E, 20°59′ N), South China Sea, 2006	5.42	65.15	5290	5275	NIWA00000000
KU6B [c]	Surface seawater, Boso Peninsula, Japan (34.9° N, 134.89° E), 2011	5.83	64.95	6009	4886	AP022337-022340
L6M1-5	Deep-sea sediment (2835 m), South Atlantic Ocean (15.18° S, 13.88° W), 2011	5.90	64.96	5821	5783	NIVZ00000000
F15	Deep-sea sediment (7118 m, 141°59.7′ E, 10°59.7′ N), Western Pacific Ocean, 2015	5.79	64.95	5694	5589	NIWB00000000

[a] Data were calculated based on the respective genome sequences. [b] CDS, coding DNA sequence. [c] The complete genome contained a circular chromosome (AP022337) and three circular plasmids (AP022338/AP022339/AP022340) [34].

2.5. Phylogenomic Characterization of Bacterial Strain LZ-28

To further infer the phylogenetic relationship among the ten selected *M. alba* strains with available genome sequences, the phylogenomic tree using an up-to-date bacterial core gene set (UBCG) consisting of 92 bacterial core genes was also performed. The constructed phylogenetic tree is shown in Figure 4. It clearly shows that, among the five marine algae-derived *M. alba* strains, strain LZ-28 clusters together with strain KD53, which is isolated from marine diatom *Phaeodactylum tricornutum* [12]. The other three strains, PBVC088, UMTAT08, and Ep20, were also clustered together in the UBCG tree. Two clusters were also observed in the phylogenomic tree. One cluster was composed of strains JLT354-W^T, JL-351, and KU6B, which were all isolated from the surface seawater, and the second cluster included two strains, L6M1-5 and F15, which were both isolated from the deep-sea sediment. It indicated that the genome-based phylogeny plays a more definitive role in the construction of a natural and objective taxonomy [43]. Additionally, the comparison of the three key phylogenomic parameters, including average nucleotide identity (ANI, Figure S1A), average amino acid identity (AAI, Figure S1B), and digital DNA–DNA hybridization (dDDH, Figure S1C) values of strain LZ-28 and the type strain of *M. alba* JLT354-W^T were 98.0%, 98.4%, and 84.3%, respectively. All the values were higher than the thresholds (95–96% for ANI, 97% for AAI, and 70% for dDDH) generally accepted for new species delineations [43]. Accordingly, it clearly confirmed that strain LZ-28 was a new member of *M. alba*, based on the obtained taxogenomic evidences.

2.6. Comparison of the Core- and Pan-Genomic Profiles among the Selected M. alba Strains

The genome sequences of the other nine selected *M. alba* strains were retrieved from the NCBI database in addition to strain LZ-28. The numbers of accessory and unique genes of the ten *M. alba* strains were created and shown in Figure 5A. It shows that 4472 core genes are shared by the ten selected *M. alba* strains, which account for 80.8% to 90.3% of the genome repertoire of each strain. Additionally, only 0.1% to 5.1% of the unique genes were distributed in individual strains, depending on the varied genomic sizes. Two functional accumulation curves were constructed according to the core- and pan-genome analyses, respectively. The pan genome of the ten *M. alba* strains was fitted into a power law function with an exponent $\gamma = 0.15$ [$Fp(x) = 5160.29 \times x^{0.15}$], and did not appear to reach saturation even with the increasing genome number. Thus, it indicated that the pan genome was in an open state. This kind of open pan genome often exists in bacterial species residing in multiple ecological environments, and has multiple ways of exchanging genetic material through horizontal gene transfer (HGT) [45]. The core genome was fitted into an exponential regression [Fc(x) = 4965.15 + $e^{-0.01x}$]. Based on the two constructed functional curves, the gene number of the pan genome increased when the species number of the *M. alba* strain accumulated. Meanwhile, the tendency of the core genome, on the contrary, implied that more strains tended to result in additional numbers of unique genes.

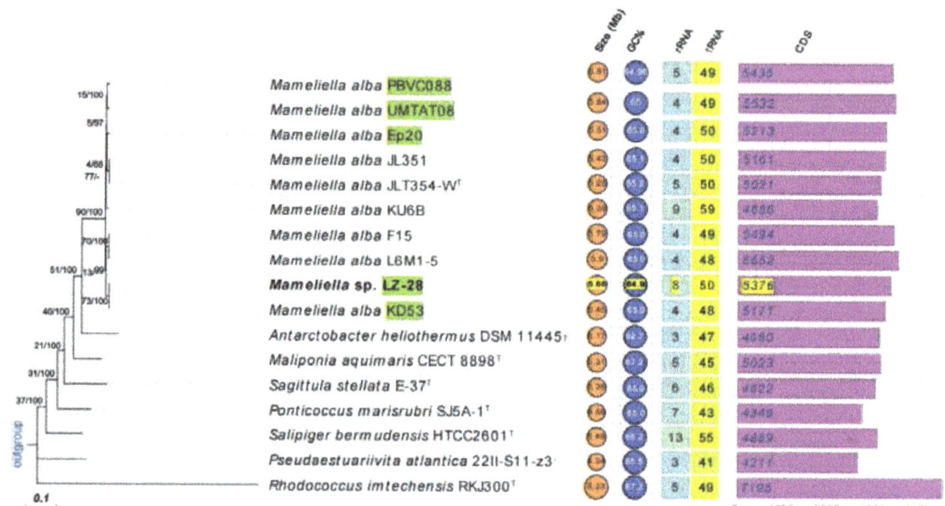

Figure 4. Phylogenomic tree constructed by up-to-date bacterial core gene (UBCG) set among the ten selected *M. alba* strains, as well as representative type species in the family *Rhodobacteraceae*, and a comparison of the five general genomic characteristics, including the genomic size, DNA G+C content (%), and numbers of predicted protein-coding genes (CDS, rRNA, and tRNA). The gene-support index (the number of individual gene trees presenting the same node for the total genes used) (**left**) and bootstrap values (**right**) were presented at the nodes of the phylogenomic tree. The five selected *M. alba* strains isolated from marine algae are marked with green, and the genome parameters for strain LZ-28 were marked with yellow. *Bar*: 0.1 nt substitutions per site.

The phylogenetic trees of the core and pan genomes of the ten selected *M. alba* were constructed and shown in Figure 6. It indicates the highly conservative evolution among the *M. alba* members investigated in this study. The obvious difference of the number of unique genes among the ten selected *M. alba* strains may be explained by their adaptions to the growth conditions and possible horizontal gene-transfer events [45]. Moreover, strain LZ-28 showed a close phylogenetic relationship with the marine algae-derived strains, either with strains KD53 and UMTAT08 in the core-genome phylogenetic tree (Figure 6A), or with strains KD53 and Ep20 in the pan-genome phylogenetic tree (Figure 6B). Thus, it indicates the genome-based phylogenetic analysis provides robust evidence for the evolutionary history of the individual *M. alba* strain [46].

2.7. Comparison of the Functional Classes of Predicted Genes among the Selected M. alba Strains

The functions of gene families in the genomes of the ten selected *M. alba* strains were evaluated by performing the COG and KEGG categories analyses. The COG distribution profile showed that most core (Figure S2A) and unique (Figure S2B) genes were both mainly related to five functional groups, including carbohydrate transport and metabolism [G], transcription [K], replication, recombination and repair [L], general function prediction only [R], and unknown function [S]. Three strains, including strain KU6B isolated from surface seawater, strain KD53 from marine diatom *Phaeodactylum tricornutum* [12], and strain UMTAT08 from marine dinoflagellate *Alexandrium tamiyavanichii* [47], harbored the largest core gene number among the ten selected *M. alba* strains during COG analysis (Figure S2A). Moreover, most unique genes were found in strains KU6B and UMTAT08 (Figure S2B). With respect to the KEGG assignment, based on the characterization of the core and unique genomes, the genes related to general function, amino acid metabolism, and carbohydrate metabolism, accounted for the major types of KEGG categories. In addition, two strains, UMTAT08 and F15, harbored the largest unique gene number among the ten selected *M. alba* strains during COG analysis (Figure S2D). Additionally, no unique gene sorted in COG and KEGG categories was observed in strain JL351 isolated from the surface seawater [41].

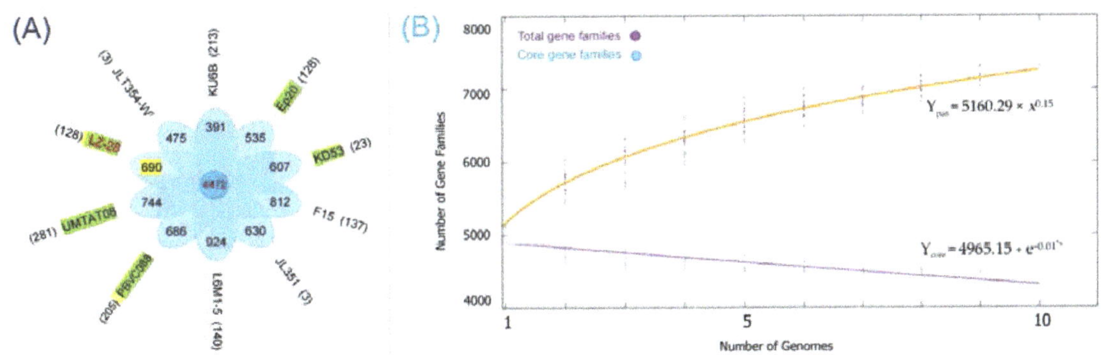

Figure 5. Core–pan-genome comparison of the ten selected *M. alba* strains based on the bacterial pan-genome analysis (BPGA) pipeline. (**A**) Petal diagram of the core, accessory, and unique genes of the pan genome. The number of core genes is displayed in the central circle. Overlapping regions represent the number of accessory genes. Strain names are marked outside each petal, along with the strain-specific gene number shown in the brackets. The five selected *M. alba* strains isolated from marine algae are marked with green. (**B**) Accumulation curves for the pan and core genomes of the ten *M. alba* strains. The purple and blue dots represent the number of pan- and core genes, respectively. The equations are illustrated in each curve.

For the core-gene groups of the COG category, the five marine algae-derived *M. alba* strains demonstrated a more scattered distribution pattern (Figure S2A). However, for the unique gene groups of COG, the principal components analysis (PCA) showed two obviously separated groups among the five marine algae-derived *M. alba* strains. One group consisted of strains Ep20 and KD53, and the second group included strains LZ-28 and PBVC088. For strain UMTAT08, isolated from *A. tamiyavanichii* [47], it was far away from the two clustered groups (Figure S2B). In addition, based on the two PCA analyses of the core- (Figure S2C) and unique-KEGG (Figure S2D) profiles, it was highly noteworthy that strain LZ-28 both showed a functional relationship similar to the other three marine algae-derived *M. alba* strains, except for strain UMTAT08. Thus, it indicated that the four selected marine algae-derived *M. alba* strains shared similar functional destinations in spite of their distinctive isolation sources.

2.8. Growth-Promoting Effects of Bacterial Strain LZ-28 and Algal Strain LZT09

Based on the microalgae growth-promoting (MGP) assay [21], strain LZ-28 demonstrated obvious growth-promoting activity when co-cultured with algal host LZT09 (Figure 7A). Interestingly, the mutual promoting effect of the algal culture extract from LZT09 on the bacterial growth of strain LZ-28 was also observed (Figure 7B). It indicated that the potential association between bacterial strain LZ-28 and algal strain LZT09 in spite of the detailed nature of action mechanism was still unclear. Furthermore, deeply unearthing the multi-omics information for both sides in co-culture circumstances is believed to offer substantial insights into the detailed mechanism governing these dynamic interactions [48–52].

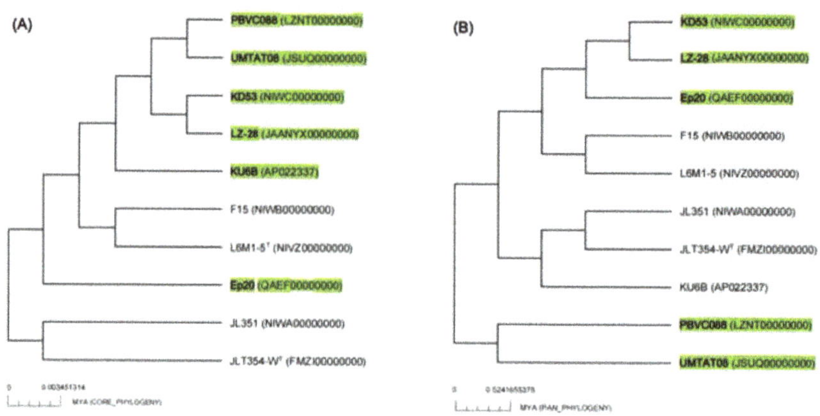

Figure 6. Phylogenetic trees constructed by core (**A**) and pan genomes (**B**) of the ten selected *M. alba* strains inferred from maximum likelihood (ML) analysis using concatenated alignment of 303 pan-orthologous genes. The estimated divergence time from present (million years ago; MYA) is given as the bar for both trees. The five bacterial strains isolated from marine algae are marked with green.

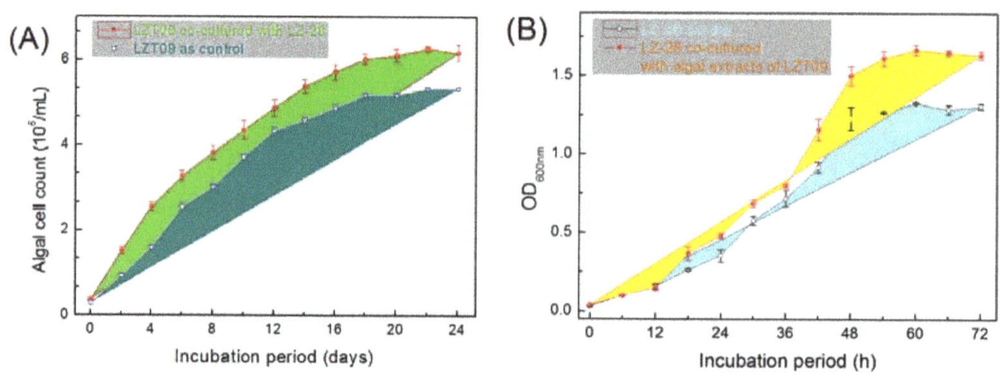

Figure 7. The microalgae growth-promoting potential of the bacterial strain LZ-28 on algal strain LZT09 (**A**), and the promoting effect of algal extract of LZT09 on the bacterial growth by the measurements of the optical changes recorded at $OD_{600\ nm}$ (**B**).

2.9. Optimization of Bacterial Growth and EPS Accumulation

Bacterial exopolysaccharides (EPSs) were revealed to serve as one essential chemical intermedia within the microscopic phycosphere niches and mediate host–microbe interactions [2–5,53]. During our previous investigations, several novel bacterial species, which produce bioactive EPSs, were also recovered from the freshwater [26] and marine phycosphere [27–31], as well as the gut microbiota of Antarctic emperor penguin [48,54,55]. In this study, the preliminary experiment showed that temperature and carbon sources were the main two factors influencing the bacterial growth and EPS accumulation. Therefore, ten carbon sources, including cellbiose, fructose, galactose, glucose, glycerol, lactose, maltose, mannose, sucrose, and trehalose, and the pH range of 5.0–9.0 were further used for the optimization of the culture conditions. Based on the obtained result, bacterial incubation at 37 °C promoted bacterial growth and shortened the incubation period of strain LZ-28, although the general trends of bacterial growth that were cultured at 28 °C and 37 °C demonstrated a similar pattern. For the pH tests, strain LZ-28 grew better in the pH values ranging from 6.0 to 9.0. Additionally, strain LZ-28 was found to achieve the fastest growth rate when the cellobiose was used as a carbon source, and cultured at 28°C or at 37 °C and at pH 7.0 (Figure 8). Based on the EPS accumulation measurements, when cultured at 28 °C, the higher EPS yield of strain LZ-28 was achieved both at pH 5.0 and pH 6.0. However, when the culture temperature was changed to 37 °C, the EPS yields were observed to gradually enhance with the increasing pH values, and reached the maximum at pH 9.0. Under the optimized conditions, the highest EPS yield of 47.7 μg/mL was obtained when sucrose (10 g/L) was used as a carbon source and cultured at 37 °C and at pH 9.0 (Figure 8).

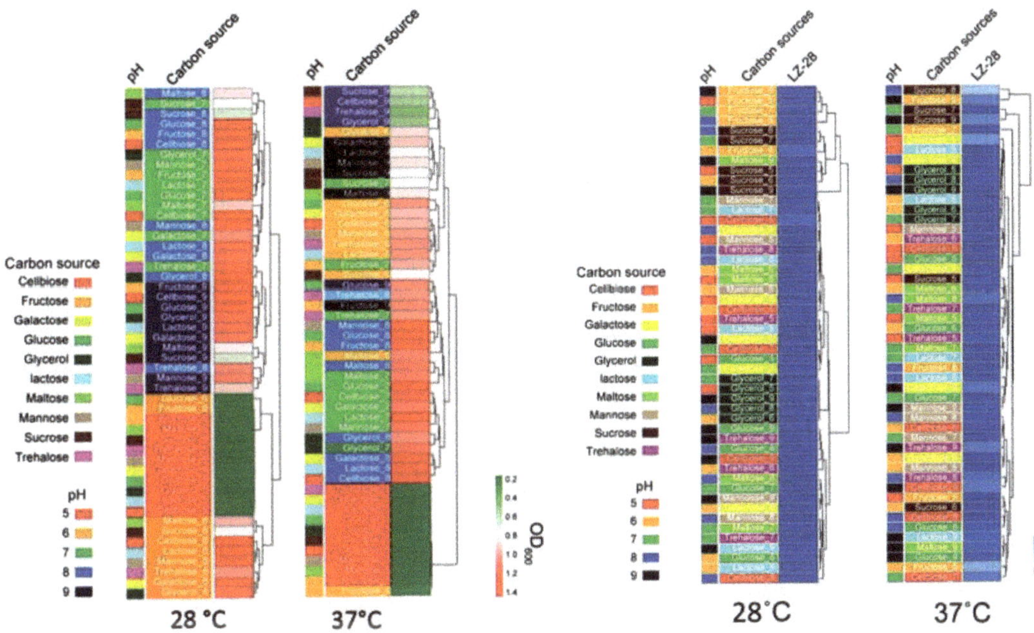

Figure 8. Comparison of the effects of the ten carbon sources and the pH range (5.0–9.0) of the media on bacterial growth (**left**; OD_{600}) and EPS accumulation (**right**; unit, μg/mL) of strain LZ-28 cultured at 28 °C and 37 °C, respectively.

2.10. Bioflocculanting-Activity Evaluation and Correlation Analysis

Based on the bacterial bioflocculanting assaying, EPS extracted from strain LZ-28 demonstrated obvious bioflocculanting activity, which showed a concentration-dependent manner (Figure 9A). The highest efficiency of the bioflocculanting rate of 95.7 ± 8.5% was achieved when 0.60 $g \cdot L^{-1}$ of EPS was applied. It exhibited a higher bioflocculanting capacity compared to the other bacterial strains, which were also recovered from the marine dinoflagellates previously reported [27–30,56–59]. To further infer whether the type and abundance of the monosaccharides of the polymer EPSs were related to the bioflocculanting activity, the characterizations of the monosaccharide compositions of the EPSs and their correlations with the bioflocculanting activity were then performed. Base on the obtained result, the relative portions of two monosaccharides, glucose and fucose, in the crude polysaccharides demonstrated significantly positive highly strong correlations (both $p < 0.05$) with the bioflocculanting capacities with the correlation coefficients of 0.9 ($p = 0.03734$) and 0.8 ($p = 0.03101$), respectively, whereas the other four monosaccharides were found to be negatively contributed without statistical significance (Figure 9B). Despite these findings, the detailed chemical structures of the EPSs produced by strain LZ-28 remain to be further elucidated. Moreover, more experimental data are still needed to dig out reliable clues to explore the structure–activity relationship of the bacterial EPSs as promising and natural microbial bioflocculants. In addition, the genomic mining also revealed the presence of several typical biosynthesis genes (wzx, exo, and muc) for bacterial EPS biosynthesis in strain LZ-28. Thus, the present study indicated that strain LZ-28 could serve as a new, fresh bacterial candidate with natural potential for the production of versatile EPS bioflocculants derived from marine microalgae with potential pharmaceutical, environmental, and biotechnological implications [27–30,55–61].

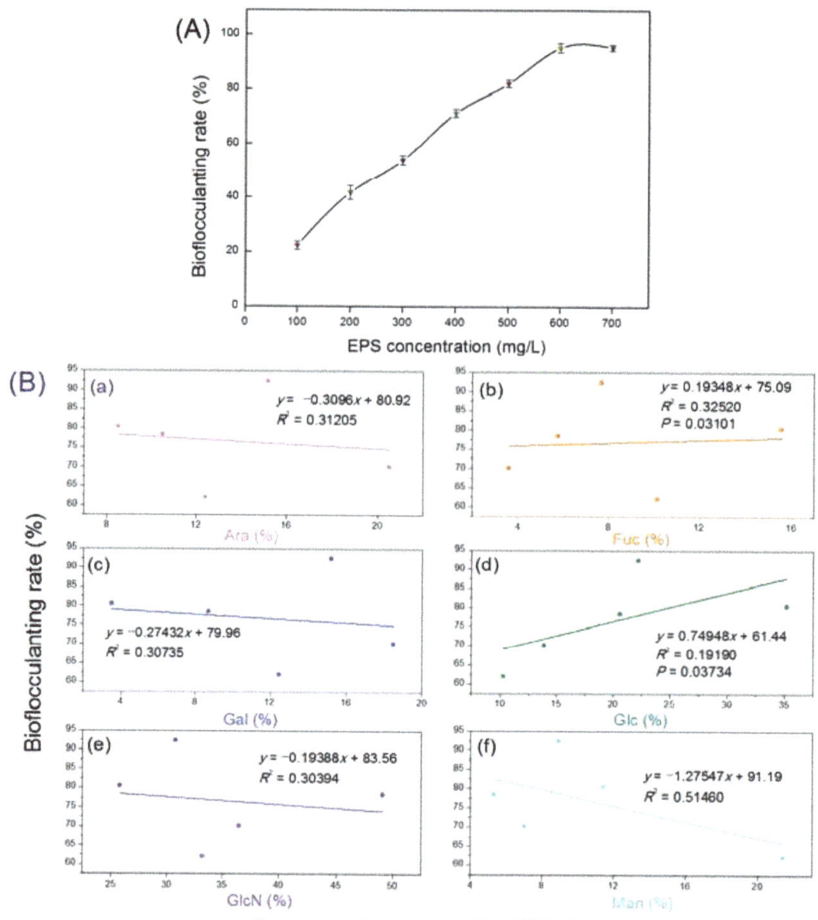

Figure 9. Bioflocculanting activities of the bacterial exopolysaccharides (EPSs) produced by bacterial strain LZ-28 (**A**), and correlation analysis (**B**) of the relative portions (%) of the six monosaccharides including arabinose (Ara, pane **a**), fucose (Fuc, pane **b**), galactose (Gal, pane **c**), glucose (Glc, pane **d**), amino-glucose (GlcN, pane **e**) and mannose (Man, pane **f**) of the bacterial EPS with the flocculanting efficiency expressed with the bioflocculanting rate (%).

3. Materials and Methods

3.1. Algal and Bacterial Strains and Culture

Alexandrium catenella LZT09 was sampled in the Zhoushan Archipelago area in the East China Sea during an algal bloom occurrence in July 2018, and then routinely cultured in our ABI laboratory [10]. LZT09 produced high levels of paralytic shellfish-poisoning (PSP) toxins [21]. LZT09 cells were cultured for 21 days in 50 mL of F/2-enriched seawater [21], and were exposed to a 12 h light–dark cycle at 25 °C (ca. 125 μmol photons m^{-2} s^{-1}). Algal cells were obtained by centrifugation (4000× *g* for 20 min) and the resultant cell pellet was stored at −20 °C until used. The axenic LZT09 was obtained by repeated washing, lysozyme/SDS and antibiotic treatment according to the procedure previously reported [62]. For comparative purposes, five strains, including *Mameliella phaeodactyli* KD53 (MCCC 1K00273), *Mameliella atlantica* L6M1-5 (MCCC 1A07531), and *Alkalimicrobium pacificum* F15 (MCCC 1A09948), purchased from the Marine Culture Collection of China (MCCC), and strains *Ponticoccus lacteus* JL351 (CGMCC 1.12986) and *Mameliella alba* JLT354-W^T (CGMCC 1.7290^T) purchased from the China General Microbiological Culture Collection Center (CGMCC) were used. All the tested strains were grown in the same conditions for the experiments.

3.2. PCR Amplification, 16S rRNA Sequencing, and Data Analysis

Total genomic DNA (gDNA) was extracted using a Fast Bacterial Genomic DNA Extraction Kit (Sangon Biotech Co., Ltd., Shanghai, China) according to the manufacturer's instructions. A general bacterial primer pair 338F/806R was used to amplify the V3–V4 variable region of the bacterial 16S rRNA gene [19,20]. The 16S rRNA sequencing libraries for Illumina library was constructed using the TruSeq® DNA PCR-Free Sample Preparation Kit (Illumina, San Diego, CA, USA), then sequenced using a NovaSeq 6000 PE250 platform utilized at BioMajor (Shanghai, China). To analyze the microbial community using the obtained Illumina sequencing data, the script from QIIME2 (https://qiime2.org, accessed on 10 May 2022) was used [63]. The paired-end reads were joined using the multiple_join_paired_ends.py script and quality filtered by the multiple_split_libraries_fastq.py script. The obtained OTUs were clustered using UCLUST with a default threshold of 0.97 [64]. The OTU phylotypes were assigned using the closed-reference OTU picking method against the QIIME modified version of the SILVA database, v128 [65]. The parameters used in the above analysis were all default settings.

3.3. Morpho-Physiological and Biochemical Characterizations

The preparation of the transmission electron microscopy (TEM) sample was performed using the negative-staining method [54,55]. Briefly, vegetative cells of strain LZ-28 were observed from a fresh, 24 h culture inoculated in MB. The cells were fixed in 2.5% glutaraldehyde in a phosphate buffer (0.1 M; pH 7.2) for 2 h at 4 °C, then washed with the same phosphate buffer post fixed with 1.0% OsO_4. The obtained samples were dehydrated in serial 30–100% ethanol, then mounted on copper grids, and observed using a JEM-1200 TEM instrument (EOL, Tokyo, Japan). Gram-staining test was followed by the our procedures previously described [56–58]. The hanging-drop technique was used for observing motility during the exponential phase of the culture growth. The growth conditions of temperature were tested using MA medium at the temperature ranges of 4, 10, 15, 20, 25, 28, 30, 33, 37, 40, 45, 50, and 55 °C. The optimal pH was investigated at different pH levels ranging from 4.0 to 11.0 with 0.5 unit intervals and cultured in MB at 28 °C for 2 days. Salt tolerance was determined by growth in an MB medium containing various NaCl concentrations (0–15.0%, w/v, with interval of 0.5%) while removed the original NaCl [47,48]. Catalase and oxidase activities were detected by adding 3% (*v/v*) H_2O_2 and using oxidase reagent (bioMérieux, Marcy-l'Étoile, France), respectively. Anaerobic growth was assessed after 1 week of cultivation in an anaerobic chamber (Bactron EZ-2; Shellab, Cornelius, OR, USA) on MA at 28 °C. Tests of phenotypic and enzymatic characterizations were conducted using API 20NE, API 20E, and API ZYM strips (bioMérieux) according to the manufacturer's instructions.

3.4. Analysis of Fatty Acid Profiles

For whole-cell fatty acid analysis, strain LZ-28 and the five reference strains were grown on MA at 28 °C for 2 days. Sufficient cells of comparable physiological ages were harvested from the third streak quadrant of the agar plates and cellular fatty acids were saponified, methylated, and extracted using the standard protocol of the Sherlock Microbial Identification (MIDI) System (Sherlock version 6.1, Newark, Delaware, USA). Fatty acid methyl esters were analyzed by an Agilent 6890N gas chromatography system (GC) and identified using the MIDI TSBA6 database [66].

3.5. Phylogenetic Analysis Based on the 16S rRNA Gene Sequences

The related 16S rRNA gene sequences of species in the genus *Mameliella* and type strains in the family *Rhodobacteraceae* were retrieved from the NCBI database, except strains PBVC088 and Ep20 [67], which were extracted from the genome sequences by PhyloSuite version 1.2.1 [68]. The identification of phylogenetic neighbors and the calculation of pairwise 16S rRNA gene sequence similarities were achieved using the EzTaxon-e server (http://eztaxon-e.ezbiocloud.net, accessed on 2 April 2022). The sequence alignments were performed with CLUSTAL_X (http://www.clustal.org, accessed on 2 April 2022) [69]. Phylogenetic trees were reconstructed with the maximum-likelihood (ML) algorithm [70] by the Tamura Nei model [71] in the software package MEGA version 7.0 (https://www.megasoftware.net, accessed on 2 April 2022) [72]. The resultant tree topologies were evaluated by bootstrap analyses [73] based on 1000 re-samplings.

3.6. Phylogenomic Calculations and UBCG Tree Construction

The draft genome sequence of LZ-28 was obtained by our previous investigation [10]. The reference genome sequences of the *Mameliella alba* strains, six *Rhodobacteraceae* strains, and one *Rhodococcus* strain were retrieved from the NCBI database. The average nucleotide identity (ANI) and the average amino acid identity (AAI) analyses were performed using the ANI/AAIMatrix genome-based distance matrix calculator (http://enve-omics.ce.gatech.edu/g-matrix, accessed on 19 March 19 2022) and the digital DNA–DNA hybridization (dDDH) analysis was conducted using the Genome-to- Genome Distance Calculator (GGDC 2.1) (http://ggdc.dsmz.de/, accessed on 19 March 2022) using Formula 2 [74]. The phylogenetic tree based on the 92 bacterial core genes was constructed by using the UBCG pipeline [75]. Further visualization and editing of the UBCG tree were accomplished by EvolView tool (https://www.evolgenius.info/evolview, accessed on 19 March 2022). Phylogenetic trees based on pan genes and 4472 bacterial core genes were conducted by the bacterial pan-genome analysis (BPGA) pipeline v.1.3 [76] using USEARCH with the default settings [77].

3.7. Pan-Genome Analysis

The BPGA pipeline was used to automate the complete pan-genome study and downstream analysis of the prokaryotic sequences [69]. All 11 protein sequences annotated by Prokka [78] were used for the BPGA (sequence identity $\geq$50%; and E value $\leq 1.0 \times 10^{-5}$) tool to perform the core–pan-genome analysis. The genomic accumulation curve was generated by the BPGA's result via gnuplot. The core, accessory, and unique genes were classified from the orthologous groups through the USEARCH clustering algorithm [77].

3.8. Comparative Analysis of Functional Genes

The online COG and KEGG databases accessed via BPGA were used to analyze the functions of the genes. The core, accessory, and unique genes were sorted, and the major concerns related to the core and the unique genes. The gene clusters were identified by using anti-SMASH 6.0.0 [79] from the reference genome sequences of *M. alba* KD53. All 11 protein sequences produced by Prokka were used for the eggnog mapper to obtain an overview of the functional genes [80]. All results were manually checked. The principal component analysis (PCA) was performed by using CANOCO 5 (http://www.canoco5.com/index.php/resources, accessed on 19 March 2022) to compare the correlations across all samples based on the relative abundance of the function genes. The Spearman's rank correlations were determined by using the SigmaPlot program version 11.0 (Systat Software, Inc., San Jose, CA, USA).

3.9. Bacterial Growth and EPS-Accumulation Analysis

For the culture optimization experiment, 1 mL of 24 h fresh bacterial culture of strain LZ-28 was obtained and mixed with 24 mL of fresh 2216 medium, then 150 μL of the mixture was added into the wells of the 96-well microplate containing different carbon sources and cultured with shaking at 60 rpm/min for 72 h. During the culture period, the bacterial growth rate was monitored by the measurement of the optical density change recorded at OD_{600nm} every 3 hours. A total of 10 carbon sources, including cellobiose, fructose, galactose, glucose, glycerol, lactose, maltose, mannose, sucrose, and trehalose, added with the final concentration of 10 g/L, and the pH range of 5.0–9.0 with 1.0 as the interval were used for the measurements cultured either at 28 °C or 37 °C. The extraction and the quantification of EPSs were performed using the phenol sulfuric acid method, as previously reported [81].

3.10. Characterization of the Monosaccharides of EPSs

The determinations of the monosaccharide of crude polysaccharides were analyzed by the HPLC method as previously described [82]. Analytical standards (HPLC purity ≥ 9%) of 6 monosaccharides, including arabinose (Ara), fucose (Fuc), galactose (Gal), glucose (Glc), amino-glucose (GlcN), and mannose (Man) were purchased from Merck (Shanghai, China). Qualitative analysis was performed by comparing the retention times and the peak areas with those of the standards.

3.11. Evaluation of Bioflocculanting and MGP Bioactivities

Bioflocculating activity evaluations of the EPSs were performed according to the procedures previously reported [27–30]. The prepared EPSs were dissolved in the distilled water for furthering the bioactivity assay. The measurements using a kaolin clay suspension flocculation assay calculated as the bioflocculation rate (%) were used and performed using 96-well microplates with at least 6 replicates [27–30]. Microalgae growth-promoting (MGP) activity of bacterial strain LZ-28 toward algal strain LZT09 in a co-culture system was evaluated, as previously reported [83]. The preparation of the algal culture extract [29,30] and the effects on bacterial growth were analyzed by measuring the optical changes of bacterial density recorded at OD_{600nm} and performed in a SpectraMax M2 model 96-well microplate reader (Molecular Devices, LLC, San Jose, CA, USA). All the results are expressed as the means ± SD. The correlation coefficients of the monosaccharide portions with the bioflocculanting activities and the statistics significance were both performed using the Spearman's rank order correlation analysis and plotted with the OriginPro (version 8.0) (OriginLab Corp., Northampton, MA, USA). A p-value of less than 0.05 was considered to be statistically significant for all analyses.

Supplementary Materials: The following are available online at https://www.mdpi.com/article/10.3390/md20050321/s1, Figure S1: Comparison of the average nucleotide identity (ANI), average amino acid identity (AAI), and digital DNA–DNA hybridization (dDDH) values based on the genomes of the ten M. alba strains. Figure S2: Comparison of the COG and KEGG categories among the ten selected M. alba strains. The functional genes of the core and unique groups distributed by COG database are shown in panes A and B, respectively. The panes C and D represent the distribution of the core and unique genes annotated by the KEGG database, respectively. The sizes of the circles in each pane are proportional to the numbers of the different COG or KEGG categories. The strains isolated from marine algae are marked with green.

Author Contributions: Conceptualization, X.-L.Z. and Q.Y.; methodology, C.-Z.R., H.-M.G., J.D., W.-Z.Z. and X.-L.Z.; software, J.D. and X.-L.Z.; validation, X.-L.Z. and Q.Y.; formal analysis, C.-Z.R., J.D., F.-F.X., Y.Y. and X.-L.Z.; investigation, X.-L.Z.; resources, Q.Y.; data curation, C.-Z.R., W.-Z.Z. and X.-L.Z.; writing—original draft preparation, C.-Z.R., X.-L.Z. and Q.Y.; writing—review and editing, X.-L.Z. and Q.Y.; visualization, H.-M.G. and X.-L.Z.; supervision, X.-L.Z. and Q.Y.; project administration, Q.Y.; funding acquisition, X.-L.Z. and Q.Y. All authors have read and agreed to the published version of the manuscript.

Funding: This research was supported by the National Natural Science Foundation of China (41876114), Specific Project for Zhejiang Ocean University from Municipal Science and Technology Bureau of Zhoushan (2022C41018), Zhejiang Province Natural Science Foundation Committee (LGF18D060001), and the National Training Program of Innovation and Entrepreneurship for Undergraduates (202110340007, 202110340015).

Data Availability Statement: All relevant data are within the manuscript.

Conflicts of Interest: The authors declare no conflict of interest.

References

1. Wayne, B.; Ralph, M. Chemotactic and growth responses of marine bacteria to algal extracellular products. *Biol. Bull.* **1972**, *143*, 265–277.
2. Samo, T.J.; Kimbrel, J.A.; Nilson, D.J.; Pett-Ridge, J.; Weber, P.K.; Mayali, X. Attachment between heterotrophic bacteria and microalgae influences symbiotic microscale interactions. *Environ. Microbiol.* **2018**, *20*, 4385–4400. [CrossRef] [PubMed]
3. Seymour, J.R.; Amin, S.A.; Raina, J.B.; Stocker, R. Zooming in on the phycosphere: The ecological interface for phytoplankton-bacteria relationships. *Nat. Microbiol.* **2017**, *2*, 17065. [CrossRef] [PubMed]
4. Ramanan, R.; Kang, Z.; Kim, B.H.; Cho, D.H.; Jin, L.; Oh, H.M.; Kim, H.S. Phycosphere bacterial diversity in green algae reveals an apparent similarity across habitats. *Algal Res.* **2015**, *8*, 140–144. [CrossRef]
5. Amin, S.A.; Parker, M.S.; Armbrust, E.V. Interactions between diatoms and bacteria. *Microbiol. Mol. Biol. Rev.* **2012**, *76*, 667–684. [CrossRef] [PubMed]
6. Kouzuma, A.; Watanabe, K. Exploring the potential of algae/bacteria interactions. *Curr. Opin. Biotechnol.* **2015**, *33*, 125–129. [CrossRef]
7. Giovannoni, S.J.; Nemergut, D. Microbes ride the current. *Science* **2014**, *345*, 1246. [CrossRef] [PubMed]
8. Buchan, A.; González, J.M.; Moran, M.A. Overview of the marine *roseobacter* lineage. *Appl. Environ. Microbiol.* **2005**, *71*, 5665–5677. [CrossRef]
9. Landry, Z.C.; Vergin, K.; Mannenbach, C.; Block, S.; Yang, Q.; Blainey, P.; Carlson, C.; Giovannoni, S. Optofluidic Single-Cell Genome Amplification of Sub-micron Bacteria in the Ocean Subsurface. *Front. Microbiol.* **2018**, *9*, 1152. [CrossRef]
10. Zhang, X.L.; Yang, X.; Wang, S.J.; Jiang, Z.W.; Xie, Z.X.; Zhang, L. Draft Genome Sequences of Nine Cultivable Heterotrophic Proteobacteria Isolated from Phycosphere Microbiota of Toxic *Alexandrium catenella* LZT09. *Microbiol. Resour. Announc.* **2020**, *9*, e00281-20. [CrossRef]
11. Zheng, Q.; Chen, C.; Yan, X.J.; Wang, Y.N.; Zeng, Y.H.; Hao, L.K.; He, W.H.; Jiao, N.Z. *Mameliella alba* gen. nov., sp. nov., a marine bacterium of the *Roseobacter* clade in the order *Rhodobacterales*. *Int. J. Syst. Evol. Microbiol.* **2010**, *60*, 953–957. [CrossRef]
12. Chen, Z.; Zhang, J.; Lei, X.; Lai, Q.; Yang, L.; Zhang, H.; Li, Y.; Zheng, W.; Tian, Y.; Yu, Z.; et al. *Mameliella phaeodactyli* sp. nov., a member of the family *Rhodobacteraceae* isolated from the marine algae *Phaeodactylum tricornutum*. *Int. J. Syst. Evol. Microbiol.* **2015**, *65*, 1617–1621. [CrossRef]
13. Liu, Y.; Zhang, X.; Lai, Q.; Shao, Z. Reclassification of *Mameliella phaeodactyli, Mameliella atlantica, Ponticoccus lacteus* and *Alkalimicrobium pacificum* as later heterotypic synonyms of *Mameliella alba* and an emended description of *Mameliella alba*. *Int. J. Syst. Evol. Microbiol.* **2018**, *68*, 1047–1051. [CrossRef]
14. Rasmussen, S.A.; Andersen, A.J.; Andersen, N.G.; Nielsen, K.F.; Hansen, P.J.; Larsen, T.O. Chemical Diversity, Origin, and Analysis of Phycotoxins. *J. Nat. Prod.* **2016**, *79*, 662–673. [CrossRef]
15. Brosnahan, M.L.; Fischer, A.D.; Lopez, C.B.; Moore, S.K.; Anderson, D.M. Cyst-forming dinoflagellates in a warming climate. *Harmful Algae* **2020**, *91*, 101728. [CrossRef]
16. McKenzie, C.H.; Bates, S.S.; Martin, J.L.; Haigh, N.; Howland, K.L.; Lewis, N.I.; Locke, A.; Peña, A.; Poulin, M.; Rochon, A.; et al. Three decades of Canadian marine harmful algal events: Phytoplankton and phycotoxins of concern to human and ecosystem health. *Harmful Algae* **2021**, *102*, 101852. [CrossRef]
17. Lewis, A.M.; Coates, L.N.; Turner, A.D.; Percy, L.; Lewis, J. A review of the global distribution of *Alexandrium minutum* (Dinophyceae) and comments on ecology and associated paralytic shellfish toxin profiles, with a focus on Northern Europe. *J. Phycol.* **2018**, *54*, 581–598. [CrossRef]
18. Thottumkara, A.P.; Parsons, W.H.; Du, B.J. Saxitoxin. *Angew. Chem. Int. Ed. Engl.* **2014**, *53*, 5760–5784. [CrossRef]
19. Zhang, X.L.; Tian, X.Q.; Ma, L.Y.; Feng, B.; Liu, Q.H.; Yuan, L.D. Biodiversity of the symbiotic bacteria associated with toxic marine dinoflagellate *Alexandrium tamarense*. *J. Biosci. Med.* **2015**, *3*, 23–28. [CrossRef]
20. Zhang, X.L.; Ma, L.Y.; Tian, X.Q.; Huang, H.L.; Yang, Q. Biodiversity study of intracellular bacteria closely associated with paralytic shellfish poisoning dinoflagellates *Alexandrium tamarense* and *A. minutum*. *Int. J. Environ. Resour.* **2015**, *4*, 23–27. [CrossRef]
21. Yang, Q.; Feng, Q.; Zhang, B.P.; Gao, J.J.; Sheng, Z.; Xue, Q.P.; Zhang, X.L. *Marinobacter alexandrii* sp. nov., a novel yellow-pigmented and algae growth-promoting bacterium isolated from marine phycosphere microbiota. *Antonie Leeuwenhoek* **2021**, *114*, 709–718. [CrossRef] [PubMed]

22. Jiang, Z.; Duan, Y.; Yang, X.; Yao, B.; Zeng, T.; Wang, X.; Feng, Q.; Qi, M.; Yang, Q.; Zhang, X.L. *Nitratireductor alexandrii* sp. nov., from phycosphere microbiota of toxic marine dinoflagellate *Alexandrium tamarense*. *Int. J. Syst. Evol. Microbiol.* **2020**, *70*, 4390–4397. [CrossRef]
23. Yang, Q.; Jiang, Z.; Zhou, X.; Zhang, R.; Xie, Z.; Zhang, S.; Wu, Y.; Ge, Y.; Zhang, X. *Haliea alexandrii* sp. nov., isolated from phycosphere microbiota of the toxin-producing dinoflagellate *Alexandrium catenella*. *Int. J. Syst. Evol. Microbiol.* **2020**, *70*, 1133–1138. [CrossRef] [PubMed]
24. Yang, X.; Jiang, Z.W.; Zhang, J.; Zhou, X.; Zhang, X.L.; Wang, L.; Yu, T.; Wang, Z.; Bei, J.; Dong, B. *Mesorhizobium alexandrii* sp. nov., isolated from phycosphere microbiota of PSTs-producing marine dinoflagellate *Alexandrium minutum* amtk4. *Antonie Leeuwenhoek* **2020**, *113*, 907–917. [CrossRef] [PubMed]
25. Yang, Q.; Jiang, Z.W.; Huang, C.H.; Zhang, R.N.; Li, L.Z.; Yang, G.; Feng, L.J.; Yang, G.F.; Zhang, H.; Zhang, X.L. *Hoeflea prorocentri* sp. nov., isolated from a culture of the marine dinoflagellate *Prorocentrum mexicanum* PM01. *Antonie Leeuwenhoek* **2018**, *111*, 1845–1853. [CrossRef] [PubMed]
26. Zhang, X.L.; Li, G.X.; Ge, Y.M.; Iqbal, N.M.; Yang, X.; Cui, Z.D.; Yang, Q. *Sphingopyxis microcysteis* sp. nov., a novel bioactive exopolysaccharides-bearing *Sphingomonadaceae* isolated from the *Microcystis* phycosphere. *Antonie Leeuwenhoek* **2021**, *114*, 845–857. [CrossRef] [PubMed]
27. Zhu, W.Z.; Ge, Y.M.; Dai, J.; Zhang, X.L.; Yang, Q. *Alexandriicola marinus* gen. nov., sp. nov., a new member of the family *Rhodobacteraceae* isolated from marine phycosphere. *Antonie Leeuwenhoek* **2022**, *115*, 473–486. [CrossRef]
28. Duan, Y.; Jiang, Z.; Wu, Z.; Sheng, Z.; Yang, X.; Sun, J.; Zhang, X.; Yang, Q.; Yu, X.; Yan, J. *Limnobacter alexandrii* sp. nov., a thiosulfate-oxidizing, heterotrophic and EPS-bearing *Burkholderiaceae* isolated from cultivable phycosphere microbiota of toxic *Alexandrium catenella* LZT09. *Antonie Leeuwenhoek* **2020**, *13*, 1689–1698. [CrossRef]
29. Yang, Q.; Ge, Y.M.; Iqbal, N.M.; Yang, X.; Zhang, X.L. *Sulfitobacter alexandrii* sp. nov., a new microalgae growth-promoting bacterium with exopolysaccharides bioflocculanting potential isolated from marine phycosphere. *Antonie Leeuwenhoek* **2021**, *114*, 1091–1106. [CrossRef]
30. Zhang, X.L.; Qi, M.; Li, Q.H.; Cui, Z.D.; Yang, Q. *Maricaulis alexandrii* sp. nov., a novel active bioflocculants-bearing and dimorphic prosthecate bacterium isolated from marine phycosphere. *Antonie Leeuwenhoek* **2021**, *114*, 1195–1203. [CrossRef]
31. Yang, Q.; Jiang, Z.; Zhou, X.; Zhang, R.; Wu, Y.; Lou, L.; Ma, Z.; Wang, D.; Ge, Y.; Zhang, X.; et al. *Nioella ostreopsis* sp. nov., isolated from toxic dinoflagellate, *Ostreopsis lenticularis*. *Int. J. Syst. Evol. Microbiol.* **2020**, *70*, 759–765. [CrossRef]
32. Simon, M.; Scheuner, C.; Meier-Kolthoff, J.P.; Brinkhoff, T.; Wagner-Döbler, I.; Ulbrich, M.; Klenk, H.P.; Schomburg, D.; Petersen, J.; Göker, M. Phylogenomics of *Rhodobacteraceae* reveals evolutionary adaptation to marine and non-marine habitats. *ISME J.* **2017**, *11*, 1483–1499. [CrossRef]
33. Attar, N. Marine microbiology. An interkingdom partnership. *Nat. Rev. Microbiol.* **2015**, *13*, 400. [CrossRef]
34. Zhang, H.; Wang, K.; Shen, L.; Chen, H.; Hou, F.; Zhou, X.; Zhang, D.; Zhu, X. Microbial Community Dynamics and Assembly Follow Trajectories of an Early-Spring Diatom Bloom in a Semienclosed Bay. *Appl. Environ. Microbiol.* **2018**, *84*, e01000-18. [CrossRef]
35. Shibl, A.A.; Isaac, A.; Ochsenkühn, M.A.; Cárdenas, A.; Fei, C.; Behringer, G.; Arnoux, M.; Drou, N.; Santos, M.P.; Gunsalus, K.C.; et al. Diatom modulation of select bacteria through use of two unique secondary metabolites. *Proc. Natl. Acad. Sci. USA* **2020**, *117*, 27445–27455. [CrossRef]
36. Bowman, J.P. Out from the Shadows—Resolution of the Taxonomy of the Family *Cryomorphaceae*. *Front. Microbiol.* **2020**, *11*, 795. [CrossRef]
37. Nakayama, T.; Nomura, M.; Takano, Y.; Tanifuji, G.; Shiba, K.; Inaba, K.; Inagaki, Y.; Kawata, M. Single-cell genomics unveiled a cryptic cyanobacterial lineage with a worldwide distribution hidden by a dinoflagellate host. *Proc. Natl. Acad. Sci. USA* **2019**, *116*, 15973–15978. [CrossRef]
38. Giovannoni, S.J.; Stingl, U. The importance of culturing bacterioplankton in the 'omics' age. *Nat. Rev. Microb.* **2007**, *5*, 820–826. [CrossRef]
39. Lewis, W.H.; Tahon, G.; Geesink, P.; Sousa, D.Z.; Ettema, T.J.G. Innovations to culturing the uncultured microbial majority. *Nat. Rev. Microbiol.* **2021**, *19*, 225–240. [CrossRef]
40. Xu, H.; Jiang, L.; Li, S.; Zeng, X.; Shao, Z. *Mameliella atlantica* sp. nov., a marine bacterium of the Roseobacter clade isolated from deep-sea sediment of the South Atlantic Ocean. *Int. J. Syst. Evol. Microbiol.* **2015**, *65*, 2255–2259. [CrossRef]
41. Yang, Y.; Sun, J.; Tang, K.; Lin, D.; Li, C.; Lin, Y. *Ponticoccus lacteus* sp. nov. of the family *Rhodobacteraceae*, isolated from surface seawater. *Int. J. Syst. Evol. Microbiol.* **2015**, *65*, 1247–1250. [CrossRef]
42. Zhang, G.; Yang, Y.; Wang, S.; Sun, Z.; Jiao, K. *Alkalimicrobium pacificum* gen. nov., sp. nov., a marine bacterium in the family *Rhodobacteraceae*. *Int. J. Syst. Evol. Microbiol.* **2015**, *65*, 2453–2458. [CrossRef]
43. Richter, M.; Rosselló-Móra, R. Shifting the genomic gold standard for the prokaryotic species defnition. *Proc. Natl. Acad. Sci. USA* **2009**, *106*, 19126–19131. [CrossRef]
44. Yamamoto, T.; Liu, Y.; Hasegawa, Y.; Iwaki, H. Complete Genome Sequence of *Mameliella alba* Strain KU6B, a Cyclohexylamine-Utilizing Marine Bacterium. *Microbiol. Resour. Announc.* **2020**, *9*, e00273-20. [CrossRef]
45. Polz, M.F.; Alm, E.J.; Hanage, W.P. Horizontal gene transfer and the evolution of bacterial and archaeal population structure. *Trends Genet.* **2013**, *29*, 170–175. [CrossRef]

46. Kumar, S.; Filipski, A.J.; Battistuzzi, F.U.; Kosakovsky Pond, S.L.; Tamura, K. Statistics and truth in phylogenomics. *Mol. Biol. Evol.* **2012**, *29*, 457–472. [CrossRef]
47. Danish-Daniel, M.; Han, M.G.; Noor, M.E.; Yeong, Y.S.; Usup, G. Draft genome sequence of *Mameliella alba* strain UMTAT08 isolated from clonal culture of toxic dinoflagellate *Alexandrium tamiyavanichii*. *Genom. Data* **2016**, *10*, 12–14. [CrossRef]
48. Geng, H.; Belas, R. Molecular mechanisms underlying roseobacter-phytoplankton symbioses. *Curr. Opin. Biotechnol.* **2010**, *21*, 332–338. [CrossRef]
49. Gao, H.M.; Xie, P.F.; Zhang, X.L.; Yang, Q. Isolation, Phylogenetic and Gephyromycin Metabolites Characterization of New Exopolysaccharides-Bearing Antarctic Actinobacterium from Feces of Emperor Penguin. *Mar. Drugs* **2021**, *19*, 458. [CrossRef]
50. Gutleben, J.; Chaib De Mares, M.; van Elsas, J.D.; Smidt, H.; Overmann, J.; Sipkema, D. The multi-omics promise in context: From sequence to microbial isolate. *Crit. Rev. Microbiol.* **2018**, *44*, 212–229. [CrossRef]
51. Stingl, U.; Tripp, H.J.; Giovannoni, S.J. Improvements of high-throughput culturing yielded novel SAR11 strains and other abundant marine bacteria from the Oregon coast and the Bermuda Atlantic Time Series study site. *ISME J.* **2007**, *1*, 361–371. [CrossRef] [PubMed]
52. Stingl, U.; Cho, J.C.; Foo, W.; Vergin, K.L.; Lanoil, B.; Giovannoni, S.J. Dilution-to-extinction culturing of psychrotolerant planktonic bacteria from permanently ice-covered lakes in the McMurdo Dry Valleys, Antarctica. *Microb. Ecol.* **2007**, *55*, 395–405. [CrossRef] [PubMed]
53. Amin, S.A.; Hmelo, L.R.; van Tol, H.M.; Durham, B.P.; Carlson, L.T.; Heal, K.R.; Morales, R.L.; Berthiaume, C.T.; Parker, M.S.; Djunaedi, B.; et al. Interaction and signalling between a cosmopolitan phytoplankton and associated bacteria. *Nature* **2015**, *522*, 98–101. [CrossRef] [PubMed]
54. Zhu, W.Z.; Ge, Y.M.; Gao, H.M.; Dai, J.; Zhang, X.L.; Yang, Q. *Gephyromycinifex aptenodytis* gen. nov., sp. nov., isolated from gut of Antarctic emperor penguin. *Antonie Leeuwenhoek* **2021**, *114*, 2003–2017. [CrossRef]
55. Zhu, W.Z.; Wang, S.H.; Gao, H.M.; Ge, Y.M.; Dai, J.; Zhang, X.L.; Yang, Q. Characterization of Bioactivities and Biosynthesis of Angucycline/angucyclinone Derivatives Derived from *Gephyromycinifex aptenodytis* gen. nov., sp. nov. *Mar. Drugs* **2022**, *20*, 34. [CrossRef]
56. Yang, Q.; Jiang, Z.; Zhou, X.; Xie, Z.; Wang, Y.; Wang, D.; Feng, L.; Yang, G.; Ge, Y.; Zhang, X. *Saccharospirillum alexandrii* sp. nov., isolated from the toxigenic marine dinoflagellate *Alexandrium catenella* LZT09. *Int. J. Syst. Evol. Microbiol.* **2020**, *70*, 820–826. [CrossRef]
57. Yang, X.; Xiang, R.; Iqbal, N.M.; Duan, Y.H.; Zhang, X.A.; Wang, L.; Yu, L.Z.; Li, J.Z.; Sun, M.F.; Yang, Q. *Marinobacter shengliensis* subsp. *alexandrii* Subsp. Nov.; Isolated from Cultivable Phycosphere Microbiota of Highly Toxic Dinoflagellate *Alexandrium catenella* LZT09 and Description of *Marinobacter shengliensis* Subsp. *shengliensis* Subsp. Nov. *Curr. Microbiol.* **2021**, *78*, 1648–1655. [CrossRef]
58. Wang, X.; Ye, Y.; Xu, F.F.; Duan, Y.H.; Xie, P.F.; Yang, Q.; Zhang, X. *Maritimibacter alexandrii* sp. nov.; a New Member of *Rhodobacteraceae* Isolated from Marine Phycosphere. *Curr. Microbiol.* **2021**, *78*, 3996–4003. [CrossRef]
59. Zhou, X.; Zhang, X.; Jiang, Z.; Yang, X.; Zhang, X.; Yang, Q. Combined characterization of a new member of *Marivita cryptomonadis*, strain LZ-15-2 isolated from cultivable phycosphere microbiota of toxic HAB dinoflagellate *Alexandrium catenella* LZT09. *Braz. J. Microbiol.* **2021**, *52*, 739–748. [CrossRef]
60. Siddharth, T.; Sridhar, P.; Vinila, V.; Tyagi, R.D. Environmental applications of microbial extracellular polymeric substance (EPS): A review. *J. Environ. Manag.* **2021**, *287*, 112307. [CrossRef]
61. Lai, H.; Fang, H.; Huang, L.; He, G.; Reible, D. A review on sediment bioflocculation: Dynamics, influencing factors and modeling. *Sci. Total Environ.* **2018**, *642*, 1184–1200. [CrossRef]
62. Su, J.; Yang, X.; Zheng, T.; Hong, H. An efficient method to obtain axenic cultures of *Alexandrium tamarense*-a PSP-producing dinoflagellate. *J. Microbiol. Methods* **2007**, *69*, 425–430. [CrossRef]
63. Hall, M.; Beiko, R.G. 16S rRNA Gene Analysis with QIIME2. *Methods Mol. Biol.* **2018**, *1849*, 113–129.
64. Prasad, D.V.; Madhusudanan, S.; Jaganathan, S. uCLUST-A new algorithm for clustering unstructured data. *ARPN J. Eng. Appl. Sci.* **2015**, *10*, 2108–2117.
65. Quast, C.; Pruesse, E.; Yilmaz, P.; Gerken, J.; Schweer, T.; Yarza, P.; Peplies, J.; Glöckner, F.O. The SILVA ribosomal RNA gene database project: Improved data processing and web-based tools. *Nucleic Acids Res.* **2013**, *41*, D590–D596. [CrossRef]
66. Sasser, M. *Identification of Bacteria by Gas Chromatography of Cellular Fatty Acids*; MIDI Technical Note 101; MIDI Inc.: Newark, DE, USA, 1990.
67. Varasteh, T.; Moreira, A.P.B.; Silva Lima, A.W.; Leomil, L.; Otsuki, K.; Tschoeke, D.; Garcia, G.; Thompson, C.; Thompson, F. Genomic repertoire of *Mameliella alba* Ep20 associated with *Symbiodinium* from the endemic coral *Mussismilia braziliensis*. *Symbiosis* **2020**, *80*, 53–60. [CrossRef]
68. Zhang, D.; Gao, F.; Jakovlić, I.; Zou, H.; Zhang, J.; Li, W.X.; Wang, G.T. PhyloSuite: An integrated and scalable desktop platform for streamlined molecular sequence data management and evolutionary phylogenetics studies. *Mol. Ecol. Resour.* **2020**, *20*, 348–355. [CrossRef]
69. Jeanmougin, F.; Thompson, J.D.; Gouy, M.; Higgins, D.G.; Gibson, T.J. Multiple sequence alignment with Clustal X. *Trends Biochem. Sci.* **1998**, *23*, 403–405. [CrossRef]
70. Thorne, J.L.; Kishino, H.; Felsenstein, J. An evolutionary model for maximum likelihood alignment of DNA sequences. *J. Mol. Evol.* **1991**, *33*, 114–124. [CrossRef]

71. Tamura, K.; Nei, M. Estimation of the number of nucleotide substitutions in the control region of mitochondrial DNA in humans and chimpanzees. *Mol. Biol. Evol.* **1993**, *10*, 512–526.
72. Kumar, S.; Stecher, G.; Tamura, K. MEGA7: Molecular evolutionary genetics analysis version 7.0 for bigger datasets. *Mol. Biol. Evol.* **2016**, *33*, 1870–1874. [CrossRef]
73. Felsenstein, J. Confidence limits on phylogenies: An approach using the bootstrap. *Evolution* **1985**, *39*, 783–791. [CrossRef]
74. Meier-Kolthoff, J.P.; Sardà Carbasse, J.; Peinado-Olarte, R.L.; Göker, M. TYGS and LPSN: A database tandem for fast and reliable genome-based classification and nomenclature of prokaryotes. *Nucleic Acid Res.* **2022**, *50*, D801–D807. [CrossRef]
75. Na, S.I.; Kim, Y.O.; Yoon, S.H.; Ha, S.M.; Baek, I.; Chun, J. UBCG: Up-to-date bacterial core gene set and pipeline for phylogenomic tree reconstruction. *J. Microbiol.* **2018**, *56*, 280–285. [CrossRef]
76. Chaudhari, N.M.; Gupta, V.K.; Dutta, C. BPGA-an ultra-fast pan-genome analysis pipeline. *Sci. Rep.* **2016**, *6*, 24373. [CrossRef]
77. Rognes, T.; Flouri, T.; Nichols, B.; Quince, C.; Mahé, F. VSEARCH: A versatile open source tool for metagenomics. *PeerJ* **2016**, *4*, e2584. [CrossRef]
78. Seemann, T. Prokka: Rapid prokaryotic genome annotation. *Bioinformatics* **2014**, *30*, 2068–2069. [CrossRef]
79. Blin, K.; Pascal Andreu, V.; de Los Santos, E.L.C.; Del Carratore, F.; Lee, S.Y.; Medema, M.H.; Weber, T. The antiSMASH database version 2: A comprehensive resource on secondary metabolite biosynthetic gene clusters. *Nucleic Acids Res.* **2019**, *47*, D625–D630. [CrossRef]
80. Huerta-Cepas, J.; Forslund, K.; Coelho, L.P.; Szklarczyk, D.; Jensen, L.J.; von Mering, C.; Bork, P. Fast Genome-Wide Functional Annottion through Orthology Assignment by eggNOG-Mapper. *Mol. Biol. Evol.* **2017**, *34*, 2115–2122. [CrossRef]
81. Masuko, T.; Minami, A.; Iwasaki, N.; Majima, T.; Nishimura, S.; Lee, Y.C. Carbohydrate analysis by a phenol-sulfuric acid method in microplate format. *Anal. Biochem.* **2005**, *339*, 69–72. [CrossRef]
82. Dai, J.; Wu, Y.; Chen, S.W.; Zhu, S.; Yin, H.P.; Wang, M.; Tang, J. Sugar compositional determination of polysaccharides from *Dunaliella salina* by modified RP-HPLC method of precolumn derivatization with 1-phenyl-3-methyl-5-pyrazolone. *Carbohydr. Polym.* **2010**, *82*, 629–635. [CrossRef]
83. Gonzalez, L.E.; Bashan, Y. Increased growth of the microalga *Chlorella vulgaris* when coimmobilized and cocultured in alginate beads with the plant-growth-promoting bacterium *Azospirillum brasilense*. *Appl. Environ. Microbiol.* **2000**, *6*, 1527–1531. [CrossRef] [PubMed]

MDPI
St. Alban-Anlage 66
4052 Basel
Switzerland
Tel. +41 61 683 77 34
Fax +41 61 302 89 18
www.mdpi.com

Marine Drugs Editorial Office
E-mail: marinedrugs@mdpi.com
www.mdpi.com/journal/marinedrugs

www.ingramcontent.com/pod-product-compliance
Lightning Source LLC
Chambersburg PA
CBHW061500210926
51850CB00008B/1775
* 9 7 8 3 0 3 6 5 5 8 2 1 9 *